AF445476

Acoustic Sensors and Their Applications

Acoustic Sensors and Their Applications

Editors

Farook Sattar
Niladri Bihari Puhan
Reza Fazel-Rezai

Basel • Beijing • Wuhan • Barcelona • Belgrade • Novi Sad • Cluj • Manchester

Editors

Farook Sattar
University of Victoria
Victoria, BC, Canada

Niladri Bihari Puhan
Indian Institute of Technology
Bhubaneswar
Bhubaneswar, India

Reza Fazel-Rezai
Americal Public University
Charles Town, WV, USA

Editorial Office
MDPI
St. Alban-Anlage 66
4052 Basel, Switzerland

This is a reprint of articles from the Special Issue published online in the open access journal *Sensors* (ISSN 1424-8220) (available at: https://www.mdpi.com/journal/sensors/special_issues/K2557Z3S38).

For citation purposes, cite each article independently as indicated on the article page online and as indicated below:

Lastname, A.A.; Lastname, B.B. Article Title. *Journal Name* **Year**, *Volume Number*, Page Range.

ISBN 978-3-0365-9847-5 (Hbk)
ISBN 978-3-0365-9848-2 (PDF)
doi.org/10.3390/books978-3-0365-9848-2

Contents

About the Editors

Farook Sattar

Farook Sattar received his Technical Licentiate and PhD degrees from Lund University, Sweden. He obtained the B.Eng and M.Eng degrees from Bangladesh University of Engineering & Technology (BUET), Dhaka in Electrical and Electronics Engineering. Farook has extensive experience and expertise as an academic as well as by collaborating across multiple disciplines. He was a faculty member in the Dept. of Information Eng., School of Electrical & Electronic Engineering, Nanyang Technological University (NTU), Singapore before moving to Canada. He has been affiliated with various universities/institutions in Canada as a research faculty as well as visiting scientist and academic advisor. His research interests include signal and image processing, speech/audio processing, bioacoustics, pattern recognition and machine learning, intelligent systems, big data, computer vision, and data hiding. He is involved in a number of funded research projects sponsored by the Swedish National Science and Technology Board (NUTEK), Singapore Academic Research Fund (AcRF), National Science and Technology Board (NSTB) of Singapore, Canada's National Research and Education Network—CANARIE, and the Natural Sciences and Engineering Research Council of Canada (NSERC). He has published over 130 research articles in refereed journals, conference proceedings, and book chapters.

Niladri Bihari Puhan

Niladri B. Puhan received the B.E. degree in electrical engineering from UCE Burla, Burla, Odisha, India, in 2000; the M.E. degree in signal processing from the Indian Institute of Science, Bangalore, India, in 2002; and the Ph.D. degree from Nanyang Technological University, Singapore, in 2007. He is currently an Associate Professor with the School of Electrical Sciences, IIT Bhubaneswar Bhubaneswar, India. His research interests include signal and image processing, biometrics, medical image analysis, and machine learning.

Reza Fazel-Rezai

Dr. Reza Fazel-Rezai is a Senior Science and Education Application Engineer at MathWorks and a faculty at the American Public University. He holds a Ph.D. degree in Biomedical Engineering with over 20 years of industry expertise as a senior research scientist and research team manager and academic experience as the founding Director and tenured Professor of Biomedical Engineering. With over 190 scientific publications, six edited books, and extensive research interests in biomedical signal and image processing, he has become an expert in pattern recognition methods, particularly in machine learning and deep learning approaches. He is passionate about utilizing and sharing his skills and knowledge to assist others in achieving their goals and objectives and helping them succeed.

Preface

Acoustic sensors are built to measure an environment and convert this information into a digital or analog data signal that can be analyzed by a computer or observer, processed, and involved in numerous applications. The signals collected by acoustic sensors contain a large amount of valid information that facilitates further processing of the collected acoustic signals.

Acoustic sensors have an extremely broad range of applications in many fields, including, engineering acoustics, underwater acoustics, environmental acoustics, architectural acoustics, physical acoustics, psychoacoustics, and so on. Acoustic sensors have extensive usage in different areas, including but not limited to automotive and transportation, robotics, healthcare, security and surveillance, entertainment, home automation, agriculture, IT & communication, commerce, space, and defense.

The acoustic sensors are designed to have some properties such as high sensitivity, linearity, low noise, high resolution, and low power consumption. There are many types of sound sensing devices and some of them highlight a couple of them. First of which is the microphones which are the most common devices to pick up sound waves and turn it into an electric signal. Like our highly sensitive ear, microphones convert our voice into waveforms. The second type is the piezoelectric sensor that also converts the pressure into small electric signals. The sound travels through the air as a wave of pressure and that's exactly what causes the piezoelectric sensor to generate in the form of an electric signal.

Acoustic sensor is one of the fastest-growing technologies in recent years. The increasing demand for acoustic sensors can be attributed to their large market serving different purposes across multiple industries. The sensor technological advances in acoustics turning heavy, non-portable, expensive hardware, labour and time-intensive methods for analysis into new small, movable, affordable, and automated systems, make acoustic sensors increasingly popular.. The future outlook of the acoustic sensor market thus looks promising and is expected to witness significant growth in the coming years. Overall, the ongoing digital transformation and increasing integration of acoustic sensors in various applications will contribute to the market's revival and long-term growth.

This Special Issue is focused on the acoustic sensors and their applications covering the design and implementation of the acoustic sensors as well as their numerous applications such as noise control, speaker recognition, leak detection, non-destructive testing, structure health monitoring, online real-time monitoring, etc. Different types of acoustic sensors including piezoelectric transducers, ultrasound sensors, and microphones, are featured here being an integral part of various issues related to ultrasonic systems, acoustic power transfer systems, sonar systems, preload measurement systems, hybrid acoustic models, aggregation acoustic models, acoustic emission, optochemical devices, optical microphones, infinity tubes, etc. The future directions and some challenges related to acoustic sensors designs and applications are further highlighted. This Special Issue supports the engineers, researchers, academics, and designers, in several interdisciplinary domains.

In conclusion, we thank all authors for their contribution to this Special Issue, and reviewers for their valuable comments. We would also like to express our sincere gratitude to the MPDI-publisher team for their immense help and support for this endeavour. A very special thanks to Mr. Kayle Liu, Managing Editor, for his super performance and invaluable support throughout our Collaboration and it has been a great pleasure working with him. Last but not least, we would like to thank our readers, anticipating that they will find this special issue as a valuable resource in their domains.

Farook Sattar, Niladri Bihari Puhan , and Reza Fazel-Rezai
Editors

Editorial

Special Issue on Acoustic Sensors and Their Applications (Vol. 1)

Farook Sattar [1],*, Niladri Bihari Puhan [2] and Reza Fazel-Rezai [3,4]

[1] Department of Electrical and Computer Engineering, University of Victoria, Victoria, BC V8P 5C2, Canada
[2] School of Electrical Sciences, Indian Institute of Technology Bhubaneswar, Khordha 752050, India; nbpuhan@iitbbs.ac.in
[3] Mathworks, Natick, MA 01760-2098, USA; rfazelre@mathworks.com
[4] School of Science, Technology, Engineering, and Math, American Public University, Charles Town, WV 64706, USA
* Correspondence: fsattar@ieee.org

Citation: Sattar, F.; Puhan, N.B.; Fazel-Rezai, R. Special Issue on Acoustic Sensors and Their Applications (Vol. 1). *Sensors* **2023**, *23*, 7726. https://doi.org/10.3390/s23187726

Received: 30 August 2023
Accepted: 4 September 2023
Published: 7 September 2023

Acoustic sensors have been in commercial use for more than 60 years. Acoustic sensing technologies have been studied extensively, and the information, transmission, reception, transformation, processing and application of acoustic signals have been developed, with acoustic sensors as a central focus. An acoustic sensor is a device that converts a sound wave signal into an electrical signal. The design and development of acoustic sensors are very important technological and scientific issues. Acoustic sensors are widely used in industrial, medical and numerous other applications including environmental and health monitoring, chemical and biochemical detection, and signal processing devices. The papers that form this Special Issue cover a variety of approaches and models related to acoustic sensors and their applications.

Liu et al. [1] aimed to fabricate a smart, nickel-based super-alloy bolt with a high-frequency (center frequency: 17.14 MHz) piezoelectric thin-film sensor by radio frequency magnetron sputtering. The proposed high-frequency probe provides a number of advantages over the commercially available piezoelectric probe in terms of a pure and broad frequency spectrum, high temperature tolerance, small preload measurement, repeatability, measured error due to thickness of couplant layer and change in measurement position.

Rong et al. [2] proposed a muffler named infinity tube with an expansion chamber (ITEC) by adopting the transfer matrix method (TMM) for noise control in the ductwork system. A thorough theoretical and numerical study was carried out for this novel sound attenuation device, ITEC, to reduce duct noise. A closed-form expression for the transmission loss of the ITEC device was derived. The advantage of ITEC over IT (infinity tube) in low-frequency noise reduction was demonstrated by comparing the transmission loss between them.

Deng et al. [3] designed an end-to-end speaker recognition system, ResSKNet-SSDP, with an improved feature extraction capability and improved adaptation to the speaker recognition task. The proposed system makes it more suitable for practical application as it is more efficient in terms of the Equal Error Rate (EER) and detection cost function (DCF), and its structure is lightweight with fewer parameters and less interference time compared to many of the existing methods.

Bente et al. [4] proposed broadband air-coupled ultrasound-emitting and -receiving transducers, demonstrating the combined use of a broadband thermoacoustic emitter and an optical microphone as receiver. They showed an initial application for simultaneous determination of thickness and sound velocity of a solid material. The sensor combination presented is simpler to use, and the transducer pair works for all samples that have a thickness resonances (TR) frequency below a critical value of 1 MHz.

Liang et al. [5] presented a secondary phase transform (PHAT) cross-correlation method to improve the performance of the acoustic methods based on cross-correlation for pipeline leakage detection in complex background noise environments, for example, in power plants or other industrial concerns. A sinc interpolation method was then introduced to automatic search for the peak value of the cross-correlation curve. An improved performance of the proposed method is shown for noise suppression and accurate time delay estimation (TDE) compared to the basic cross-correlation method, which can be beneficial in engineering applications.

Yao et al. [6] investigated the capabilities of a commercial AE (Acoustic Emission) sensor in decomposing AE wave modes generated by pencil lead breaks (PLBs) on a thin metal plate incorporating the LAMDA (linear array for modal decomposition and analysis) sensor and finite element analysis (FEA). It is shown that a transverse PLB produces a dominant A0 mode, while a longitudinal PLB produces a combination of A0 and S0 modes. This work can further initiate the application of LAMDA sensors to identify the wave modes generated by the different damage mechanisms in composite panels.

Vetrab and Gosztolya [7] used hybrid HMM/DNN embedding extractor models in computational para-linguistic tasks. The proposed HMM/DNN hybrid acoustic-model-based feature extraction technique was then found efficient at extracting features from different para-linguistic tasks. In order to perform classification through SVM models, different aggregation methods are used to convert acoustic frame-level features into utterance-level features. The proposed scheme can be considered a competitive and resource-efficient approach for various computational para-linguistic tasks.

Liu and Abdulla [8] introduced an improved acoustic power transfer (APT) system by proposing a 3D-printable and cost-efficient piezoelectric transducer equilateral triangular peripheral clamp. This study integrates an impedance matching circuit into the Mason circuit and investigates the impact of fixed constraints on the piezoelectric transducer's sound pressure and output voltage. The novel findings of this study can aid researchers and practitioners in various fields that employ APT systems to improve their performance in air.

Scheuer and DeCorby [9] applied an ultrasensitive, broadband optomechanical air-coupled ultrasound sensor to investigate the acoustic signals produced by pressurized nitrogen escaping from a variety of small syringes. Harmonically-related MHz-range jet tones were generated by passing pressurized nitrogen gas through a collection of small syringes and shown the extension of the previously established theory to this range. The results from the constructed compact probe for all-optical detection in the MHz range could have practical implications for the non-contact monitoring and detection of early-stage leaks in pressured fluid systems.

In the tenth study, Han et al. [10] developed a sophisticated embedded ultrasonic system to monitor the mechanical properties of ice in real-time and online under various temperature conditions. With the help of the numerical models, the wave propagation in ice was investigated, and the influence of varying temperatures on the wave propagation was also discussed. The proposed system provides a platform to continuously obtain the response of the ice. With this system, the ice properties under specific temperature conditions are then identified based on the intrinsic relationships between the wave propagation velocities and the mechanical properties of ice.

Finally, in the last paper, Jang et al. [11] introduced an electrical equivalent circuit for sonar sensors based on their impedance characteristics by proposing an estimation approach to derive the equivalent circuit. In this study, a particle swarm optimization (PSO) algorithm is employed for parameter estimation of high-degree electrical equivalent circuits. The proposed approach maintained high precision of the derived equivalent circuit even with variations in the number of resonances and the sensor's impedance characteristics. It is expected to provide accurate estimations of the load characteristics when designing amplifiers for sonar operation.

Author Contributions: writing—original draft preparation, F.S.; writing—review and editing, N.B.P., R.F.-R. and F.S. All authors have read and agreed to the published version of the manuscript.

Acknowledgments: We sincerely appreciate the time, effort, and contribution of the authors and esteemed reviewers in improving the quality of the papers. Special thanks to Managing Editor, and all supporting team of the journal *Sensors* for their continuous help in promoting and publishing this Special Issue.

Conflicts of Interest: The authors declare no conflicts of interest.

References

1. Liu, S.; Sun, Z.; Ren, G.; Liao, C.; He, X.; Luo, K.; Li, R.; Jiang, W.; Zhan, H. Ultrasonic Measurement of Axial Preload in High-Frequency Nickel-Based Superalloy Smart Bolt. *Sensors* **2023**, *23*, 220. [CrossRef]
2. Xue, R.; Mak, C.M.; Cai, C.; Ma, K.W. An Infinity Tube with an Expansion Chamber for Noise Control in the Ductwork System. *Sensors* **2023**, *23*, 305. [CrossRef] [PubMed]
3. Deng, F.; Deng, L.; Jiang, P.; Zhang, G.; Yang, Q. ResSKNet-SSDP: Effective and Light End-To-End Architecture for Speaker Recognition. *Sensors* **2023**, *23*, 1203. [CrossRef] [PubMed]
4. Bente, K.; Rus, J.; Mooshofer, H.; Gaal, M.; Grosse, C.U. Broadband Air-Coupled Ultrasound Emitter and Receiver Enable Simultaneous Measurement of Thickness and Speed of Sound in Solids. *Sensors* **2023**, *23*, 1379. [CrossRef] [PubMed]
5. Liang, H.; Gao, Y.; Li, H.; Huang, S.; Chen, M.; Wang, B. Pipeline Leakage Detection Based on Secondary Phase Transform Cross-Correlation. *Sensors* **2023**, *23*, 1572. [CrossRef] [PubMed]
6. Yao, X.; Vien, S.; Rajic, N.; Rosalie, C.; Rose, L.R.F.; Davies, C.; Chiu, W.K. Modal Decomposition of Acoustic Emissions from Pencil-Lead Breaks in an Isotropic Thin Plate. *Sensors* **2023**, *23*, 1988. [CrossRef] [PubMed]
7. Vetrab, M.; Gosztolya, G. Using Hybrid HMM/DNN Embedding Extractor Models in Computational Paralinguistic Tasks. *Sensors* **2023**, *23*, 5208. [CrossRef] [PubMed]
8. Liu, L.; Abdulla, W. Improving APT Systems' Performance in Air via Impedance Matching and 3D-Printed Clamp. *Sensors* **2023**, *23*, 5347. [CrossRef] [PubMed]
9. Scheuer, K.G.; DeCorby, R.G. All-Optical, Air-Coupled Ultrasonic Detection of Low-Pressure Gas Leaks and Observation of Jet Tones in the MHz Range. *Sensors* **2023**, *23*, 5665. [CrossRef] [PubMed]
10. Han, H.; Wei, L.; Alkayem, N.F.; Cao, M. Embedded Ultrasonic Inspection on the Mechanical Properties of Cold Region Ice under Varying Temperatures. *Sensors* **2023**, *23*, 6045. [CrossRef] [PubMed]
11. Jang, J.; Choi, J.; Lee, D.; Mok, H. Estimation Method of an Electrical Equivalent Circuit for Sonar Transducer Impedance Characteristic of Multiple Resonance. *Sensors* **2023**, *23*, 6636. [CrossRef]

Article

Ultrasonic Measurement of Axial Preload in High-Frequency Nickel-Based Superalloy Smart Bolt

Shuang Liu [1,*], Zhongrui Sun [1,2], Guanpin Ren [1,2], Cheng Liao [2], Xulin He [2], Kun Luo [2], Ru Li [2], Wei Jiang [2,*] and Huan Zhan [2,*]

[1] Department of Applied Physics, College of Mathematics and Physics, Chengdu University of Technology, Chengdu 610059, China
[2] Chengdu Development Center of Science and Technology of CAEP, Chengdu 610299, China
* Correspondence: liushuang19@cdut.edu.cn (S.L.); jiangwei.phys@aliyun.com (W.J.); zhanhuanhunan@163.com (H.Z.)

Citation: Liu, S.; Sun, Z.; Ren, G.; Liao, C.; He, X.; Luo, K.; Li, R.; Jiang, W.; Zhan, H. Ultrasonic Measurement of Axial Preload in High-Frequency Nickel-Based Superalloy Smart Bolt. *Sensors* 2023, 23, 220. https://doi.org/10.3390/s23010220

Academic Editors: Farook Sattar, Niladri Bihari Puhan and Reza Fazel-Rezai

Received: 4 December 2022
Revised: 18 December 2022
Accepted: 22 December 2022
Published: 25 December 2022

Abstract: A high-frequency, piezoelectric thin-film sensor was successfully deposited on a nickel-based superalloy bolt by radio frequency magnetron sputtering to develop a smart, nickel-based superalloy bolt. Ultrasonic response characterization, high accuracy, and repeatability of ultrasonic measurement of axial preload in nickel-based superalloy smart bolts are reported here and were fully demonstrated. The axial preload in the nickel-based superalloy smart bolt was directly measured by the bi-wave method (TOF ratio between transverse and longitudinal-mode waves) without using the traditional integration of a longitudinal and shear transducer. A model concerning the bolt before and after tensioning was established to demonstrate the propagation and displacement distribution of the ultrasonic waves inside a nickel-based superalloy smart bolt. The measured A-scan signal presented significantly favorable features including a mixture of transverse and longitudinal mode waves, a pure and broad frequency spectrum which peaked at 17.14 MHz, and high measurement accuracy below 3% for tension of 4 kN–20 kN. For the temporal ultrasonic signal, the measurement envelopes were narrower than for the counterpart of the simulation, justifying the 'filtration' advantage of the high-frequency sensor. Both the TOF change of the single longitudinal-mode wave and the TOF ratio between transverse- and longitudinal-mode waves increased linearly with preload force in the range of 0 kN to 20 kN. Compared with the commercial piezoelectric probe, the proposed probe, based on the combination of a high-frequency, piezoelectric thin-film sensor and a magnetically mounted transducer connector, exhibited high tolerance to temperatures as high as 320 °C and high repeatability free from some interference factors such as bolt detection position change and couplant layer thickness. The results indicate that this system is a promising axial preload measurement system for high-temperature fasteners and connectors, and the proposed sensor is a practical, high-frequency ultrasonic sensor for non-destructive testing.

Keywords: axial preload; ultrasonic measurement; nickel-based superalloy; bolt preload; high-frequency piezoelectric thin-film sensor; non-destructive testing

1. Introduction

Nickel-based superalloys are attractive materials for the components used in the hot zones of jet turbine engines and other industrial applications where exceptional resistance to high-temperature working conditions is required [1–3]. As bolts are critical, high-temperature components, nickel-based superalloy bolts are a credible choice for top-notch, rugged applications in aircraft, aerospace, oil and gas, and various industries [1]. The accurate measurement and monitoring of the axial preload of bolts are critical for maintaining important infrastructure performance and preventing premature failure accidents [2–6]. On this basis, a lot of attention has recently been paid to research in the field of structural health monitoring and the measuring of bolts and bolted connections under the influence of loss of load capacity and high-temperature working conditions [2–10]. There are many ways to directly or indirectly measure bolt preload [2,4–15]. Direct measurement methods are mainly strain gauges [14] and torque control [15]. These methods are employed less due to their low accuracy and significant error, and they do not allow for the online monitoring of bolted structures. This has pushed detection and investigation methods toward indirect methods [7–17]. Many indirect methods such as impedance-based methods [16], piezo active sensing methods [17], acoustoelastic-based methods [7–10], etc., have been intensely investigated in recent years and have been proved to evaluate bolt preload force. Some detection methods can basically realize unmanned online monitoring. However, most indirect methods have different intrinsic shortcomings. For instance, impedance-based methods [16] suffer from relatively high cost, large size of setup, and sensitivity to thermal and load fluctuations, although they are highly sensitive and suitable for in situ monitoring.

Among these indirect methods, acoustoelastic-based methods, i.e., ultrasonic preload measurement, combine the attributes of high precision, excellent real-time performance, and strong sensing penetration and, therefore, have been the most widely used for the detection of axial preload in bolts [2,4,7–13]. Ultrasonic preload measurement methods can be roughly classified into the mono-wave method and bi-wave method according to the nature of the measurement [2,4]. The mono-wave method, proposed earlier, refers to the use of a single transverse wave or longitudinal wave to measure the axial stress in a bolt [2,4,7,11]. This method is generally used to control and verify the preload applied during the assembly of the bolt [2,4]. For comparison, the bi-wave method relies on a combination of transverse- and longitudinal-mode waves to measure the axial preload in bolts [18]. The method provides easier detection of the axial preload in an already tightened bolt [2,4,18,19]. Currently, to achieve the bi-wave method for bolt measurement, a transverse- or longitudinal-mode wave probe is adopted to collect ultrasonic wave signals [18,19]. Another way is to integrate the transverse- and longitudinal-mode wave probe together for the bi-wave method. These ways not only have intricate measurement and large-size probe fabrication processes, but they also easily cause additional measurement errors in terms of axial preload during the measurement process.

Prior to measurement, to avoid error caused by the repeated coupling of transducers, the end face of the subject bolts is processed by some methods such as turning and polishing [7]. However, it not only damages the bolt body but also consumes a great deal of time and labor. These limitations can be overcome by moving to an electromagnetic acoustic transducer (EMAT) [2] and a permanently mounted transducer system (PMTS) [20]. They have been gradually replacing the piezoelectric transducer in recent years. However, the two transducers have not been widely used in engineering applications relating to bolt preload or stress. For example, the use of an EMAT for bolt preload measurement is limited by low conversion efficiency and difficulty in exciting the longitudinal wave on the end faces of ferromagnetic bolts [2,4,20]. The special handling process of the bolt and lead content frustrates the wide use of PMTS [2]. Similar to PMTS, an ultrasonic sensor deposited or bonded on the bolt surface has been demonstrated recently [7]. The common mechanism of the ultrasonic sensor action is described in Figure 1. Based on the inverse piezoelectric effect, the excitation electric pulse signal is easily converted into an ultrasonic wave signal which propagates along the axial direction of the bolt and generates the echo at the bottom; subsequently, the echo propagates along the axial reverse direction and reaches the ultrasonic sensor, and the echo is converted into an electrical pulse signal with preload- or stress-related information through the piezoelectric effect. Based on a smart piezoelectric sensor bolt with a frequency constant of 1.89 MHz, Q. Sun et al. [7] recently reported precise bolt preload measurement by only the mono-wave method. However, the piezoceramic transducer was fixed in the center of the bolt head by low-temperature-tolerance (150 °C) epoxies. Currently, most ultrasonic-related sensors and probes are low frequency and have low temperature tolerance due to the couplant requirement and single-mode wave transducers below 10 MHz [2,4,7–20]. To the best of our knowledge, little literature has so far reported on the axial preload measurement of non-ferromagnetism or weak-ferromagnetism smart bolt specimens such as nickel-based superalloy and titanium alloy bolts. There have been almost no reports on high-frequency, piezoelectric ultrasonic sensors for the axial preload of bolts before.

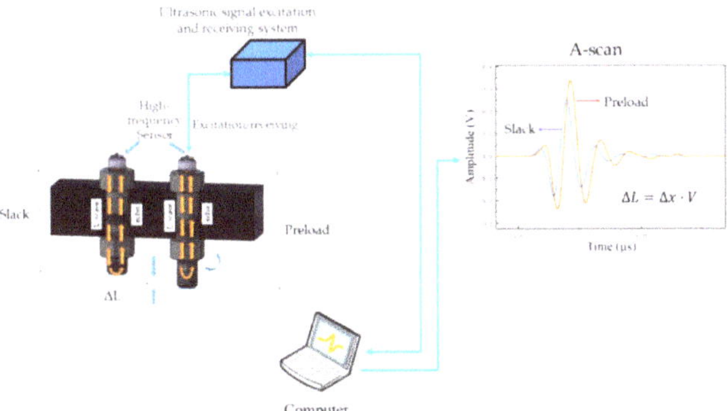

Figure 1. Experimental setup schematic and fundamental principle diagram of measuring bolt axial preload by the mono-wave method. (L-wave: longitudinal wave, V: velocity of longitudinal wave, ΔL, Δx: length and time change of the smart bolt induced by the preload compared with the slack bolt).

In this work, smart, nickel-based superalloy bolts with a high-frequency, piezoelectric thin-film sensor were successfully fabricated by radio frequency magnetron sputtering. High accuracy and repeatability of ultrasonic measurement of axial preload for the smart, nickel-based superalloy bolts were fully demonstrated. Thanks to the proposed probe, i.e., the combination of a high-frequency, piezoelectric thin-film sensor and a magnetically mounted transducer connector, the collected ultrasonic wave signals possessed pure and broad frequency centered at 17.14 MHz and included both transverse- and longitudinal-mode waves. These facts enabled us to measure the axial preload in the nickel-based superalloy by either the mono-wave method or the bi-wave method. Compared with the simulated ultrasonic response, the measured ultrasonic response showed narrower envelopes due to the 'filtration' character of the proposed sensor. The pure ultrasonic response allowed a small preload of 800 N to be successfully measured with the introduction of a spline interpolation algorithm. The proposed high-frequency probe showed obvious advantages over the commercial piezoelectric probe in terms of frequency spectrum, high temperature tolerance, small preload measurement, repeatability, measured error caused by couplant layer thickness, and measurement position change.

2. Research Methods

Measurement and Experiment Process

Ultrasonic pulse-echo technology based on acoustoelasticity theory was used here to measure the axial stress in a smart bolt [2]. For the mono-wave method, axial preload in the nickel-based superalloy bolt was calculated by the TOF change measurement of the pulse echo before and after tightening [2,4]. The relationship between the TOF change and axial preload force can be expressed as follows [2]:

$$F = \frac{E \cdot S \cdot \Delta L}{L} \tag{1}$$

where F is the axial preload force, $\Delta L = \Delta x \cdot V$ is the change of bolt length, Δx is the change in TOF from before to after tightening, and V is the propagation speed of the ultrasonic longitudinal wave in the material; E is the elastic modulus of bolt material, S is the effective cross-sectional area, and L is the clamping length of the bolt.

An integrated excitation and collection ultrasonic system based on a high-frequency piezoelectric sensor was constructed and is depicted in Figure 1. This proposed system consists of an ultrasonic signal excitation and acquisition integrated module, a computer with measurement software, and an ultrasonic probe. The independently developed excitation and acquisition integrated unit could not only launch the exciting voltage in the range of 0~400 V but could also collect ultrasonic wave signals with a frequency from 0.2 MHz to 25 MHz. The corresponding sampling frequency was 100 MHz, and the maximum gain value was as high as 89 dB. The analog signal filter in the frequency domain was from 2 MHz to 25 MHz in this work. The influence of temperature on the preload measure was not considered here.

Different from the commercial piezoelectric probe shown in Figure 2a, the ultrasonic probe consists of two disconnected parts: a high-frequency, piezoelectric thin-film sensor and a magnetically mounted transducer connector, as shown in Figure 2b. To address the disadvantages of the ultrasonic probe, the high-frequency, piezoelectric thin-film sensor was deposited on the slightly polished surface of the nickel-based superalloy head, as shown in Figure 2c. The smart bolts that were the specimens in the experiment were M8 nickel-based superalloy bolts with four thin-film, multi-function layers: an electrode layer, isolated layer, high-frequency, piezoelectric thin-film layer, and transition layer. The composition of the four layers was GH4 169, ZnO, Cr_2O_3, and Sn, respectively. The ZnO ceramic target was a high-frequency piezoelectric material with a high purity of 99.99% in this work. The deposition processes were performed inside radio frequency magnetron sputtering [21], as shown in Figure 2d. The fabrication procedure was as follows: Through a nanosecond pulsed laser polishing technique [22], the surface roughness of the nickel-based superalloy bolt head was controlled below 0.4 μm, and the parallelism between the top and the bottom of the bolt was controlled at 0.01 mm. Then, the nickel-based superalloy sample was consecutively cleaned by an acetone, ethanol, and deionized water solution through an ultrasound cleaner and dried by high-purity nitrogen; to remove the surface pollutants of the target, it was required to pre-sputter the high-frequency piezoelectric target for 20 min. After that, the sputtering experiment was started; by radio frequency magnetron sputtering method, four thin-film, multi-function layers were gradually deposited on the polished surface of the nickel-based superalloy bolt head. The process parameters for preparing the high-frequency thin films on the surface of the nickel-based superalloy bolt were as follows: the base vacuum pressure was 8.4×10^{-4} Pa, the sputtering pressure was 1.0 Pa, the sputtering power was 250 W, the ratio of argon to oxygen flow rate was 10:2, the distance between the target and the substrate was 7 cm, the rotation rate of the sample stage was 5 r/min, annealing for 1 h in oxygen atmosphere, the annealing temperature was 400 °C, and the sputtering time was 40 min. To avoid the introduction of additional stress, the thin-film deposition was performed at room temperature. The thickness of the whole thin film was ~14 μm. The thickness of the electrode layer, isolated layer, high-frequency, piezoelectric thin-film layer, and transition layer was 5.2 μm, 6.0 μm, 2.1 μm, and 1.1 μm, respectively. To the protect the high-frequency thin-film layer from corrosion and pollution, an isolated metallic oxide layer was introduced between the electrode layer and the high-frequency thin-film layer. The material property of the thin-film sensor is shown and discussed in the following sections.

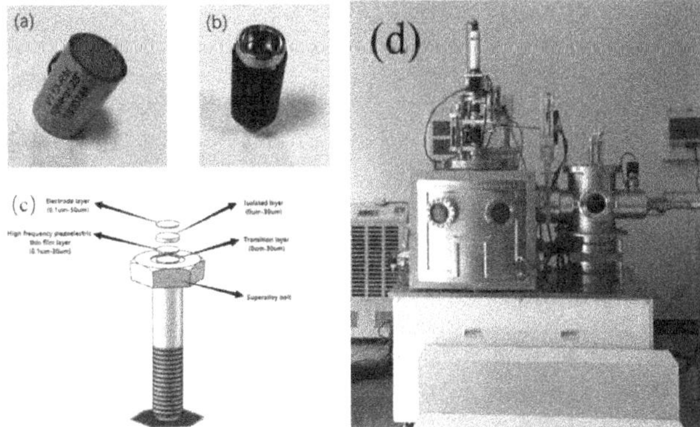

Figure 2. Photographs of (**a**) a commercial piezoelectric probe of 10 MHz and (**b**) a magnetically mounted transducer connector; (**c**) schematic diagram of a high-frequency M8 smart nickel-based superalloy bolt; (**d**) photograph of radio frequency magnetron sputtering.

3. Result and Discussion

3.1. Theoretical Model and Simulation Results

To simulate the propagation process of ultrasound inside the nickel-based superalloy bolt, numerical calculation was carried out thoroughly with the well-known finite element method that is reported elsewhere [23,24]. The expression of the analytical function of the electrically excited bolt is [25]:

$$an1(t) = V_0 * sin(A * t) * gp1(t) \tag{2}$$

where V_0 is the excitation voltage, A is the amplitude, $gp1(t)$ is the Gaussian pulse, and the function expression of the Gaussian pulse is [26]:

$$gp1(t) = A * e^{-\frac{2(t-2T_0)}{T_0}} \tag{3}$$

where A is the amplitude, and T_0 is the period.

In the piezoelectric effect, the deformation of the piezoelectric film is caused by the voltage applied by the electric excitation. The mechanism of the piezoelectric effect can be summarized as the following governing equation [25]:

$$\rho \frac{\partial^2 u}{\partial t^2} = \nabla \cdot S + Fv \tag{4}$$

where ρ is the material density, and u is the displacement field. S is the stress tensor, and Fv is the deformation gradient.

Table 1 gives the input parameters of the acoustoelasticity model for the smart bolt in the work. Figure 3 shows the displacement distribution of the ultrasonic wave along the whole bolt in the axial direction for cases of 0 N and 20 kN. To simplify the simulation result, we only considered elongation of the bolt length induced by the axial preload, and, therefore, the length of the bolt mode lengthened by 0.233 mm according to the experiment assessment, as shown in Figure 3. An ultrasonic wave was produced on the upper surface of smart bolt through 60 V voltage excitation and directly transmitted toward the bottom of the bolt at time 2 μs. The trailing waves could obviously be seen at time 2–6 μs, and part of the ultrasonic wave was reflected by the boundary of the bolt side. The ultrasonic wave reached the boundary of the bolt bottom at time 6 μs and then completed the reflection behavior on the boundary at time 8 μs. One part of reflection wave energy transmitted in the opposite axial direction. It was observed, from 12 μs to 14 μs, that the ultrasonic wave was gradually detected at the bolt head to generate the first longitudinal wave. Compared with the slack bolt, the bolt to which 20 kN preload had been applied presented a similar ultrasonic propagation process with a corresponding time decay. Figure 4a shows the simulation results of the temporal ultrasonic signal before and after tensioning at 20 kN. With the introduction of 20 kN tension, the whole temporal ultrasonic signal presented a delay phenomenon, and a TOF change for the first longitudinal wave was observed. The absence of a shear wave in the temporal signal is attributed to the pressure acoustic module of the used COMOSOL Multiphysics software, which only contains longitudinal wave mode. The displacement field distribution of the upper surface for the slack bolt and the preload bolt with 20 kN was calculated and presented in Figure 4b. Two peak components of the displacement field were found to correspond to the first and second longitudinal wave. The axial preload of 20 kN had no remarkable influence on the displacement field intensity of the longitudinal wave, which led to a time delay.

Table 1. Input parameters of acoustoelasticity model for smart bolt.

Parameter	Unit	Value
Longitudinal wave speed	m/s	5788
Frequency, f0	MHz	17
Function, A	$2 * \pi * f0$ (MHz)	10.6814
Period, T0	s	5.8824×10^{-8}
Wavelength, λ0	m	3.4047×10^{-4}
Exciting voltage, V0	V	60

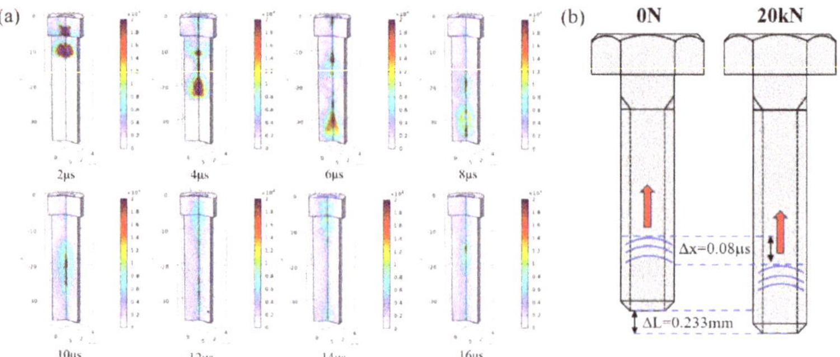

Figure 3. (a) Evolution of the ultrasonic wave inside the nickel−based superalloy bolt, (b) schematic diagram of ultrasonic wave propagation difference between the slack bolt and the bolt with 20 kN applied.

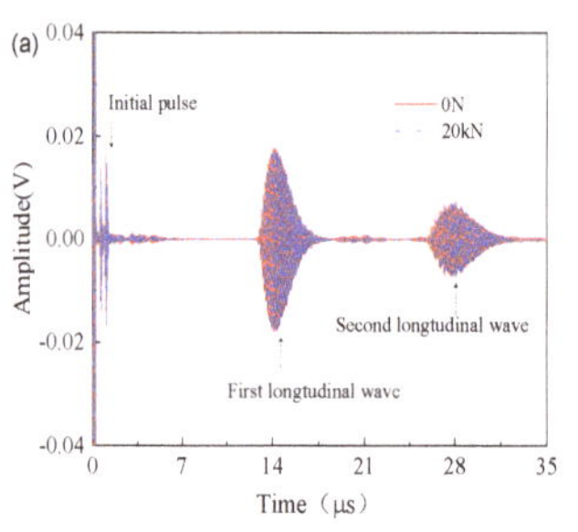

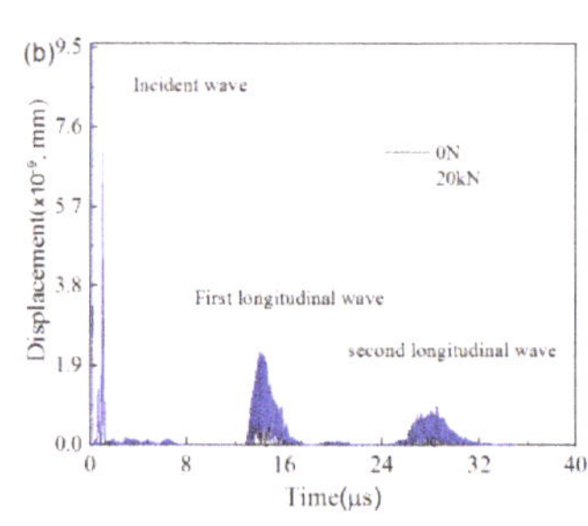

Figure 4. (a) The simulated temporal ultrasonic signal inside the nickel−based superalloy bolt, (b) displacement field contrast of the upper surface between the bolt with 20 kN axial preload applied and the slack bolt.

3.2. Measured Ultrasonic Wave Properties

Spline function interpolation was adopted to increase data points and improve the resolution of the propagation time difference [26]. Figure 5a shows the measured A-scan waveform collected from the nickel-based superalloy bolt head. The first longitudinal wave was the envelope, which peaked at 14.10 μs, corresponding to the time of one round trip of the longitudinal wave propagating along the whole bolt in the axial direction. Likewise, the second longitudinal wave corresponded to the time of two round trips of longitudinal wave propagation, and the envelope was centered at 27.95 μs with a reduction of voltage amplitude. The intense envelope between the first and second longitudinal waves was the first shear wave, peaking at 26.85 μs. The unique feature of the coexisting longitudinal and shear wave made us directly use the bi-wave method for the preload measurement. Until now, rare ultrasonic transducers have been reported to generate longitudinal and transverse waves simultaneously. In addition to this, it was noted that the voltage amplitude of the first shear wave was more intense than the counterpart of the longitudinal wave. The reason may be attributed to the piezoelectric thin-film sensor possessing a more intense poling phenomenon at the plane direction compared with the perpendicular direction when the thin-film layer was excited by 60 V voltage. Limited by the experimental conditions, however, the corresponding microscopic mechanism could not be effectively obtained in this work. Figure 5b compares the TOF of the pulse echo before and after applying tension at 20 kN. The use of tension led to an obvious TOF change and a reduction of amplitude intensity. Nevertheless, the profile of the curve presented no change. In this case, the cross-correlation algorithm [9] was used to obtain the TOF change before and after tension. Compared with the simulation results (see the Figure 4), the experimental results (see Figure 5a,b) presented more narrow envelopes. As the sensor intrinsic material property could not be considered during the simulation result, the reflected waves and the scattering waves from other surfaces were mixed into the echo signal. However, the measured narrow envelopes could be filtered by the high-frequency sensor. The intrinsic reason was unclear. For the outline and time position, the experimental results basically coincided with theoretical simulation result. To check the high temperature tolerance, the smart nickel-based superalloy bolt was inserted into a heating furnace with a temperature of 320 °C and kept at a continuous heat for 1 h. As shown in Figure 5c, the whole temporal ultrasonic signal at 320 °C also presented time decay comparable to its counterpart at 22 °C, while the amplitude intensity presented no significant reduction. It suggests that the smart bolt can tolerate a high temperature of 320 °C without failure. The further high-temperature test was limited by the tolerance temperature ($\leq$250 °C) of the available BNC connector line. For comparison, ultrasonic measurement of the bolt was carried out by replacing the magnetically mounted transducer connector with a commercial piezoelectric probe centered at 10 MHz. As indicated in Figure 5d, the temporal ultrasonic wave signal presented some obvious background noise and no shear mode wave between the first and second longitudinal waves. The envelope of the first peaked longitudinal wave of the commercial probe was obviously wider than the counterpart of the smart bolt. It validated the 'filtration' advantage of the high-frequency piezoelectric sensor with weak interference from the reflected waves and the scattering waves from other surfaces.

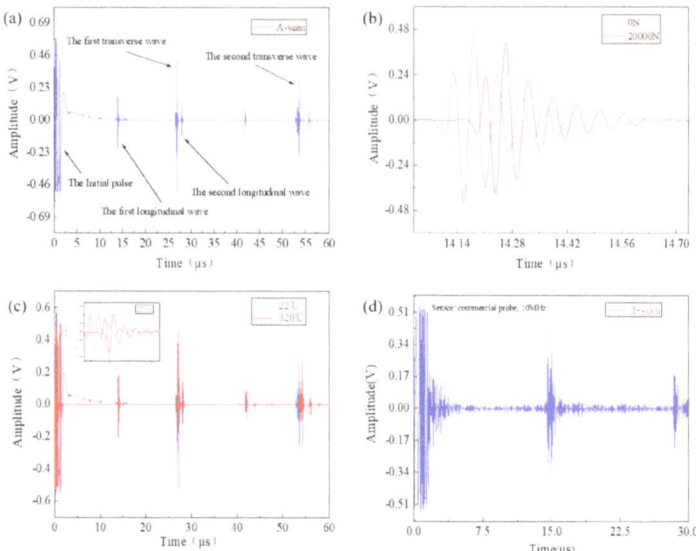

Figure 5. (**a**) The measured A−scan waveform collected from the high−frequency, piezoelectric thin−film sensor, (**b**) comparing TOF change of the pulse echo before and after applying tension at 20 kN, (**c**) A−scan waveforms of the smart bolt at 22 °C and 320 °C (the insert compares the waveform of the first longitudinal wave for the bolt at 22 °C and 320 °C), (**d**) the measured A−scan waveform collected from the commercial piezoelectric probe.

Frequency spectra of the first longitudinal waves collected from the commercial probe and high-frequency piezoelectric sensor were obtained by the Fastest Fourier Transform in the West (FFTW) algorithm. The results are shown in Figure 6. For the commercial probe, in addition to the frequency centered at 8.32 MHz, several other harmonic waves could be easily observed, as shown in Figure 6a. These harmonic waves originated from nonlinear ultrasonic effects, including sum-frequency and difference-frequency effects [27], and the piezoelectric material intrinsic property [28]. An increase in tension led to a more intense inhomogeneous stretch along the axial direction of the bolt. Owing to nonlinear ultrasonic effects, energy transfer between low- and high-frequency signals occurred continually with the increase in the tension. In this case, the similarity of the ultrasonic wave signals before and after straining was reduced, so the calculated results by the cross-correlation algorithm were difficult to keep at high accuracy. By comparison, the first longitudinal-mode wave collected by the piezoelectric sensor corresponded to a pure and broad spectrum centered at 17.14 MHz, as shown in Figure 6b. Neither a low-frequency peak nor a high-frequency peak was observed. For the smart bolt reported in [7], the center frequency of the piezoelectric sensor was 1.89 MHz, and no frequency spectrum was presented. As reported in [11], the frequency spectra presented obvious higher-harmonic-related components. With the increase in the tension, the spectrum center showed no obvious shift, although the amplitude intensity presented a slight reduction. It justifies the advantages of the piezoelectric sensor with a high nonlinear suppression feature and almost no energy transfers of different frequency signals. This ensures that mono-wave methods based on the cross-correlation algorithm are capable of achieving a precise measure of the bolt preload.

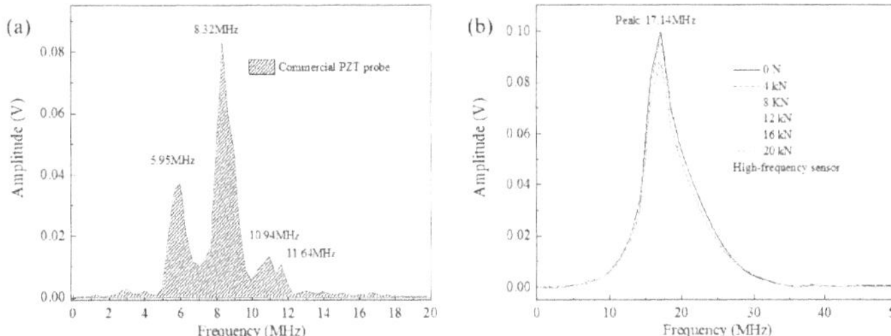

Figure 6. Frequency spectra of the measured ultrasonic wave collected from (**a**) the commercial PZT probe and (**b**) the high−frequency piezoelectric sensor.

3.3. Axial Preload Measurement

During the measurement experiment, a specially designed clamp was used to connect the nickel-based superalloy smart bolt specimens together with a magnetic connector to form a whole body. After that, axial tension was applied from 0 to 20 kN with an interval of 4 kN by the calibrated electrical universal material testing machine. In the experiment, the repeatability error for 4 kN, 8 kN, 12 kN, 16 kN, and 20 kN was calculated to be 0.42%, 0.37%, 0.36%, 0.27%, and 0.22%, respectively. Synchronously, the ultrasonic measurement system excited and collected ultrasonic wave signals by computer software. For the five bolt samples, axial preload was measured experimentally and linearly fitted as a function of the TOF change, as shown in Figure 7a. For each sample, actual measured tension values and fitted tension values had a very high fitting precision, i.e., a good linear relationship between the TOF change of the first longitudinal wave and tension. Neither a nonlinear curve nor some deviation points were observed. This validates the effectiveness of the TOF change calculated by the cross-correlation algorithm. To examine the feasibility of the bi-wave method, the TOF ratio of the first longitudinal wave to the first shear wave as a function of the tension was measured and is shown in Figure 7b. A good linear relationship between the tension and TOF ratio of the first longitudinal wave to the first shear wave was also observed. It means the in-service bolt preload could be directly measured by the calibration curve function shown in Figure 7b. For the in-service superalloy smart bolt, the proposed measurement system with only a magnetic connector directly measured its axial preload only one time due to the simultaneous generation of the longitudinal and shear waves. Compared with the previously reported bi-wave methods [18,19], the bi-wave method based on the high-frequency sensor is more effective and accurate for measuring axial preload of in-service bolts. The coefficient of determination R^2 for the bi-wave method was 0.99878, which is in agreement with that of the mono-wave wave method. Therefore, axial preload of the installing, already tightened, or in-service superalloy smart bolts is capable of being effectively and accurately measured by the proposed ultrasonic measurement system based on a high-frequency sensor.

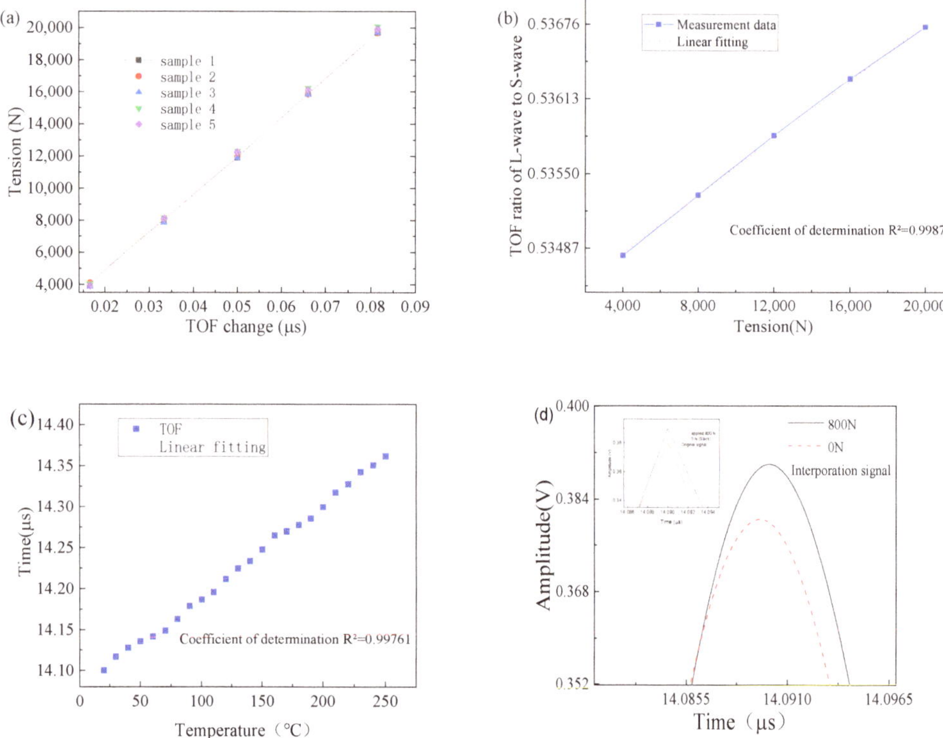

Figure 7. (**a**) Relationship between the TOF difference and the tension; (**b**) TOF ratio of the first longitudinal wave to the first shear wave as a function of the tension; (**c**) relationship between the temperature and TOF in an unstressed bolt; (**d**) TOF change before and after the applied axial preload of 800 N after interpolation. The inset is the original temporal signal of the first longitudinal wave for the cases of 0 N and 800 N.

In the experiment, a heat module installed on an electrical universal material testing machine was inconvenient and prohibited for safety. As the change in temperature from room temperature to 250 °C was large, many parameter calibrations were complicated and limited by the presented experiment condition. Considering these, the relationship between the temperature and TOF in an unstressed bolt was measured as shown in Figure 7c. TOF in unstressed sample 5 was firstly recorded at 20 °C. Sample 5 was inserted into a heap of sandy soil sustainably heated by the furnace and kept for half an hour at each measurement temperate. TOF of the heated sample 5 was quickly measured and recorded. This process was repeated until the temperature reached 250 °C. The temperature increased in 10 °C steps. It showed a nearly linear relationship between the temperature and TOF of sample 5. The coefficient of determination R^2 was 0.99761. The results could be used as compensation for the actual high-temperature tension measurement [2] in the following work. To examine the ability to measure the small preload below 1 kN, a small axial preload of 800 N was applied to sample 5 by an electrical universal material testing machine. The measured temporal ultrasonic signal curves are shown in Figure 7d. As shown in the inset of Figure 7d, no TOF change was observed for the original temporal signal for the slack and small preload due to the limited sampling frequency of 100 MHz. The spline interpolation algorithm was used to increase time resolution. After that, the preload value measured by the proposed system fluctuated in the range of 0.75~0.84 kN, corresponding to an error of less than ±7%.

The ultrasonic signal amplitude of the slack bolt was slightly lower than that of the bolt with 800 N applied, as described in Figure 7d. Both the value fluctuation and the lower amplitude of the slack smart bolt may be attributed to the small inner noise interference of the acquisition unit and different contact pressure between the magnetically mounted transducer connector and the smart bolt.

3.4. High Accuracy and Repeatability Property

To identify the advantage of the high-frequency sensor for preload measurement, five nickel-based superalloy bolts were chosen as experimental specimens and loaded with tension from 4 kN to 20 kN by the calibrated testing machine. The measurement error was obtained from a function (measure error $\sigma = 100\% * (F_{measure} - F_{apply})/F_{apply}$, where $F_{measure}$ is the tension value measured by the software, and F_{apply} is the tension value applied by the testing machine). The calculation results of measurement error are shown in Figure 8a. It was noted that the absolute error of this system was below ± 0.3 kN, and the relative error was less than $\pm 3\%$. The error results were in agreement with the counterpart reported in [7]. As a nickel-based superalloy bolt is a kind of extremely weak magnetic material, insufficient contact between the magnetic connector and the smart bolt was easily induced during the stretching test process. The low ultrasonic preload measured error for the weak-magnetism bolt justifies the advantage of the high measurement accuracy of the proposed measurement system. Compared with a carbon-steel-based bolt [7], the superalloy smart bolt with weak magnetism easily led to deficient contact and was excited by external voltage. As a result, the excited position of the sensor easily changed the required uniform piezoelectric property around the center region of the sensor. Additionally, the contact position change easily caused the angle deviation between the excitation and collection of ultrasonic wave. For a carbon-steel-material smart bolt, the absolute error of the system was less than ± 0.2 kN (corresponding to $<\pm 1\%$) in the experiment because the magnetic connection contributed to the fixed contact position point and both the excitation and collection of the ultrasonic wave. These justify the high accuracy of the ultrasonic measurement system based on a high-frequency piezoelectric sensor.

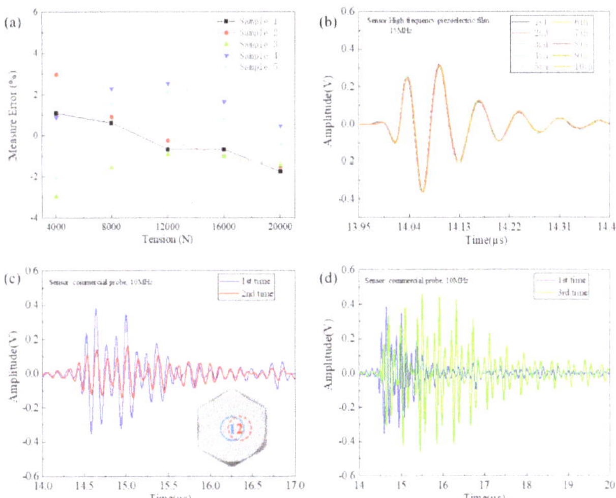

Figure 8. (**a**) The relative error of ultrasonic measurement based on the high-frequency sensor; (**b**) temporal ultrasonic wave signs for ten different contact positions between the magnet connector and smart bolt; (**c**) temporal ultrasonic wave signs for two different contact positions between the commercial piezoelectric probe and smart bolt (inset shows schematic diagram of two different detection positions on the surface of the bolt head.); (**d**) temporal ultrasonic wave signs for two different couplant layer thicknesses.

To examine the high repeatability of the proposed sensor, the magnetic connector was mounted at ten different positions on the nickel-based superalloy bolt head surface. The corresponding ten ultrasonic measurements were carried out. The measurement results are shown in Figure 8b. All of the first longitudinal mode signals almost overlapped together. This means the proposed ultrasonic system is insensitive to the contact position. This is due to the vibration–electrical energy conversion occurring between the four thin-film layers, as shown in Figure 2c. For comparison, the commercial piezoelectric probe could be used to measure ultrasonic wave signal at two different contact positions and coupling layer thicknesses. These measurement results are presented in Figure 8c,d. Obvious deviation phenomena between the ultrasonic wave signals could be easily seen in the cases of different contact positions and coupling layer thicknesses. This drawback makes it difficult for the commercial piezoelectric probe to measure the axial preload in actual engineering applications.

4. Conclusions

In summary, newly developed nickel-based superalloy smart bolts with a high-frequency (center frequency: 17.14 MHz), piezoelectric thin-film sensor were fabricated by radio frequency magnetron sputtering. The ultrasonic temporal and spectrum domain response and axial preload measurement of the nickel-based superalloy bolt were fully demonstrated. The proposed piezoelectric sensor possesses a pure and broad frequency spectrum from 10 MHz to 30 MHz, high-temperature-tolerance ability up to 320 °C, and a mixture of transverse- and longitudinal-mode waves. Compared with the commercial piezoelectric probe, the proposed sensor presents stable ultrasonic properties with high suppression of the nonlinear ultrasonic effect. The measured temporal signal presented narrower signal envelopes than the simulation results, indicating the 'filtration' advantage of the proposed sensor. Both the mono-wave method and the bi-wave method were applied to measure the axial tension for small preloads from 800 N to 20 kN. TOF presented a linear increase trend with temperature increasing from room temperature to 250 °C. The ultrasonic system based on the smart bolt showed excellent characteristics of high repeatability, low measurement error of axial preload below $\pm 3\%$ (comparable to the counterpart of the previously reported carbon-steel bolt), small preload measurement ability, and high stability. As a fixed contact position point was difficult to keep long enough for weakly magnetic smart bolts, high-accuracy ultrasonic measurement of the axial preload in this work indirectly demonstrated the uniform piezoelectric distribution in the whole high-frequency sensor. Future research work will be transferred to monitor the axial preload of the high-temperature, in-service bolt and establish a high-precision and -frequency laser ultrasonic system based on the high-frequency piezoelectric sensor.

Author Contributions: Conceptualization, S.L., G.R., Z.S., R.L. and H.Z.; methodology, G.R., Z.S. and S.L.; software, S.L., G.R. and H.Z.; validation, G.R., R.L. and W.J.; formal analysis, H.Z. and S.L.; investigation, C.L., X.H., K.L., R.L. and W.J.; writing—original draft preparation, S.L.; writing—review and editing, H.Z. and S.L.; visualization, S.L., G.R. and Z.S.; supervision, H.Z. and S.L.; project administration, S.L. All authors have read and agreed to the published version of the manuscript.

Funding: This research received no external funding.

Institutional Review Board Statement: Not applicable.

Informed Consent Statement: Not applicable.

Data Availability Statement: Data is contained within the article.

Acknowledgments: This work was supported by grants from the public service platform of innovation achievements industrialization, Chengdu Development Center of Science and Technology, China Academy of Engineering Physics (CAEP). Great thanks are given to Xia Ding (Teclab (CHINA) LIMITED) for experimental instruction and results discussion.

Conflicts of Interest: The authors declare no conflict of interest.

References

1. Andrew, D.L.; Carlson, S.S.; Macha, J.H.; Pilarczyk, R.T. Investigating and interpreting failure analysis of high strength nuts made from nickel-base superalloy. *Eng. Fail. Anal.* **2017**, *74*, 35–53. [CrossRef]
2. Pan, Q.; Pan, R.; Shao, C.; Chang, M.; Xu, X. Research Review of Principles and Methods for Ultrasonic Measurement of Axial Stress in Bolts. *Chin. J. Mech. Eng.* **2020**, *33*, 11. [CrossRef]
3. Wang, J.J.; Wen, Z.X.; Pei, H.Q.; Gu, S.N.; Zhang, C.J.; Yue, Z.F. Thermal damage evaluation of nickel-based superalloys based on ultrasonic nondestructive testing. *Appl. Acoust.* **2021**, *183*, 108329. [CrossRef]
4. Nikravesh, S.M.Y.; Goudarzi, M. A Review Paper on Looseness Detection Methods in Bolted Structures. *Lat. Am. J. Solids Struct.* **2017**, *14*, 2153–2176. [CrossRef]
5. Wang, F.; Song, G. Bolt-looseness detection by a new percussion-based method using multifractal analysis and gradient boosting decision tree. *Struct. Health Monit.* **2020**, *19*, 2023–2032. [CrossRef]
6. Tran, D.Q.; Kim, J.W.; Tola, K.D.; Kim, W.; Park, S. Artificial Intelligence-Based Bolt Loosening Diagnosis Using Deep Learning Algorithms for Laser Ultrasonic Wave Propagation Data. *Sensors* **2020**, *20*, 5329. [CrossRef] [PubMed]
7. Sun, Q.; Yuan, B.; Mu, X.; Sun, W. Bolt preload measurement based on the acoustoelastic effect using smart piezoelectric bolt. *Smart Mater. Struct.* **2019**, *28*, 055005. [CrossRef]
8. Kim, N.; Hong, M. Measurement of axial stress using mode-converted ultrasound. *NDT E Int.* **2009**, *42*, 164–169. [CrossRef]
9. Pan, Q.; Pan, R.; Chang, M.; Xu, X. A shape factor based ultrasonic measurement method for determination of bolt preload. *NDT E Int.* **2020**, *111*, 102210. [CrossRef]
10. Song, C.; Kim, Y.J.; Cho, C.B.; Chin, W.J.; Park, K.-Y. Estimation on Embedment Length of Anchor Bolt inside Concrete Using Equation for Arrival Time and Shortest Time Path of Ultrasonic Pulse. *Appl. Sci.* **2020**, *10*, 8848. [CrossRef]
11. Zhou, Y.; Yuan, C.; Sun, X.; Yang, Y.; Wang, C.; Li, D. Monitoring the looseness of a bolt through laser ultrasonic. *Smart Mater. Struct.* **2020**, *29*, 115022. [CrossRef]
12. Liu, E.; Liu, Y.; Wang, X.; Ma, H.; Chen, Y.; Sun, C.; Tan, J. Ultrasonic Measurement Method of Bolt Axial Stress Based on Time Difference Compensation of Coupling Layer Thickness Change. *IEEE Trans. Instrum. Meas.* **2021**, *70*, 1009212. [CrossRef]
13. Xingliang, H.; Ping, C. Ultrasonic Measurement of Bolt Axial Stress Using the Energy Ratio of Multiple Echoes. *IEEE Sens. J.* **2022**, *22*, 3928–3936. [CrossRef]
14. Wang, Y.Q.; Wu, J.K.; Liu, H.B.; Kang, K.; Liu, K. Geometric accuracy long-term continuous monitoring using strain gauges for CNC machine tools. *Int. J. Adv. Manuf. Technol.* **2018**, *98*, 1145–1153. [CrossRef]
15. Tsuji, H.; Nakano, M. Bolt preload control for bolted flange joint. In Proceedings of the ASME Pressure Vessels and Piping Conference, Vancouver, BC, Canada, 5–9 August 2002; pp. 163–170.
16. Huo, L.; Chen, D.; Liang, Y.; Li, H.; Feng, X.; Song, G. Impedance based bolt pre-load monitoring using piezoceramic smart washer. *Smart Mater. Struct.* **2017**, *26*, 057004. [CrossRef]
17. Wang, T.; Song, G.; Wang, Z.; Li, Y. Proof-of-concept study of monitoring bolt connection status using a piezoelectric based active sensing method. *Smart Mater. Struct.* **2013**, *22*, 087001. [CrossRef]
18. Liu, E.; Liu, Y.; Chen, Y.; Wang, X.; Ma, H.; Sun, C.; Tan, J. Measurement method of bolt hole assembly stress based on the combination of ultrasonic longitudinal and transverse waves. *Appl. Acoust.* **2022**, *189*, 108603. [CrossRef]
19. Walaszek, H.; Bouteille, P. Application of ultrasonic measurements to stress assesment on already tightened bolts. *NDT* **2014**, *1*, 2.
20. Ding, X.; Wu, X.; Wang, Y. Bolt axial stress measurement based on a mode-converted ultrasound method using an electromagnetic acoustic transducer. *Ultrasonics* **2014**, *54*, 914–920. [CrossRef]
21. Lobe, S.; Bauer, A.; Uhlenbruck, S.; Fattakhova-Rohlfing, D. Physical Vapor Deposition in Solid-State Battery Development: From Materials to Devices. *Adv. Sci.* **2021**, *8*, e2002044. [CrossRef]
22. Annamaria, G.; Massimiliano, B.; Francesco, V. Laser polishing: A review of a constantly growing technology in the surface finishing of components made by additive manufacturing. *Int. J. Adv. Manuf. Technol.* **2022**, *120*, 1433–1472. [CrossRef]
23. Ren, G.; Xu, L.; Zhan, H.; Liu, S.; Jiang, W.; Li, R. Quantifying the shape effect of plasmonic gold nanoparticles on photoacoustic conversion efficiency. *Appl. Opt.* **2022**, *61*, 4567–4570. [CrossRef] [PubMed]
24. Ren, G.; Sun, Z.; Dai, X.; Liu, S.; Zhang, X.; Chen, X.; Yan, M.; Liu, S. Laser ultrasonic nondestructive evaluation of sub-millimeter-level crack growth in the rail foot weld. *Appl. Opt.* **2022**, *61*, 6414–6419. [CrossRef]
25. Manbachi, A.; Cobbold, R.S.C. Development and Application of Piezoelectric Materials for Ultrasound Generation and Detection. *Ultrasound* **2011**, *19*, 187–196. [CrossRef]
26. Stockwell, R.G.; Mansinha, L.; Lowe, R.P. Localization of the complex spectrum: The S transform. *IEEE Trans. Signal Process.* **1996**, *44*, 998–1001. [CrossRef]
27. Castellano, A.; Fraddosio, A.; Piccioni, M.D.; Kundu, T. Linear and Nonlinear Ultrasonic Techniques for Monitoring Stress-Induced Damages in Concrete. *J. Nondestruct. Eval. Diagn. Progn. Eng. Syst.* **2021**, *4*, 041001. [CrossRef]
28. Ren, G.; Zhan, H.; Liu, Z.; Jiang, W.; Li, R.; Liu, S. Evaluation of Axial Preload in Different-Frequency Smart Bolts by Laser Ultrasound. *Sensors* **2022**, *22*, 8665. [CrossRef] [PubMed]

 *sensors*

Article

An Infinity Tube with an Expansion Chamber for Noise Control in the Ductwork System

Rong Xue [1], Cheuk Ming Mak [1,*], Chenzhi Cai [2] and Kuen Wai Ma [1]

[1] Department of Building Environment and Energy Engineering, The Hong Kong Polytechnic University, Hung Hom, Kowloon, Hong Kong, China

[2] School of Civil Engineering, Central South University, Changsha 410017, China

* Correspondence: cheuk-ming.mak@polyu.edu.hk

Abstract: This paper proposes a muffler with simple geometry to effectively reduce low-frequency noise in ductwork systems. A muffler named infinity tube with an expansion chamber (ITEC) is developed from the infinity tube (IT). Theoretical and numerical analyses of wave propagation in the ITEC have been conducted in this paper. The transfer matrix method is adopted to predict transmission loss theoretically. The theoretical results are validated by the finite element method simulation. The comparison of the transmission loss between the IT and ITEC illustrates that the ITEC has an advantage in low-frequency noise reduction. The transmission loss results of the ITEC are compared with the Helmholtz resonator system to assess the potential for industrial application. Finally, the geometric parameters of the proposed ITEC on its noise attenuation performance have been analyzed. The proposed ITEC can effectively reduce low-frequency noise, and it is suitable for ductwork systems in constrained spaces.

Keywords: infinity tube; transfer matrix method; transmission loss; noise control

Citation: Xue, R.; Mak, C.M.; Cai, C.; Ma, K.W. An Infinity Tube with an Expansion Chamber for Noise Control in the Ductwork System. *Sensors* **2023**, *23*, 305. https://doi.org/10.3390/s23010305

Academic Editors: Farook Sattar, Niladri Bihari Puhan and Reza Fazel-Rezai

Received: 7 December 2022
Revised: 20 December 2022
Accepted: 24 December 2022
Published: 28 December 2022

1. Introduction

The ductwork system is of vital importance to modern buildings [1]. It is an essential part of the HVAC system to supply fresh air, recycle exhaust, and maintain a comfortable indoor environment. However, the ductwork system always encounters noise problems [2,3]. Researchers have invented many kinds of well-designed mufflers to reduce duct-borne noise in ductwork systems. In industry, dissipative and reactive silencers have a wide range of applications [4]. The principle of dissipative silencers is using sound-absorbing materials to convert sound energy into heat. However, subject to the dimensions of porous material, the dissipative silencers are not suitable for attenuating low-frequency noise [5]. The reactive silencers are composed of acoustic elements which could alter the impedance and reflect the incident acoustic waves. The Helmholtz resonator [6], quarter-wavelength tube [7], and expansion chamber [8] are typical reactive silencers used in duct systems. For the Helmholtz resonator and quarter-wavelength tube, the noise attenuation band is near the resonance frequency and relatively narrow. The expansion chamber needs a large expansion ratio to maintain the noise attenuation ability, which leads the expansion chamber to be cumbersome [9]. A novel muffler is required to avoid the disadvantages of the existing silencers.

The Herschel-Quincke tube (hereafter HQ) consists of two pipes parallelly mounted along with arbitrary length and cross-section area [10,11]. Stewart discussed the sound transmission characteristic of HQ tube devices theoretically [12]. After years of research, the HQ tube system has been shown to be an effective silencer for low-frequency noise attenuation. Selamet et al. [13] conducted an experimental, theoretical, and computational investigation on HQ tube transmission loss (TL). Thereafter, Selamet et al. [14] eliminated the geometric restrictions of HQ and altered the HQ tube to an N-duct configuration. Kim et al. [15] designed a virtual HQ tube system to achieve the desired transmission loss

---

performance under a required frequency range. Wang et al. [16] combined the HQ tube with micro-perforated panels and developed a new noise control device. Ahmadian et al. [17] developed a genetic algorithm to optimize the parameters of a HQ tube. Mazzaro et al. [18] numerically investigated air flow movement inside the HQ tubes. The disadvantage of the HQ tube is clear. HQ needs two length parameters to locate the position on the main duct and determine the resonant frequency [19]. Therefore, compared with other reactive silencers, HQ appears to be cumbersome. In addition, the research by Torregrosa et al. [20] showed that the HQ tube would cause flow repartition between the main duct and HQ device.

Lato et al. [19] developed the traditional HQ tube into the infinity tube (hereafter IT) by combining the previous two connecting points of the HQ in different duct cross-sections into one cross-section. With a simpler geometry, the IT is easier to manufacture and install than the HQ tube. In addition, IT could avoid flow repartition. The research showed that IT is an innovative muffler and would have the potential for industrial applications. To further improve the noise attenuation ability of IT, an expansion chamber muffler would be used to replace the side branch of the IT device. The improved IT device would be referred to as the infinite tube with an expansion chamber (ITEC) throughout the whole study.

In this paper, the transfer matrix method (TMM) is adopted to predict the noise attenuation ability of the ITEC. TMM and the statement of pressure equality and conservation of volumetric flow are performed to solve the transmission loss of the ITEC analytically. The finite element method (FEM) simulation of the ITEC has been conducted to validate the theoretical prediction results. Then, the transmission loss results of the ITEC are compared with infinity tubes and other silencers to examine the noise attenuation performance. The most frequently used reactive silencer in industry, the Helmholtz resonator, is chosen for comparison. Finally, the effects of the geometric parameter of the proposed ITEC on the noise attenuation performance are investigated.

2. Analytical Model of the ITEC

2.1. Sound Propagation Inside the Duct System and Transfer Matrix Method

Considering only that the plane wave exists inside a duct system, the sound wave propagation along the X-direction would be governed by the classical acoustic wave equation as:

$$\frac{\partial^2 p}{\partial x^2} = \frac{1}{c_0^2} \frac{\partial^2 p}{\partial t^2} \tag{1}$$

where p is the acoustic pressure, $c_0 = 343$ m/s represents the sound speed in the air, and t is the time. Assuming that the wave is harmonic in time, the sound pressure and particle velocity could be solved as:

$$p(x,t) = p_I e^{i(\omega t - kx)} + p_R e^{i(\omega t + kx)} \tag{2}$$

$$u(x,t) = \frac{p_I}{\rho_0 c_0} e^{i(\omega t - kx)} - \frac{p_R}{\rho_0 c_0} e^{i(\omega t + kx)} \tag{3}$$

where $i = \sqrt{-1}$ is the imaginary unit, $\rho_0 = 1.204$ kg/m^3 represents the air density, p_I and p_R are complex pressure amplitudes indicating acoustic waves that propagate along with two opposite directions, ω is the angular frequency, and $k = \omega/c_0$ is the wave number.

The transfer matrix method (TMM) has been widely used to evaluate the noise attenuation performance of the mufflers. The transfer matrix of a circular duct of uniform cross-sectional area and length, e.g., from point C to point D in Figure 1d, is given by:

$$\begin{bmatrix} p(0,t) \\ \rho_0 c_0 u(0,t) \end{bmatrix} = \begin{bmatrix} T_{11} & T_{12} \\ T_{21} & T_{22} \end{bmatrix} \begin{bmatrix} p(L_{EC},t) \\ \rho_0 c_0 u(L_{EC},t) \end{bmatrix} = \mathbf{T_{CD}} \begin{bmatrix} p(L_{EC},t) \\ \rho_0 c_0 u(L_{EC},t) \end{bmatrix} \tag{4}$$

where L_{EC} is the length from point C to point D. Sound pressure $p(0,t)$ and $p(L_{EC},t)$ as well as volume velocity $u(0,t)$ and $u(L_{EC},t)$ can be solved via Equations (2) and (3):

$$
\begin{aligned}
p(0,t) &= [p_I + p_R]e^{i\omega t}\\
p(L_{EC},t) &= [p_I e^{-ikL_{EC}} + p_R e^{ikL_{EC}}]e^{i\omega t}\\
&= [(p_I + p_R)\cos kL_{EC} - i(p_I - p_R)\sin kL_{EC}]e^{i\omega t}
\end{aligned}
\tag{5}
$$

$$
\begin{aligned}
u(0,t) &= \tfrac{1}{\rho_0 c_0}[p_I - p_R]e^{i\omega t}\\
u(L_{EC},t) &= \tfrac{1}{\rho_0 c_0}[p_I e^{-ikL_{EC}} - p_R e^{ikL_{EC}}]e^{i\omega t}\\
&= \tfrac{1}{\rho_0 c_0}[(p_I - p_R)\cos kL_{EC} - i(p_I + p_R)\sin kL_{EC}]e^{i\omega t}
\end{aligned}
\tag{6}
$$

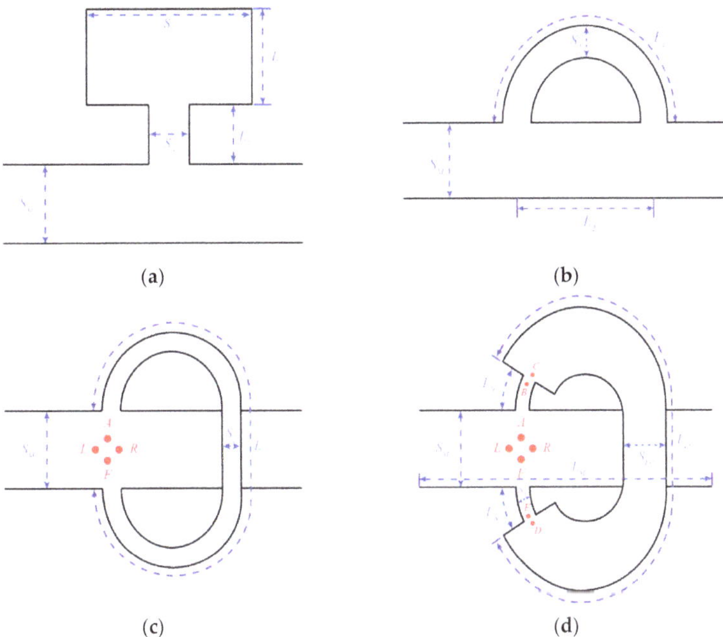

Figure 1. Schematic of the Helmholtz resonator, HQ tube, infinity tube, and infinity tube with an expansion chamber. (**a**) Helmholtz resonator (HR); (**b**) HQ tube; (**c**) infinity tube (IT); (**d**) infinity tube with an expansion chamber (ITEC).

By ignoring the time-harmonic terms, Equations (5) and (6) could be re-arranged to a matrix form:

$$
\begin{bmatrix} p(L_{EC}) \\ \rho_0 c_0 u(L_{EC}) \end{bmatrix} = \begin{bmatrix} \cos kL_{EC} & -i\sin kL_{EC} \\ -i\sin kL_{EC} & \cos kL_{EC} \end{bmatrix} \begin{bmatrix} p(0) \\ \rho_0 c_0 u(0) \end{bmatrix}
\tag{7}
$$

Equation (7) could be used to determine the sound pressure and particle velocity transmitted through length L inside a uniform cross-section duct. Then, the transfer matrix $\mathbf{T_{CD}}$ of Equation (4) could be obtained by inverting Equation (7):

$$
\begin{bmatrix} p(0) \\ \rho_0 c_0 u(0) \end{bmatrix} = \begin{bmatrix} \cos kL_{EC} & i\sin kL_{EC} \\ i\sin kL_{EC} & \cos kL_{EC} \end{bmatrix} \begin{bmatrix} p(L_{EC}) \\ \rho_0 c_0 u(L_{EC}) \end{bmatrix} = \mathbf{T_{CD}} \begin{bmatrix} p(L_{EC}) \\ \rho_0 c_0 u(L_{EC}) \end{bmatrix}
\tag{8}
$$

TMM could also solve the transfer matrix at conjunction points. According to the statement of pressure equality and conservation of volumetric flow, the transfer matrix from point B to point C in Figure 1b is given by:

$$\begin{bmatrix} p(B) \\ \rho_0 c_0 u(B) \end{bmatrix} = \begin{bmatrix} 1 & 0 \\ 0 & \frac{S_{EC}}{S_N} \end{bmatrix} \begin{bmatrix} p(C) \\ \rho_0 c_0 u(C) \end{bmatrix} = \mathbf{T_{BC}} \begin{bmatrix} p(C) \\ \rho_0 c_0 u(C) \end{bmatrix} \tag{9}$$

where S_{EC} is the area of duct CD, and S_N is the area of neck AB.

By calculating the product of the transfer matrix in each subsystem, the sound pressure and particle velocity between points A and F in Figure 1b are given by:

$$\begin{bmatrix} p(A) \\ \rho_0 c_0 u(A) \end{bmatrix} = \mathbf{T_T} \begin{bmatrix} p(F) \\ \rho_0 c_0 u(F) \end{bmatrix} = \begin{bmatrix} T_{T11} & T_{T12} \\ T_{T21} & T_{T22} \end{bmatrix} \begin{bmatrix} p(F) \\ \rho_0 c_0 u(F) \end{bmatrix}$$

$$= \mathbf{T_{AB}T_{BC}T_{CD}T_{DE}T_{EF}} \begin{bmatrix} p(F) \\ \rho_0 c_0 u(F) \end{bmatrix} \tag{10}$$

where $\mathbf{T_T}$ represents the transfer matrix for the side-branch tube of the ITEC, and $\mathbf{T_{AB}}$ to $\mathbf{T_{EF}}$ represents the transfer matrix of each cascaded subsystem. The transfer matrix from $\mathbf{T_{AB}}$ to $\mathbf{T_{EF}}$ could be easily obtained by referring to Equations (8) and (9).

2.2. Transfer Matrix of the ITEC

The sound transmission characteristics inside the side-branch tube are pre-requisites to acquiring the sound pressure and particle velocity of the whole duct system. This indicates that the transfer matrix between points L and R of the main duct is required. The continuous conditions of pressure equilibrium and conservation of volume velocity at the junction position yield:

$$\begin{cases} p(L) = p(R) = p(A) = p(F) \\ S_M u(L) + S_N u(F) = S_M u(R) + S_N u(A) \end{cases} \tag{11}$$

Re-arranging Equation (11), we can obtain:

$$u(L) = \frac{S_N}{S_M}(u(A) - u(F)) + u(R) \tag{12}$$

The relationship between $U(A)$ and $U(F)$ in Equation (12) could be derived from Equations (10) and (11):

$$\begin{cases} p(A) = p(F) \\ p(A) = T_{T11} p(F) + T_{T12} \rho_0 c_0 u(F) \\ \rho_0 c_0 u(A) = T_{T21} p(F) + T_{T22} \rho_0 c_0 u(F) \end{cases} \tag{13}$$

Equation (13) along with Equations (12) and (11) could be solved to determine the transfer matrix between points L and R:

$$\begin{bmatrix} p(L) \\ \rho_0 c_0 u(L) \end{bmatrix} = \mathbf{T_M} \begin{bmatrix} p(R) \\ \rho_0 c_0 u(R) \end{bmatrix} = \begin{bmatrix} T_{M11} & T_{M12} \\ T_{M21} & T_{M22} \end{bmatrix} \begin{bmatrix} p(R) \\ \rho_0 c_0 u(R) \end{bmatrix}$$

$$= \begin{bmatrix} 1 & 0 \\ \frac{S_N}{S_M} \frac{T_{T12}T_{T21}+(T_{T22}-1)(1-T_{T11})}{T_{T12}} & 1 \end{bmatrix} \begin{bmatrix} p(R) \\ \rho_0 c_0 u(R) \end{bmatrix} \tag{14}$$

Finally, the transmission loss of the whole duct system could be expressed as:

$$TL = 20 \log_{10} |\frac{p(L)}{p(R)}| = 20 \log_{10} |\frac{1}{2}(T_{M11} + T_{M12} + T_{M21} + T_{M22})| \tag{15}$$

3. Results and Discussion

3.1. Validation of Theoretical Prediction

The finite element method (FEM) simulations are performed by commercial software COMSOL Multiphysics to validate the accuracy of the transmission loss results of the analytical model. The parameters of the ITEC model in this validation case are listed in Table 1. The transmission loss results of the frequency domain between 1 and 1000 Hz are shown in Figure 2. It can be seen that the ITEC has three resonance frequencies in the selected frequency domain. The TMM results match the FEM results well, especially near the first transmission loss peak. The frequencies corresponding to transmission loss peaks are listed in Table 2. It could be seen that the transmission loss peaks have a maximum error of 10 Hz near the second peak and a minimum error of 4 Hz near the first peak. In general, the analytical model has relatively high accuracy.

Table 1. Parameters of the ITEC model for simulation.

	Length (mm)	Area (mm^2)
Main duct	$L_M = 1000$	$S_M = 5674.5$
Neck	$L_N = 95.91$	$S_N = 1418.6$
Expansion Chamber	$L_{EC} = 958.19$	$S_{EC} = 5674.5$

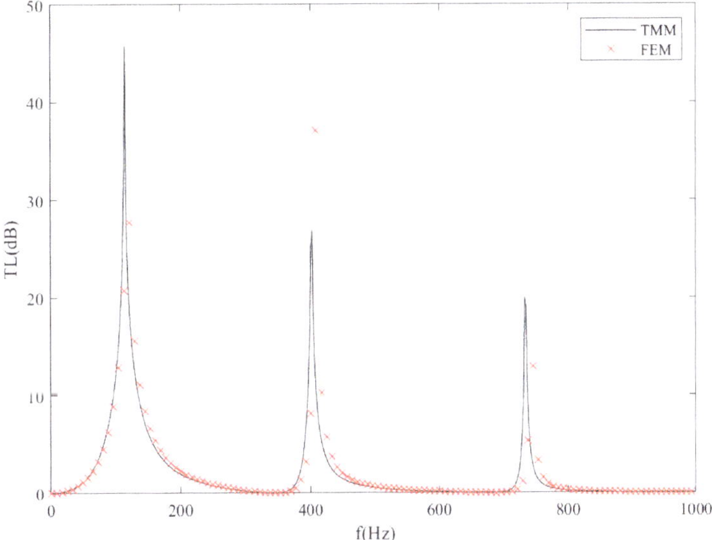

Figure 2. Comparison between the theoretical and numerical results of transmission loss spectra of the ITEC tube; the solid lines represent theoretical predictions by TMM, and the dotted crosses represent simulation results.

Table 2. The frequencies correspond to the transmission loss peak.

	TMM	FEM
1st peak	115 Hz	119 Hz
2nd peak	403 Hz	409 Hz
3rd peak	733 Hz	743 Hz

The analytical model could also be used to predict the transmission loss of the infinity tube. According to Lato et al. [19], for an infinity tube, as shown in Figure 1c, the transmission loss is:

$$TL = 20 \log_{10} \left| \frac{2(S_M - S_2 + S_M e^{ikL_2} + S_2 e^{ikL_2})}{1 + 2S_M e^{ikl_2}} \right| \tag{16}$$

where S_M represents the cross-section area of the main duct, S_2 denotes the cross-section area of IT, and L_2 represents the length of IT. In Figure 1b, if S_{EC} has the same value as S_N, the ITEC would become an infinity tube. Therefore, the analytical model of the ITEC could also be used to predict the transmission loss of IT. In this case, matrices $\mathbf{T_{BC}}$ and $\mathbf{T_{DE}}$ in Equation (10) turn into identity matrices, which indicates that Equation (10) becomes:

$$\mathbf{T_T} = \begin{bmatrix} T_{T11} & T_{T12} \\ T_{T21} & T_{T22} \end{bmatrix} = \mathbf{T_{AB}T_{CD}T_{EF}} \tag{17}$$

Using the expression of $\mathbf{T_T}$ in Equation (17), the transmission loss of the infinity tube could be deduced. In the following research, the analytical model based on Equation (15) is used to calculate the transmission loss of IT and then is compared with the results from Equation (16).

As shown in Figure 3, the solid lines represent the transmission loss of ITs with $L_2 = 1.15$ m and various S_2/S_M ratios. At the same time, ITs with the same geometries are used to examine the transmission loss by Equation (15). Figure 3 illustrates a good agreement between the analytical model from the second part and from the research conducted by Lato et al. [19]. The results indicate that Equation (15) could predict the IT transmission loss.

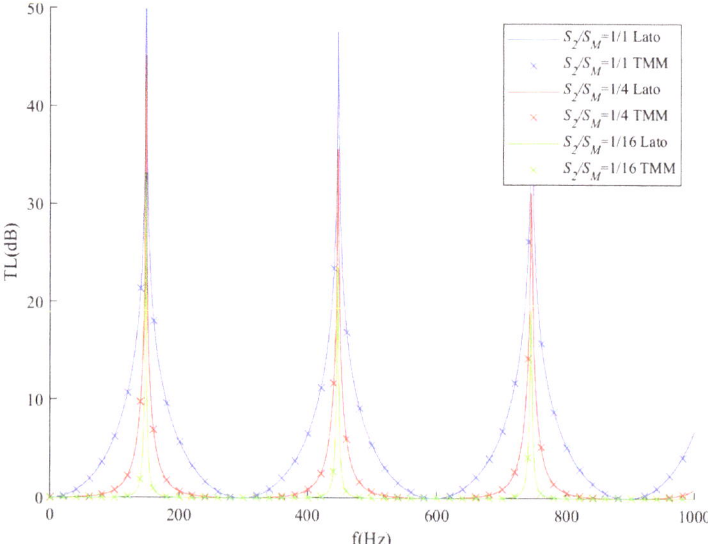

Figure 3. Comparison between the transmission loss of IT based on the analytical model from Section 2 and Equation (16); the solid lines are transmission loss results extracted from Lato et al. [19], and the dotted crosses are calculated by TMM from Section 2.

3.2. Noise Attenuation Ability of the ITEC

A comparison of the transmission loss between the IT and ITEC is carried out to examine the noise attenuation ability of the ITEC. The parameters of the ITEC are the same as the geometric model of Table 1, and the IT parameters are selected as the $S_2/S_M = 1/4$

case in Figure 3, which indicates that IT has the same cross-section area as the neck of the ITEC.

Figure 4 shows the analytical transmission loss between the IT and ITEC. The transmission loss peaks of the ITEC are 115, 403, and 733 Hz, while the transmission loss peaks of IT are 149, 447, and 745 Hz. The results show that an expansion chamber could lead to a decrease of 34, 46, and 12 Hz in resonance frequency. On the other hand, compared with IT, the noise attenuation bands of the ITEC under three transmission loss peaks are non-uniform. In the lower frequency (1–350 Hz), the attenuation band of the ITEC is significantly wider than IT. In medium frequency (350–650 Hz), they are approximately close to each other. In the higher frequency (650–1000 Hz), the attenuation band of the ITEC is narrower than IT. This feature indicates that the ITEC is more efficient in reducing low-frequency noise. In addition, since the ITEC has decreased resonance frequency, it has an advantage in low-frequency noise control compared with IT.

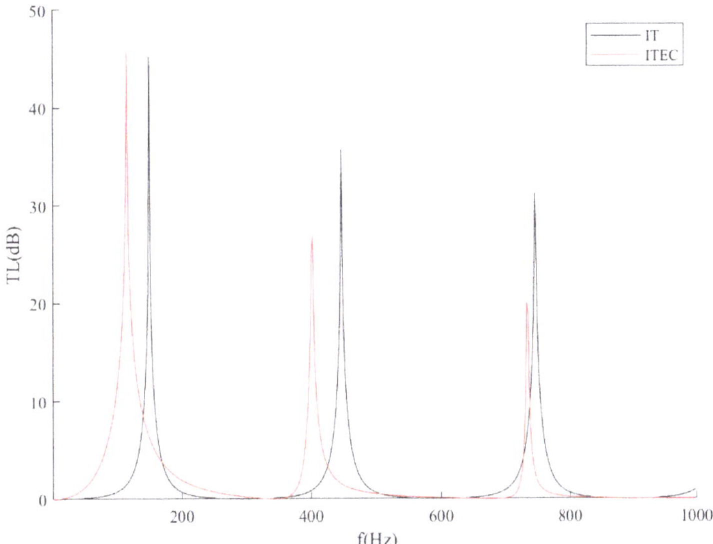

Figure 4. Comparison of transmission loss between the IT and ITEC; the black lines represent IT results, and red lines represent ITEC results.

The Helmholtz resonator is widely used as a muffler for ductwork systems in industry [8]. To further examine the noise attenuation ability and assess the potential in the industrial application of the ITEC, a comparison of the transmission loss between the ITEC and Helmholtz resonator system is conducted here. As illustrated in Figure 5a, if a sharable sidewall is placed at the midpoint of the ITEC, the whole system could be regarded as two curved Helmholtz resonators mounted on the same cross-section of the main duct. Cai and Mak [21] have examined the transmission loss of parallel HRs system, which is shaped as shown in Figure 5b. Figure 5a,b indicate that curved HRs are geometrically similar to the parallel HRs system. However, the ductwork system is always located in a limited space. A curved HRs system could save more space if it has the same cavity volume as a straight HRs system.

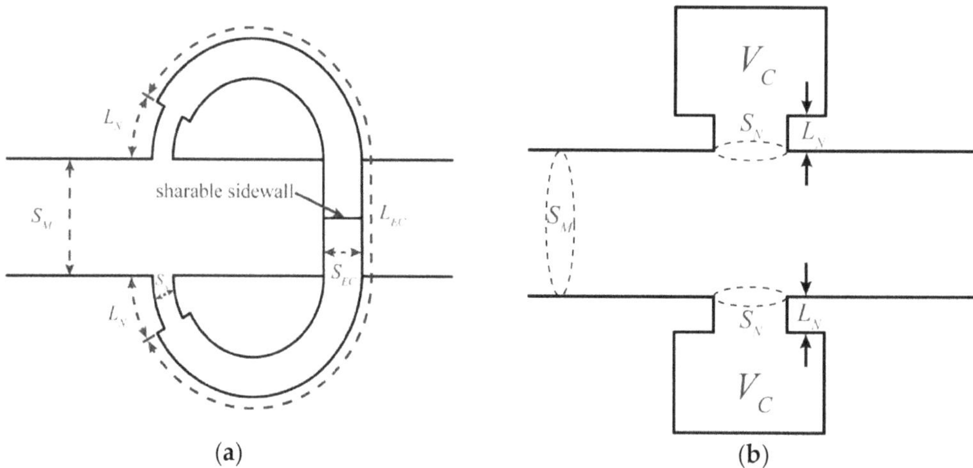

Figure 5. Configuration of curved HRs and parallel HRs systems. (**a**) Curved HRs; (**b**) parallel HRs.

In this study, the parameters of curved HRs are the same as the ITEC in Table 1. The neck parameters of parallel HRs are L_N = 95.91 mm and S_N = 1418.6 mm^2. The cavity volume of parallel HRs is the same as half of the chamber volume of curved HRs, which is easy to obtain from Table 1. As shown in Figure 6, the transmission loss of curved HRs is the same as the ITEC, with the same resonance frequencies and noise attenuation bandwidths. This indicates that the sharable sidewall has no impact on the noise attenuation mechanism of the ITEC. However, the transmission loss of the parallel HRs system is different. It has only two resonance frequencies from 1 to 800 Hz, while the ITEC and curved HRs have three. In addition, under the lower-frequency domain (1–350 Hz), the resonance frequency of parallel HRs is 133 Hz, which has an increase of 14 Hz compared with the ITEC and curved HRs; under the moderate frequency domain (350–650 Hz), the resonance frequency of parallel HRs is 473 Hz, which has an increase of 64 Hz compared with the ITEC and curved HRs. The characteristic of resonance frequencies shows that the ITEC and curved HRs are entirely different from the parallel HRs system, although they are geometrically similar. The ITEC has a lower resonance frequency than parallel HRs, which indicates that the ITEC is more suitable for reducing low-frequency noise. Furthermore, the curved shape of the ITEC and curved HRs system has an advantage in a constrained space. These characteristics indicate that the ITEC would have potential in ductwork systems.

3.3. Parametric Study of the ITEC

In this section, ITECs with different geometric parameters are analyzed to discuss the influence of geometrics on noise attenuation performance. Figure 7 shows the transmission loss results of ITECs with different length ratios. The total length of ITECs (L_{EC} + 2L_N) is fixed (1150 mm), while the neck and expansion chamber lengths have different values. The length values are shown in Table 3. Both FEM simulation and TMM analysis are conducted to validate the accuracy of transmission loss results. According to research in the previous parts, the ITEC would have three peaks of 1–350 Hz, 350–650 Hz, and 650–1000 Hz, respectively. Therefore, the frequency domains are divided into three sub-domains to distinguish 1st, 2nd, and 3rd TL peaks. Under the lower-frequency domain, the peaks of ITECs with different length ratios are close, while they have more significant differences under moderate and higher-frequency domains. We could summarize the following principles for ITECs with different length ratios:

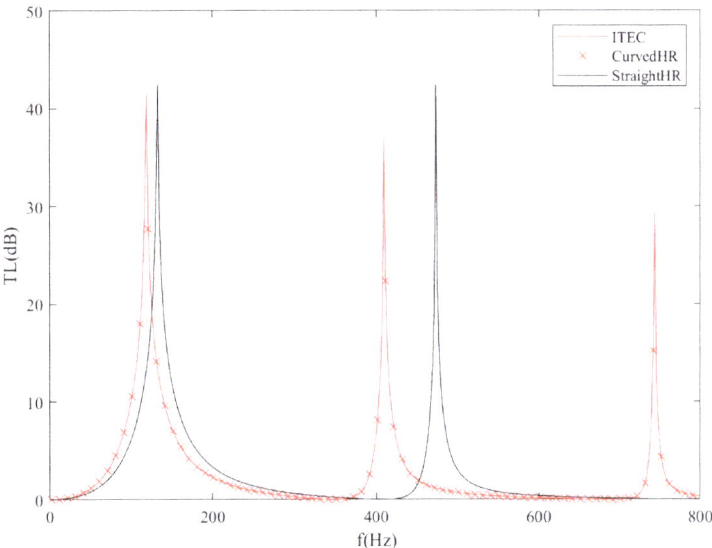

Figure 6. Comparison of the simulated transmission loss results between the ITEC and the curved HRs and parallel HRs systems; the red lines represent the ITEC results, black lines represent parallel HRs system results, and the red dotted crosses represent curved HRs system results.

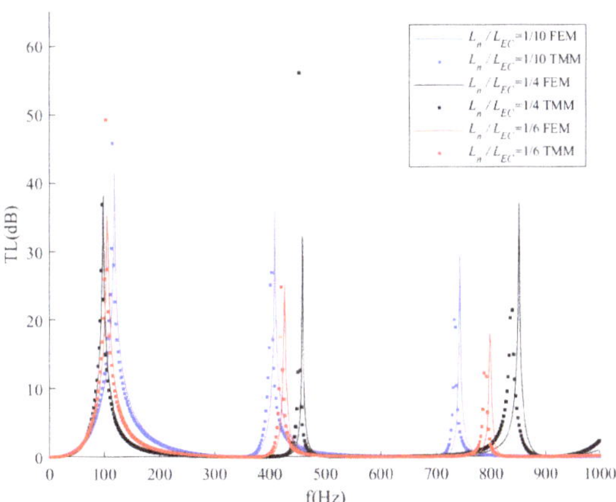

Figure 7. Comparison of transmission loss with respect to different length ratios of the expansion chamber.

Table 3. Parameters of L_N and L_{EC} of ITECs in Figure 7.

L_N/L_{EC}	L_N	L_{EC}
1/10	95.91 mm	958.19 mm
1/6	143.75 mm	862.5 mm
1/4	191.65 mm	766.67 mm

(1) Under the lower-frequency domain, length ratios have little influence on resonance frequency and attenuation bandwidth. ITECs with a higher length ratio would slightly decrease transmission loss peaks and have slightly narrower attenuation bands.

(2)   Under the moderate frequency domain, the length ratio significantly influences transmission loss performance. ITECs with higher length ratios have a higher resonance frequency and narrower attenuation bands. Compared with the low-frequency condition, ITECs have significantly narrower attenuation bands under the moderate frequency domain, which indicates that ITECs with higher length ratios are not suitable for medium-frequency noise attenuation.

(3)   Length ratio has a significant influence on transmission loss under the higher-frequency domain. With the length ratio changing from 1/10 to 1/4, the transmission loss peak has increased by nearly 100 Hz. In addition, ITECs with a length ratio equal to 1/4 have a significantly wider bandwidth than the other length ratios, which indicates that the increase in neck length of the ITECs would be good for high-frequency noise attenuation.

Furthermore, the influence of the cross-section area ratio is investigated. As listed in Table 4, three S_N/S_{EC} ratios correspond to three different expansion chamber radii and a fixed neck radius. The transmission loss results are shown in Figure 8. It could be obtained from Figure 8 that a higher cross-section area ratio would be better for low-frequency noise attenuation. The ITEC with $S_N/S_{EC} = 1/4$ has the lowest peak frequency and widest noise attenuation bandwidth under the lower-frequency domain. On the contrary, the lower cross-section area ratio would be better for high-frequency noise attenuation. The ITEC with $S_N/S_{EC} = 1/1.44$ has the highest peak frequency and widest noise attenuation band under the higher-frequency domain.

Table 4. Parameters of S_N and S_{EC} of the ITECs in Figure 8.

S_N/S_{EC}	R_N	R_{EC}
1/1.44	21.25 mm	25.5 mm
1/2.25	21.25 mm	31.875 mm
1/4	21.25 mm	42.5 mm

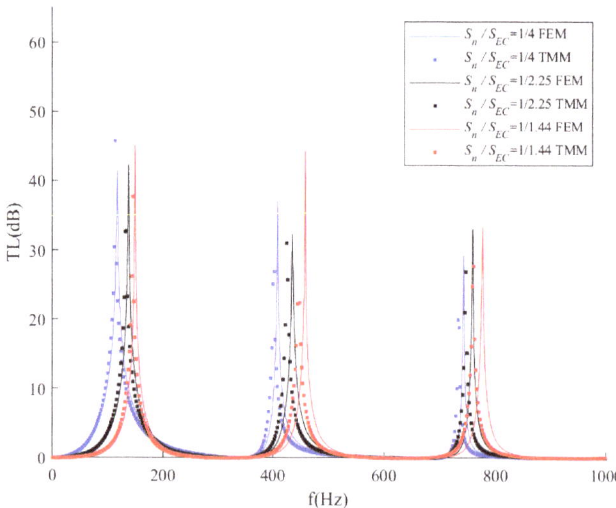

Figure 8. Comparison of transmission loss with respect to different cross-section area ratios of the expansion chamber.

In Figure 9, we perform the transmission loss results of the ITECs with different total lengths. The total length of the ITECs is changed from 0.6 and 0.75 times to the original length ($L = L_{EC} + 2L_n = 1.15$ m); L_{EC} and L_n are also scaled down simultaneously, whereas the radii of the neck and expansion chamber have remained unchanged. As shown in Figure 9, the shorter total length would have a broader sound attenuation bandwidth. At the same time, the transmission loss peak would shift to the higher-frequency domain, even exceeding 1000 Hz, the upper limit of this research. In addition, the TMM results of 0.6 L would lead to a more significant error than the FEM results, which is due to the fact that the neck length of 0.6 L is very short. According to Ingard [22], an end correction is non-negligible for the aperture neck to improve transmission loss accuracy. For this reason, TMM in this study is not suitable for the short L_n case. Both TMM and FEM results show that the ITECs with shorter lengths would have a broader noise attenuation band. Meanwhile, the transmission loss peaks tend to shift to the higher-frequency domain. Therefore, the ITECs with shorter lengths would have better noise attenuation performance but are not suitable for low-frequency noise reduction. The transmission loss peak (*TLmax*) and the resonance frequency (f_0) of ITECs with different geometric parameters are summarized in Table 5.

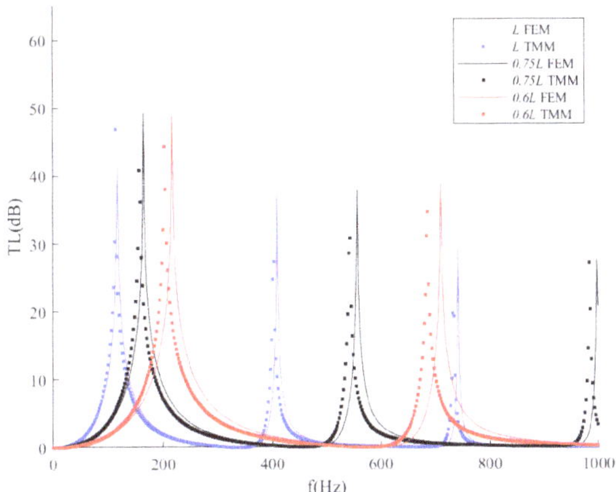

Figure 9. Comparison of transmission loss with respect to different lengths of the ITEC.

Table 5. Summary of the transmission loss peak and the resonance frequency of ITEC with different geometric parameters.

	Peak 1		Peak 2		Peak 3	
	TLmax (dB)	f_0 (Hz)	*TLmax* (dB)	f_0 (Hz)	*TLmax* (dB)	f_0 (Hz)
$L_n/L_{EC} = 1/10$	41.40	119.00	37.06	409.00	29.27	743.00
$L_n/L_{EC} = 1/6$	35.20	105.00	25.20	427.00	18.01	797.00
$L_n/L_{EC} = 1/4$	38.10	99.00	32.20	459.00	36.90	851.00
$S_n/S_{EC} = 1/4$	41.43	119.00	37.06	409.00	29.27	743.00
$S_n/S_{EC} = 1/2.25$	42.30	139.00	32.30	435.00	33.10	759.00
$S_n/S_{EC} = 1/1.44$	45.10	151.00	44.40	459.00	33.30	777.00
L	41.40	119.00	37.06	409.00	29.27	743.00
0.75 L	49.22	165.00	37.86	557.00	27.60	997.00
0.60 L	48.97	217.00	38.79	711.00	-	-

Figure 10 illustrates the influence of different geometric parameters of ITEC on the *TLmax*. It can be seen that changing the length ratio leads to a change of 19 dB in the *TLmax* in higher-frequency domain and a change of 11.8 dB in the *TLmax* in moderate frequency domain. Changing the cross-section area ratio has a change of 12.1 dB in higher-frequency domain. Changing the total length has a change of 7.8 dB in lower-frequency domain.

Therefore, adjusting the length ratio and the cross-section area ratio are beneficial for improving the higher- and medium-frequency noise attenuation ability, and adjusting the total length is useful for improving the lower-frequency noise attenuation ability. Figure 11 illustrates the influence of different geometric parameters of ITEC on f_0. Changing the total length has more of a significant impact on the resonance frequency than changing the length ratio and the cross-section area ratio. The f_0 under three peaks has an increase of 98, 302, and 254 Hz. This indicates that adjusting the total length is an effective way to control the frequency of noise reduction.

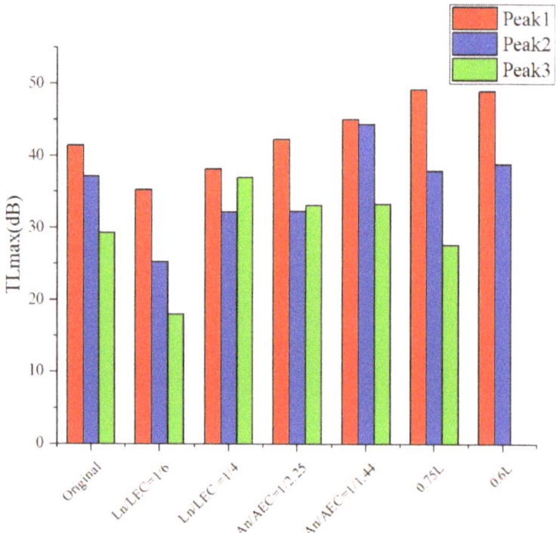

Figure 10. Variation of the transmission loss peak of the ITEC with different geometric parameters.

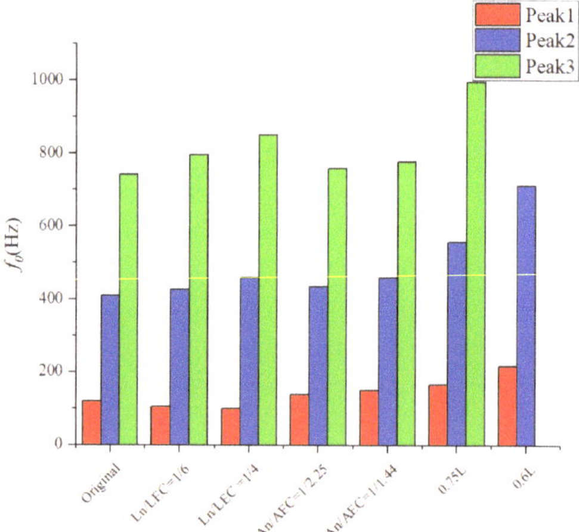

Figure 11. Variation of the resonance frequency of the ITEC with different geometric parameters.

4. Conclusions

This paper conducts a thorough theoretical and numerical investigation of an innovative noise attenuation device, the ITEC. The conclusions are summarized as follows:

A closed-form equation for the transmission loss of the ITEC device has been derived. The analytical model is compared with the FEM model to validate the accuracy of the transmission loss results. In addition, the analytical model is used to predict the transmission loss of the IT device. Transmission loss results indicate that IT could be regarded as a particular case of the ITEC.

The transmission loss results of the ITEC are compared with those of IT, which shows that the ITEC is more suitable for reducing low-frequency noise than IT devices. The transmission loss results of the ITEC are compared with those of curved HRs and parallel HRs systems. The results show that the ITEC has the same transmission loss results as those of curved HRs system, which indicates that the sharable sidewall does not affect the noise attenuation characteristics of the ITEC device. The ITEC has 14 and 64 Hz resonance frequency reduction than parallel HRs, which indicates that the ITEC is more suitable for reducing low-frequency noise. In addition, the geometry of the ITEC shows that it has an advantage in a constrained space, which indicates that the ITEC would have potential in ductwork systems.

A parametric study is conducted to investigate the influence of geometric parameters on the noise attenuation performance of the ITEC. Transmission loss results of the ITECs with different length ratios, cross-section area ratios, and total length are conducted by TMM and FEM. The transmission loss results could provide guidance on choosing the geometric parameters of the ITECs to reduce duct noise.

Author Contributions: R.X., C.C., C.M.M. and K.W.M. conceived this study; R.X. simulated and analyzed the data; C.M.M. contributed analysis tools; C.C. and C.M.M. provided advice for the preparation and revision of the paper; R.X. wrote the paper; C.M.M. and K.W.M. reviewed the manuscript for scientific contents. All authors have read and agreed to the published version of the manuscript.

Funding: The work described in this article was fully supported by grants from the Research Grants Council of the Hong Kong Special Administrative Region, China (Project No.15207820), the National Natural Science Foundation of China (Grant No. 51908554), and Hunan Provincial Natural Science Foundation of China (Project No. 2020JJ5712).

Institutional Review Board Statement: Not applicable.

Informed Consent Statement: Not applicable.

Data Availability Statement: Not applicable.

Conflicts of Interest: The authors declare no conflict of interest.

Nomenclature

The following abbreviations are used in this manuscript.

A	Cross-section area of an acoustical element
c_0	Sound speed of air
f_0	The resonance frequency of the Helmholtz resonator
k	Wave number
L_{EC}	Length of the expansion chamber
L_N	Neck length of the ITEC
p	Acoustic pressure
p_I	Sound pressure of incident plane wave
p_R	Sound pressure of reflected plane wave
S_{EC}	Area of the expansion chamber
S_N	Area of the ITEC neck
$\mathbf{T}$	Transfer matrix of an acoustical element
TL	Transmission Loss
u	Particle velocity
ρ_0	Air density
ω	Angular frequency

References

1. Munjal, M.L. *Acoustics of Ducts and Mufflers*; Wiley: New York, NY, USA, 2014.
2. Cai, C.; Mak, C.M. Generalized flow-generated noise prediction method for multiple elements in air ducts. *Appl. Acoust.* **2018**, *135*, 136–141. [CrossRef]
3. Mori, M.; Masumoto, T.; Ishihara, K. Study on acoustic, vibration and flow induced noise characteristics of T-shaped pipe with a square cross-section. *Appl. Acoust.* **2017**, *120*, 137–147. [CrossRef]
4. Perrey-Debain, E.; Maréchal, R.; Ville, J.M. Side-branch resonators modelling with Green's function methods. *J. Sound Vib.* **2014**, *333*, 4458–4472. [CrossRef]
5. Shi, X.; Mak, C.-M. Sound attenuation of a periodic array of micro-perforated tube mufflers. *Appl. Acoust.* **2017**, *115*, 15–22. [CrossRef]
6. Rayleigh, L. The Theory of the Helmholtz Resonator. *Proc. R. Soc. London. Ser. A Contain. Pap. A Math. Phys. Character* **1916**, *92*, 265–275.
7. Davis, D.D.; Stokes, G.M.; Moore, D.; Stevens, G.L.; Aeronautics, U.S. Theoretical and Experimental Investigation of Mufflers with Comments on Engine-Exhaust Muffler Design. Patent Application No. NACA-TR-1192, 1 January 1954.
8. Munjal, M.L. *Acoustics of Ducts and Mufflers with Application to Exhaust and Ventilation System Design*; John Wiley & Sons: New York, NY, USA, 1987.
9. Rossing, T.D. *Springer Handbook of Acoustics*; Springer: New York, NY, USA, 2014.
10. Herschel, J.F.W. On the absorption of light by coloured media, viewed in connexion with the undulatory theory. *Lond. Edinb. Dublin Philos. Mag. J. Sci.* **1833**, *3*, 401–412. [CrossRef]
11. Quincke, G. Ueber Interferenzapparate für Schallwellen (On an interfer-ence device for sound waves). *Ann. Der Phys.* **1866**, *128*, 177–192. [CrossRef]
12. Stewart, G.W. The Theory of the Herschel-Quincke Tube. *Phys. Rev.* **1928**, *31*, 696–698. [CrossRef]
13. Selamet, A.; Dickey, N.; Novak, J. The Herschel-Quincke tube: A theoretical, computational, and experimental investigation. *J. Acoust. Soc. Am.* **1994**, *96*, 3177–3185. [CrossRef]
14. Selamet, A.; Easwaran, V. Modified Herschel-Quincke tube: Attenuation and resonance for n-duct configuration. *J. Acoust. Soc. Am.* **1997**, *102*, 164–169. [CrossRef]
15. Kim, D.-Y.; Ih, J.-G.; Åbom, M. Virtual Herschel-Quincke tube using the multiple small resonators and acoustic metamaterials. *J. Sound Vib.* **2020**, *466*, 115045. [CrossRef]
16. Wang, Z.B.; Chiang, Y.K.; Choy, Y.S.; Wang, C.Q.; Xi, Q. Noise control for a dipole sound source using micro-perforated panel housing integrated with a Herschel–Quincke tube. *Appl. Acoust.* **2019**, *148*, 202–211. [CrossRef]
17. Ahmadian, H.; Najafi, G.; Ghobadian, B.; Reza Hassan-Beygi, S.; Bastiaans, R.J.M. Analytical and numerical modeling, sensitivity analysis, and multi-objective optimization of the acoustic performance of the herschel-quincke tube. *Appl. Acoust.* **2021**, *180*, 108096. [CrossRef]
18. Siqueira Mazzaro, R.; de Morais Hanriot, S.; Jorge Amorim, R.; Américo Almeida Magalhães Júnior, P. Numerical analysis of the air flow in internal combustion engine intake ducts using Herschel-Quincke tubes. *Appl. Acoust.* **2020**, *165*, 107310. [CrossRef]
19. Lato, T.; Mohany, A.; Hassan, M. A passive damping device for suppressing acoustic pressure pulsations: The infinity tube. *J. Acoust. Soc. Am.* **2019**, *146*, 4534–4544. [CrossRef] [PubMed]
20. Torregrosa, A.; Broatch, A.; Payri, R. A study of the influence of mean flow on the acoustic performance of Herschel-Quincke tubes. *J. Acoust. Soc. Am.* **2000**, *107*, 1874–1879. [CrossRef] [PubMed]
21. Cai, C.; Mak, C.M. Acoustic performance of different Helmholtz resonator array configurations. *Appl. Acoust.* **2018**, *130*, 204–209. [CrossRef]
22. Ingard, U. On the Theory and Design of Acoustic Resonators. *J. Acoust. Soc. Am.* **1953**, *25*, 1037–1061. [CrossRef]

Article

ResSKNet-SSDP: Effective and Light End-To-End Architecture for Speaker Recognition

Fei Deng [1], Lihong Deng [1,*], Peifan Jiang [1], Gexiang Zhang [2,3] and Qiang Yang [3]

[1] College of Computer Science and Cyber Security (Oxford Brookes College), Chengdu University of Technology, Chengdu 610059, China
[2] Artificial Intelligence Research Center, Chengdu University of Technology, Chengdu 610059, China
[3] School of Control Engineering, Chengdu University of Information Engineering, Chengdu 610059, China
* Correspondence: denglihongneo@stu.cdut.edu.cn

Abstract: In speaker recognition tasks, convolutional neural network (CNN)-based approaches have shown significant success. Modeling the long-term contexts and efficiently aggregating the information are two challenges in speaker recognition, and they have a critical impact on system performance. Previous research has addressed these issues by introducing deeper, wider, and more complex network architectures and aggregation methods. However, it is difficult to significantly improve the performance with these approaches because they also have trouble fully utilizing global information, channel information, and time-frequency information. To address the above issues, we propose a lighter and more efficient CNN-based end-to-end speaker recognition architecture, ResSKNet-SSDP. ResSKNet-SSDP consists of a residual selective kernel network (ResSKNet) and self-attentive standard deviation pooling (SSDP). ResSKNet can capture long-term contexts, neighboring information, and global information, thus extracting a more informative frame-level. SSDP can capture short- and long-term changes in frame-level features, aggregating the variable-length frame-level features into fixed-length, more distinctive utterance-level features. Extensive comparison experiments were performed on two popular public speaker recognition datasets, Voxceleb and CN-Celeb, with current state-of-the-art speaker recognition systems and achieved the lowest EER/DCF of 2.33%/0.2298, 2.44%/0.2559, 4.10%/0.3502, and 12.28%/0.5051. Compared with the lightest x-vector, our designed ResSKNet-SSDP has 3.1 M fewer parameters and 31.6 ms less inference time, but 35.1% better performance. The results show that ResSKNet-SSDP significantly outperforms the current state-of-the-art speaker recognition architectures on all test sets and is an end-to-end architecture with fewer parameters and higher efficiency for applications in realistic situations. The ablation experiments further show that our proposed approaches also provide significant improvements over previous methods.

Keywords: speaker recognition; end-to-end; selective kernel convolution; aggregation model

Citation: Deng, F.; Deng, L.; Jiang, P.; Zhang, G.; Yang, Q. ResSKNet-SSDP: Effective and Light End-To-End Architecture for Speaker Recognition. *Sensors* **2023**, *23*, 1203. https://doi.org/10.3390/s23031203

Academic Editors: Farook Sattar, Niladri Bihari Puhan and Reza Fazel-Rezai

Received: 21 December 2022
Revised: 10 January 2023
Accepted: 12 January 2023
Published: 20 January 2023

1. Introduction

Speaker recognition is intended to identify the speaker by obtaining identity information from the audio. With the widespread use of voice commands, speaker recognition has become a necessary measure to protect user security and privacy. However, the real-world recording environment can be too noisy for the audio to contain the speaker's identity information. Intrinsic factors such as age, emotion, and intonation may also have an impact. Therefore, speaker recognition remains a challenging task. The key to achieving effective speaker recognition is to extract the fixed-dimensional and discriminative features from the audio.

In the past few decades, the traditional i-vector system with probabilistic linear discriminant analysis (PLDA) [1,2] was the principal method of speaker recognition. However, with the development of deep learning, deep neural networks (DNNs) have brought substantial improvements to speaker recognition. Compared with the traditional i-vector system, the DNN architecture does not require manual feature extraction. Instead, it can directly process noisy datasets to extract frame-level features and then aggregate the variable-length frame-level features into fixed-dimensional utterance-level features for end-to-end training. The DNN-based end-to-end speaker recognition system [3,4] achieved a better performance than the i-vector system and occupies a dominant position with excellent feature extraction capabilities. In recent years, researchers have been attempting to build more effective speaker recognition architectures to obtain more discriminative features. Specifically, these attempts can be divided into two categories: (1) more efficient network architectures; and (2) better aggregation models.

Convolutional neural networks (CNNs) are the most popular feature extractors for speaker recognition tasks. CNN-based [5,6] feature extractors are widely used due to their strong feature extraction capabilities. These feature extractors can be classified into two classes: (1) one-dimensional convolution-based structures [7–12]; and (2) two-dimensional convolution-based structures [13–16]. The one-dimensional convolution generates two-dimensional outputs with time and channel dimensions. The time-delay neural network (TDNN) [7–9] is a typical one-dimensional convolution structure. The two-dimensional convolutional structure treats the input acoustic features as an image with three dimensions—time, frequency, and channel—and uses two-dimensional convolution to produce three-dimensional outputs. In the two-dimensional convolutional structures, the time and frequency dimensions decrease, and the channel dimensions increase with the downsampling operation. ResNet is a representative two-dimensional convolutional structure [5]. However, CNNs have their imperfections. Convolutional operations use a fixed-size convolutional kernel to capture the time and frequency information of the audio. The size of the convolution kernel limits the receptive fields. As a result, the feature extractor is also limited in its capabilities, which leads to its inability to capture essential global information and model long-term contexts [17,18]. Some researchers have used deeper, wider, and more complex network structures to solve these problems. Although these methods can expand the receptive fields, they lead to a significant increase in parameters and inference time.

Another challenge is the aggregation of frame-level features. In speaker recognition, the length of the input audio is variable, and an aggregation model is needed to aggregate the variable-length features into fixed-dimensional utterance-level features after the feature extractor has acquired the frame-level features [19,20]. The most common approach is to use global average pooling (GAP) directly to aggregate frame-level features into utterance-level features. However, global average pooling causes the frame-level features to lose time and frequency information, and the aggregated utterance-level features are not discriminative. In addition, audio sometimes changes or pauses while speaking, and the global average pooling cannot focus on these significant parts. To address this problem, researchers have proposed attention-based aggregation models to aggregate the frame-level features [21–23]. However, these methods also use global average pooling on the frame-level features for pre-processing. They also lose some of the information and do not take full advantage of the audio's time-frequency information.

On the other hand, with the widespread use of mobile devices [24–27], the design of speaker recognition systems tends to be light and efficient. However, existing models cannot be lighter, and the performance decreases drastically when made lighter. Therefore, we also need to design more light models for mobile devices.

To address these problems, we propose a more effective and lighter CNN-based end-to-end speaker recognition architecture, ResSKNet-SSDP. We propose a residual selective kernel network (ResSKNet), which can capture the neighborhood and global information of the features and can more efficiently model long-term contexts, resulting in more discriminative features. We propose self-attentive standard deviation pooling (SSDP). This avoids the pooling operation and preserves the time and frequency information, allowing a more accurate selection of information-rich frame-level features which contain more speaker identify information. Self-attentive standard deviation pooling can also capture the short- and long-term variations of the frame-level features, thus aggregating the variable-length frame-level features into fixed-length, robust, and more discriminative utterance-level features. To prove the effectiveness of our proposed method, we performed experiments with various settings in realistic environments. The main contributions of this study are as follows:

1. We propose ResSKNet. This is a more efficient and lighter network structure that can capture global information and effectively model long-term contexts;
2. We propose self-attentive standard deviation pooling (SSDP). This can capture short- and long-term changes in the frame-level features, aggregating the variable-length frame-level features into fixed-length, more distinctive utterance-level features;
3. We build a light and efficient end-to-end speaker recognition system. Extensive experiments conducted on two popular public datasets demonstrate the effectiveness of the proposed architectures.

In the rest of this paper, we primarily present the most popular and advanced network structures and aggregation methods and their disadvantages in Section 2. In Section 3, we present the details of the proposed method. In Section 4, we introduce the dataset used, training details, testing details, and testing methods. In Section 5, we discuss and analyze the experimental results. Finally, we summarize the work conducted and explain the limitations of the current work and future research directions in Section 6.

2. Related Works

This section introduces the current research related to network structures and aggregation models for speaker recognition systems. In the network structure, we present the most common and state-of-the-art methods based on one-dimensional convolution and two-dimensional convolution at present. In the aggregation models, we describe the statistical-based approach, the attention-based approach, and the dictionary-based approach. In addition, we summarize the advantages and disadvantages of these methods.

2.1. Network Structure

To obtain a larger receptive field and improve the system performance, researchers usually improve the network structure, such as Snyder et al., who proposed a x-vector speaker recognition system [7]. It extracts frame-level features by overlaying TDNN layers and then uses statistical pooling to aggregate the frame-level features into a fixed-dimensional speaker vector. The x-vector has become the most commonly used method for speaker recognition due to its excellent performance and light structure. The structural design of the x-vector also laid the foundation for most of the speaker recognition systems, such as E-TDNN [7], F-TDNN [8], DTDNN [18], and ECAPA-TDNN [12]. In addition, many works have used famous CNN structures, such as ResNet [5], ResNeXt, and Res2Net [16], in extracting the frame-level features. ResNet [5] introduced short connections to the neural network, thus alleviating the problem of gradient disappearance, and obtaining a deeper network structure, simultaneously. ResNet is also the most popular feature extractor in current speaker recognition systems. ResNeXt [16] adopts the repetitive layer strategy of VGG/ResNets while exploiting the split-transform-merge strategy in a simple and scalable way. A module in ResNeXt performs a set of transformations, each on a low-dimensional feature, whose output is aggregated by summation. ResNeXt reduces the parameters and inference time of the model, but it has yet to improve its modelling of long-term contexts. Res2Net [16] focuses on the revision of ResNet blocks, which build hierarchical residual block-like connections in a single residual block. It splits the features within one block into multiple channel groups and designs residual-like connections across the different channel groups. This residual-style connection increases the coverage of the receptive fields and yields many different combinations of the receptive fields for the improved modeling of long-term contexts. Res2Net has a smaller parameter, while achieving a better performance improvement. Although these methods achieved better performance improvement, they require more computation resources and are difficult to lighten the structure. The light structure makes it difficult to capture global information and model long-term contexts.

Therefore, we designed ResSKNet. It combines the advantages of both regular and dilation convolution and uses an attention mechanism to adjust the weights between the two convolutions according to the input features. Thus, it can capture global information and adaptively adjust the weights between the short-term and long-term contexts of the input audio to better model long-term contexts.

2.2. Aggregation Model

In speaker recognition, the length of the input acoustic features is variable. Therefore, a flexible processing method should have the ability to accept audio of any duration and obtain the fixed-dimensional features. In speaker recognition systems, the GAP aggregation model [10,20,24] is the most commonly used method for aggregating the frame-level features into fixed-dimensional utterance-level features. The reference [7] employs statistical pooling (SP) to aggregate the features, which computes the mean vector of the frame-level features and the standard deviation vector of the second-order statistics; then, it stitches them together as an utterance-level feature. However, our voices sometimes change, and there are brief pauses in the audio. Hence, researchers have proposed a self-attentive pooling (SAP) aggregation model [3,21,22] based on the attention mechanism to solve this problem and select frames that contain speaker information more effectively. Okabe et al. constructed the attention statistics pooling (ASP) aggregation model [23] based on the SP aggregation model by introducing an attention mechanism. It is calculated in the same way as the SP aggregation model, but has better performance than SP. They also introduced the NetVLAD aggregation model in computer vision that aggregates features into fixed dimensions via clustering [20,28]. NetVLAD assigns each frame-level feature to a different cluster center and encodes the residuals as output features. However, attention-based pooling methods are weakly robust and can exhibit a lower performance than GAP in different experimental settings. In addition, these methods usually use pooling operations to process features, resulting in a loss of information in the time and frequency dimensions and weakening the discriminative character of utterance-level features. Chung et al. conducted experiments using the NetVLAD method and its variants on the Voxceleb dataset and achieved the best results for speaker recognition [21]. It is a more robust and effective aggregation model, but the number of parameters increases as the cluster center increases. Therefore, we propose self-attentive standard deviation pooling. We avoid the pooling operation and retain the time and frequency information of the features, which allows self-attentive standard deviation pooling to select information-rich frames and capture the short- and long-term changes of the frame-level features more accurately. Thus, it aggregates more discriminative utterance-level features. In addition, it has stronger robustness and fewer parameters.

3. Materials and Methods

In this section, we introduce the proposed architecture, ResSKNet-SSDP, as shown in Figure 1. It consists of four parts: (1) feature extraction network. We use ResSKNet to extract frame-level features, which can acquire local and global information more efficiently to model both short-term and long-term contexts, as shown in the red box in Figure 1. The structure of ResSKNet is shown in Table 1; (2) feature aggregation. The aggregation model is a bridge between the frame-level features and the utterance-level features. We aggregate the variable-length frame-level features into the fixed-dimensional utterance-level features through self-attentive standard deviation pooling, as shown in the blue box in Figure 1; (3) speaker recognition loss function. AM-softmax loss is used during training to classify speakers more accurately [29]; (4) similarity metric. This is used to identify speakers, after the training, by calculating the distance of a pair of utterance-level features to determine whether the audio comes from the same speaker.

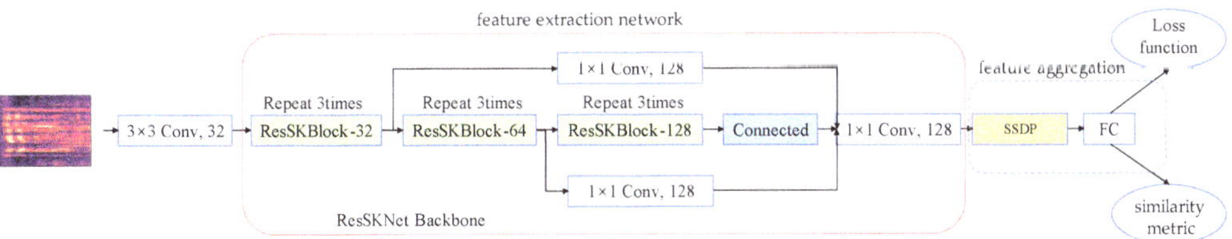

Figure 1. Overview of the ResSKNet-SSDP speaker recognition system.

Table 1. The structure of ResSKNet.

Layer	Blocks ($T \times 40 \times 1$)	Channels	Output Size ($T \times F \times C$)
Conv 1	Conv2d, 3×3, stride 1	32	$T \times 40 \times 16$
Stage 1	ResSK Block$\times$3, stride 2	32	$T \times 40 \times 32$
Stage 2	ResSK Block$\times$3, stride 2	64	$T/2 \times 20 \times 66$
Stage 3	ResSK Block$\times$3, stride 2	128	$T/4 \times 10 \times 128$

3.1. ResSKNet

Dilated convolution is a special convolutional operation that skips the input values by a certain step and uses filters in the region over its convolution kernel. Compared to standard convolution, dilated convolution can produce a larger receptive field by skipping and can more effectively model variable-length audio with long-term contexts. However, it is a challenge to use dilation convolution in deep convolutional neural networks. Performing lots of dilated convolutional layers in a deep structure will lead to a gridding problem that causes the complete loss of the local information and makes the information at a distance unrelated. In addition, it is problematic to determine in which layers to use dilation convolution for deep structures.

To solve the above problem, we designed a ResSKNet block, which consists of a residual selective kernel convolution (ResSKConv) module and 1×1 convolution with residual connections. We improved and introduced the selective kernel convolution (SKConv) module [18] into the end-to-end speaker recognition architecture, as shown in Figure 2. It performs standard and dilated convolutional layers on two parallel paths to capture the local and global information to better model short- and long-term contexts. Then, we use a self-attentive module to obtain the global information and adaptively adjust the weight between the short-term and long-term contexts.

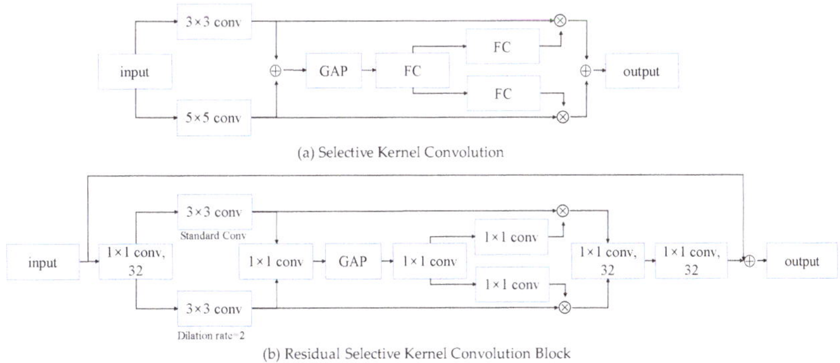

(a) Selective Kernel Convolution

(b) Residual Selective Kernel Convolution Block

Figure 2. Residual selective kernel convolution structure.

Using $x \in R^{T \times F \times C}$ to denote the input features, T, F, and C are the time dimensions, frequency dimensions, and channel dimensions, respectively. SKConv performs 3×3 convolutional F' and 3×3 dilated convolutional F'', respectively. The dilation rate is set to 2.

$$u = \delta(B(F'(x))) \tag{1}$$

$$u' = \delta(B(F''(x))) \tag{2}$$

In Equations (1) and (2), B and δ denote the batch normalization (BN) and ReLU activation functions, respectively. We do not use addition to fuse the features on the two parallel paths. Instead, we connect the features of the two branch outputs along the channel dimension and use a 1×1 convolution to fuse them, where $U \in R^{T \times F \times 2C}$ is the global feature obtained from the two paths. Compared to the addition method, we use convolution to fuse the features of two parallel paths along the channel dimension, thus increasing the depth and making full use of the features. Then, we generate a global channel feature $U_C \in R^{1 \times 1 \times C}$ using global average pooling and perform a feature transformation on it to generate channel attention $w_{CF} \in R^{1 \times 1 \times C}$ and $w_{CF'} \in R^{1 \times 1 \times C}$ on the two paths. Finally, after connecting the weighted elements of the features on both paths along the channel dimension, the features $y \in R^{T \times F \times C}$ are then output by using a 1×1 convolution.

$$U = conv_1(u, u') \tag{3}$$

$$U_C = GAP(U) \tag{4}$$

$$w_{CF} = \text{softmax}(conv_3(\delta(conv_2(U_C)))) \tag{5}$$

$$w_{CF'} = \text{softmax}(conv_4(\delta(conv_2(U_C)))) \tag{6}$$

$$y = conv_5(u \times w_{CF}, u' \times w_{CF'}) \tag{7}$$

where $conv_1 \in R^{1 \times 1 \times C}$, $conv_2 \in R^{1 \times 1 \times C/r}$, $conv_3 \in R^{1 \times 1 \times C}$, $conv_4 \in R^{1 \times 1 \times C}$, and $conv_5 \in R^{1 \times 1 \times C}$ are the convolution operations, r is the scale factor used to reduce the parameters and to obtain the dependencies between channels. Usually, we set r to 16. We do not use a fully connected layer for feature transformation because a fully connected layer is inefficient and useless. We use a 1×1 convolution, which captures the dependencies between channels more efficiently and thus produces more accurate channel attention [30].

Current speaker recognition systems only use the features of the last convolutional layer to obtain the frame-level features. Considering the hierarchical character of CNNs, these deeper features are the most complex and should be closely related to the speaker's identity. However, according to references [31,32], we believe that shallow features are also helpful in making utterance-level features more discriminative. Therefore, in our proposed system, we connect the output features of each stage of the ResSKNet block. We use 1×1 convolution to transform the output of each stage to the same size as the output of the last stage. The final architecture is shown in Figure 1.

3.2. Self-Attentive Standard Deviation Pooling

Attention-based aggregation models, such as SAP and ASP, can select frame-level features based on their importance. However, they can exhibit a similar or lower performance than GAP in different experimental settings. This indicates that these attention-based aggregation models are not accurate in selecting more informative frame-level features and have weak robustness. In addition, these methods usually use pooling operations to process features, resulting in the loss of time and frequency information and reducing the utterance-level feature differentiation. Therefore, we propose self-attentive standard deviation pooling. Self-attentive standard deviation pooling avoids pooling operations and preserves the time and frequency information of the frame-level features, allowing a more accurate selection of information-rich frames. At the same time, the self-attentive standard deviation pooling computation can capture the short- and long-term changes of frame-level features, aggregating variable-length frame-level features into fixed-length, more differentiated utterance-level features. It is also more robust than the previous attention-based aggregation models.

Use $x \in R^{T \times F \times C}$ to denote the frame-level features extracted by the convolutional neural network (T is the time dimension, F is the frequency dimension, and C is the channel dimension). The existing aggregation methods typically use an average pooling layer along the frequency axis of the features to generate a time feature descriptor matrix $X \in R^{T \times C}$ and $h = [x_1, x_2, \dots, x_T]$, where $x_t \in R^{1 \times C}$. However, this causes the features to lose information in the frequency dimension and limits the performance of the aggregation model. We transform x to obtain the time-frequency feature description matrix $H \in R^{N \times C}$ ($N = T \times F$) and $h = [h_1, h_2, \dots, h_N]$, where $h_t \in R^{1 \times C}$. We retain the time and frequency information of the features, and then use it to generate the importance weight $w_t \in R^{1 \times C}$ corresponding to each frame-level feature. Considering that the frame-level features extracted by the neural network already have great screening, the non-linear activation function is not applied to change the distribution of the features during aggregation, causing feature distortion. Therefore, we use a linear attention mechanism with a softmax function in the self-attentive standard deviation aggregation model to generate the weights, as shown in Equation (8). The $f_{SL}()$ denotes the linear attention mechanism and it can enhance the significant parts of the frame-level features and suppress the unessential parts without causing feature distortion. The non-linear activation functions change the information in the frame-level features and the feature distribution, resulting in feature distortion and the inaccurate selection of frame-level features. Thus, self-attentive standard deviation pooling retains more speaker information and can select frame-level features more accurately.

$$e_t = f_{SL}(h_t) = w^T h_t + b \tag{8}$$

$$w_t = \frac{exp(e_t)}{\sum_{i=1}^{T} exp(e_i)} \tag{9}$$

The mean vector $\mu \in R^{1 \times C}$ and the standard deviation vector $\sigma \in R^{1 \times C}$ are calculated as shown in Equations (10) and (11). However, statistical methods cannot identify which part of the audio contains the speaker identity information. Therefore, we use attention to calculate the mean vector and the standard deviation vector.

$$\mu = \frac{1}{T} \sum_{t}^{T} h_t \tag{10}$$

$$\sigma = \sqrt{\frac{1}{T} \sum_{t}^{T} h_t \odot h_t - \mu \odot \mu} \tag{11}$$

where h_t denotes the time-frequency feature descriptions, and $\odot$ denotes the Adama product.

In self-attentive standard deviation pooling, we obtain the mean vector by weighted summation, as shown in Equation (12): w_t is the importance weight, and $\alpha \in R^{1 \times C}$ is a learnable vector. The average vector generated by attention can be trained along with the neural network. It is learnable. In addition, it can suppress noise, thus filtering out part of the interference information to retain more effective time-frequency information. Finally, the features h_t and weights w_t are concatenated and reduced with the mean vector μ to obtain the self-attentive standard deviation vector $e \in R^C$, as shown in Equation (13). The self-attentive standard deviation vector is the utterance-level features obtained from the aggregating frame-level features. As the standard deviation contains other speaker features, in terms of time variability over long-term contexts, the self-attentive standard pooling model captures features' long-term variation.

$$\mu = \sum_{t=1}^{N} w_t \alpha \tag{12}$$

$$e = \frac{\sum_{t=1}^{N} w_t h_t - \mu}{N} \tag{13}$$

We did not use the same calculation as the statistical method because the frame-level features obtained by the convolutional neural network are first-order information, and there is an information mismatch between them and the higher-order information obtained by the statistical method [33]. In contrast, self-attentive standard deviation pooling uses first-order information for calculation throughout the process, and the output utterance-level features are also first-order information, avoiding this mismatch. At the same time, in the audio features, the difference in frequencies between different time-frequency locations is very large, and there is a coincidence in the direct addition of the results. The self-attentive standard deviation calculation eliminates this chance and makes the produced utterance-level features more discriminative. It goes without saying that self-attentive standard deviation pooling is also differentiable. Therefore, self-attentive standard deviation pooling can also be trained, along with the speaker recognition system, based on backpropagation.

4. Experimental Setups

4.1. Dataset

The experimental speaker datasets were adopted from the CN-Celeb [34] and Voxceleb [19] datasets, which have been commonly used for speaker recognition tasks in recent years.

There are 3000 speakers and 11 different genres in CN-Celeb, which include various real-world noises, cross-channel mismatches, and speaking styles in wild speech utterances. The training set has more than 600,000 utterances from 2800 speakers and 18,024 utterances from 200 speakers. There are 3,604,800 pairs in the test trials. Moreover, domain mismatch between the enrollment and test in the trials makes this dataset a very challenging one in speaker verification.

Voxceleb is a large text-independent speaker recognition dataset containing the Voxceleb1 and the Voxceleb2 dataset, and Voxceleb2 contains more than 1 million audio clips of 5994 speakers extracted from YouTube videos. The average time duration was 7.8 s, from different acoustic environments, making speaker identification more challenging. Voxceleb1 contains over 100,000 pieces of audio from 1251 speakers. There are three test sets: Voxceleb1-O, Voxceleb1-E, and Voxceleb1-H. Voxceleb1-O is a test set that includes 40 speakers independent of Voxceleb1 and does not overlap with the speakers in Voxceleb1. The Voxceleb1-E test set uses the entire Voxceleb1 dataset, while the Voxceleb1-H test set is specific. It contains samples from the same country of nationality and the same gender. The Voxceleb2 dataset is an extended version of the Voxceleb1 dataset, but the two datasets are mutually exclusive. As mentioned in reference [31], Voxceleb2 contains some flaws in its annotation. Therefore, it is not recommended for testing models. It is widely used for training. As with most existing references, we use Voxceleb2 for training and Voxceleb1 as the test set.

4.2. Training Details

We selected the 40-dimensional Filter Banks as the input features to the deep convolutional neural network without voice activity detection (VAD) and data augmentation [35]. The acoustic features have a frame length of 25 ms and a frame shift of 10 ms. Compared with the other acoustic features [36,37], Filter Banks are more in line with the nature of the sound signal and suitable for the reception characteristics of the human ear. We also do not use complex processing at the back-end, such as PLDA. During the training, we cut out a three-second clip from the audio.

The loss function of the system adopts AM-softmax [29] with margin = 0.1 and scale = 30. Compared with the softmax loss function, AM-softmax improves the verification accuracy by introducing a boundary in the angular space. It is calculated as follows:

$$L_i = -\log \frac{e^{s(\cos\theta_{yi}-m)}}{e^{s(\cos\theta_{yi}-m)} + \sum_{j \neq y_i} e^{s\cos(\theta_j)}} \tag{14}$$

where L_i is the cost of classifying the sample correctly, and $\theta_y = arccos\,(w^T x)$ refers to the angle between the sample feature and the decision hyperplane (w), with both vectors normalized by L2. Therefore, the angle is minimized by making $\cos(\theta_{yi})-m$ as large as possible, where m is the angle boundary. The hyperparameter s controls the "temperature" of the loss function, producing higher gradients for well-separated samples and further reducing the intra-class variance.

The Adam optimizer [38] with an initial learning rate of 0.001 was used to optimize the network parameters and attenuate 0.1 times every five cycles for training.

4.3. Testing and Testing Standards

During the test phase, we used the same settings as [19], extracted ten 3-s segments from each test audio as samples, and then sent them to the system to extract the utterance-level features of each segment and to calculate the distance between all of the combinations ($10 \times 10 = 100$) of pairs of segments. Then, the average of 100 distances denotes the score.

This study adopts the commonly used equal error rate (EER) and minimum detection cost function 2010 (DCF10) [39] as the evaluation indices to objectively evaluate the performance of different aggregation models. They indicate that the smaller the value, the better the performance. The calculation formula of the minimum detection cost function is:

$$DCF = C_{FR}F_{FR}P_{target} + C_{FA}F_{FA}(1 - P_{target}) \tag{15}$$

where C_{FR} and C_{FA} are the weights of false rejection rate F_{FR} and false acceptance rate F_{FA}, respectively, and P_{target} and $1-P_{target}$ are the prior probability of real speaking and impersonation tests. We use the parameters $C_{FR} = 1$, $C_{FA} = 1$, and $P_{target} = 0.01$ (DCF10) set by NIST SRE2010. DCF not only considers the different costs of false rejection and false reception but also considers the prior probability of the test. Therefore, MinDCF is more informative than EER in model performance evaluation.

5. Results

5.1. Ablation Experiments

5.1.1. Evaluating the Residuals Selective Kernel Convolution

To validate the effectiveness of the proposed ResSKNet-SSDP speaker recognition system, we first performed a series of ablation experiments. In addition, we counted their parameters and inference time. The inference time is the time required by the speaker recognition system to convert the audio into the utterance-level features. We both used the same experimental setups, simple training methods, and direct aggregation of frame-level features using global average pooling (GAP). Table 2 shows the test results of the regular convolution, SK conv, with our proposed ResSK conv on voxceleb1-O. As shown in Table 2, the EER/DCF of the regular convolution is 4.17%/0.3882. After using SK conv, the EER/DCF decreases to 3.42%/0.3522. It significantly improves the system performance, although the parameters and inference time increased slightly. It also shows that using SK conv improves the modelling of the long-term contexts. Testing again using our proposed ResSK conv as shown in Table 2, the EER/DCF further decreased to 2.85%/0.3126. This indicates that our ResSK conv obtained a larger perceptual field than the SK conv by using dilated convolution. In addition, we captured the dependencies between channels more effectively by using 1×1 convolution in the ResSK conv than the fully connected layer in the SK conv, which produces more accurate channel attention. These improvements allow

ResSK conv to capture local and global information more effectively and thus better model short- and long-term contexts. Compared to the regular conv, it increases the parameters by 0.3 M and inference time by 4.1 ms, but the performance is improved by 31.7%. Compared with SK conv, it increases the inference time by 1.7 ms with 0.1 M more parameters, but the performance is improved by 16.7%.

Table 2. Results of residual selective kernel convolution on Voxceleb1-O.

Model	Method	Aggregation	Loss	Dims	EER (%)	DCF	Params (M)	Time (ms)
ResSKNet	regular conv	GAP	AM-softmax	512	4.17	0.3882	2.9	27.8
ResSKNet	SK conv	GAP	AM-softmax	512	3.42	0.3522	3.1	30.2
ResSKNet	ResSK conv	GAP	AM-softmax	512	2.85	0.3126	3.2	31.9

5.1.2. Evaluating the Dilation Rate of the Residuals Selective Kernel Convolution

As shown in Table 3, when the dilation rate is 0, the EER/DCF is 3.29%/0.3417, and the perceptual field of the convolution at this time is consistent with that of the conventional convolution. When the dilation rate is 1, the EER/DCF decreases to 3.16%/0.3253. At this time, the ResSK conv has a larger receptive field. When the dilation rate is 2, the EER/DCF further decreases to 3.16%/0.3253, the receptive field also increases further and achieves the best performance. However, as the dilation rate increases, the performance of ResSK conv gradually decreases. The increase in the dilation rate leads to the loss of a lot of local information and makes the information discontinuous and irrelevant. In addition, we found that the inference time gradually increased with an increase in the dilation rate. This is because, as the dilation rate increases, the convolutional kernel of ResSK conv also gradually increases, and it also limits the bottleneck of the inference speed due to the GPU access memory bandwidth. Therefore, in this study, the best performance of ResSK conv is achieved when the dilation rate is 2, which can capture local and global information more effectively and thus better model the long-term contexts.

Table 3. Results of dilation rate on Voxceleb1-O.

Model	Dilation Rate	Aggregation	Loss	Dims	EER (%)	DCF	Params (M)	Time (ms)
ResSKNet	0	GAP	AM-softmax	512	3.29	0.3417	3.2	30.9
ResSKNet	1	GAP	AM-softmax	512	3.16	0.3253	3.2	31.3
ResSKNet	2	GAP	AM-softmax	512	2.85	0.3126	3.2	31.9
ResSKNet	3	GAP	AM-softmax	512	3.39	0.3614	3.2	32.4
ResSKNet	4	GAP	AM-softmax	512	3.68	0.3871	3.2	33.2

5.1.3. Evaluating the ResSKNet

Table 4 shows the results of ResSKNet with other networks on the Voxceleb1-O test, set under the same experimental conditions. As shown in Table 4, the worst performers were x-vector [7] and ResNet-34 [19], with EER/DCF of 4.39%/0.3726 and 4.47%/0.3909, respectively. However, compared to ResNet-34, the x-vector has the lightest structure and shorter inference time with fewer parameters. Due to its deeper and wider network structure, ResNet-50 [19] exhibits a significant performance improvement, with the EER/DCF falling to 3.89%/0.3710. However, both the inference time and its parameters increase dramatically. We then tested with ResNeXt [16] and found that ResNeXt achieved a similar performance to ResNet-50 with an EER/DCF of 3.85%/0.3822. Compared to ResNet-50, ResNeXt has 12.2 M fewer parameters and 55.2 ms less inference time. This indicates that the structure of ResNeXt can effectively reduce the parameters and inference time of the model, but it still cannot better model the long-term context. Testing Res2Net [16] again, we found that Res2Net achieves the lowest EER/DCF, of 3.32%/0.3442, compared to the previously tested methods. Its parameters and inference time are close to ResNeXt but still much higher than the x-vector structure. This suggests that Res2Net is a more efficient network structure that can better model long-term contexts, but is still not applicable in practice due to its parameters and inference time. Finally, we tested the proposed ResSKNet. ResSKNet further improves its performance compared to Res2Net, achieving the lowest EER/DCF of 2.85%/0.3126. This indicates that ResSKNet has better feature extraction capability and can more effectively model long-term contexts and our designed ResSKNet structure makes full use of both shallow and deep features, resulting in more informative frame-level features with more speaker identity information. Meanwhile, it has 2.1 M fewer parameters and 36.1 ms less inference time compared to the x-vector, but a 35.1% improvement in performance. This proves that the ResSKNet we constructed has a more efficient and light structure. To further visualize the effectiveness of the proposed architecture, we plotted detection error trade-off (DET) curves for all comparable models, as shown in Figure 3. We found that ResSKNet maintained a great performance advantage and was always below all of the curves.

Table 4. Results of networks on Voxceleb1-O.

Model	Aggregation	Loss	Dims	EER (%)	DCF	Params (M)	Time (ms)
x-vector [7]	GAP	AM-softmax	512	4.39	0.3726	5.3	63.5
ResNet-34 [19]	GAP	AM-softmax	512	4.47	0.3909	21.4	117.6
ResNet-50 [19]	GAP	AM-softmax	512	3.89	0.3701	36.5	237.6
ResNeXt [16]	GAP	AM-softmax	512	3.85	0.3822	24.3	182.4
Res2Net [16]	GAP	AM-softmax	512	3.32	0.3442	26.0	204.2
ResSKNet	GAP	AM-softmax	512	2.85	0.3126	3.2	31.9

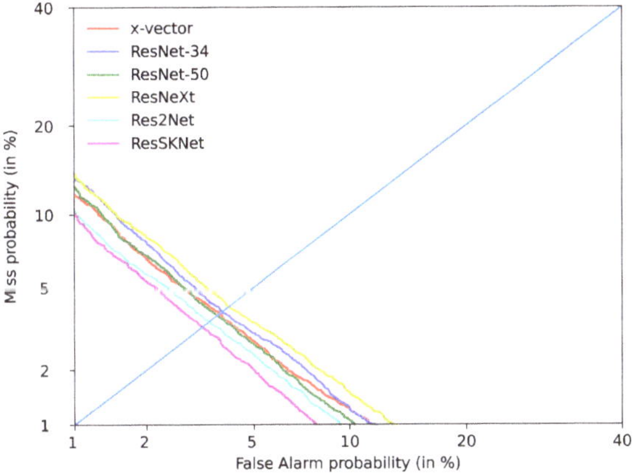

Figure 3. DET curve of different networks on Voxceleb1-O.

5.1.4. Evaluating the SSDP Aggregation Model

Table 5 shows the performance of our proposed self-attentive standard deviation pooling with other aggregation models on the Voxceleb1-O test set. Similarly, both the training and testing were performed under the same experimental conditions, and we did not introduce the attention method into the network. We directly use ResSKNet as the backbone network. As shown in Table 5, the performance of the baseline system can also be significantly improved using the aggregation model. The SP aggregation model [7] reduces the EER/DCF of the system to 2.64%/0.3883 by calculating the mean and standard deviation vector of the frame-level features. The SAP aggregation model [21] further improves its performance by focusing on more significant frame-level features, achieving a lower EER/DCF than the statistical approach of 2.62%/0.3599. The ASP aggregation model [23] combines attention and statistics to further reduce the EER/DCF to 2.57%/0.3563. However, it is not a significant improvement over the SAP aggregation model. This indicates that the ASP aggregation model does not effectively combine attentional and statistical methods. NetVLAD [28] still shows the best performance with an EER/DCF of 2.42%/0.3391. Finally, we tested the SSDP proposed in this paper. We found that SSDP achieves the lowest EER/DCF of 2.33%/0.2451, which is better than NetVLAD. This proves that the utterance-level features obtained from SSDP are more discriminative. It can more accurately select frame-level features and capture short- and long-term changes in the frame-level features, resulting in robust and more discriminative utterance-level features. Compared to ASP, it combines attention and statistics methods more effectively. We estimated the parameters and inference time of these aggregation models. The proposed SSDP has 0.2 M more parameters and 4.4 ms more inference time than the GAP, with the simplest structure and the least number of parameters, but the performance is improved by 18%. On the other hand, compared to NetVLAD, self-attentive standard deviation pooling has fewer parameters, a smaller inference time, and better performance. We similarly plotted the detection error trade-off (DET) curves for all of the comparable aggregation models, as shown in Figure 4. Our proposed SSDP also significantly outperforms the NetVLAD, which had previously achieved the best results, and is always below the NetVLAD curve.

Table 5. Results of aggregation models on Voxceleb1-O.

Model	Aggregation	Loss	Dims	EER (%)	DCF	Params (M)	Time (ms)
ResSKNet	GAP [19]	AM-softmax	512	2.85	0.3126	3.2	31.9
ResSKNet	SP [7]	AM-softmax	512	2.64	0.2883	3.2	33.1
ResSKNet	SAP [21]	AM-softmax	512	2.67	0.2699	3.4	35.3
ResSKNet	ASP [23]	AM-softmax	512	2.57	0.2563	3.6	37.6
ResSKNet	NetVLAD [28]	AM-softmax	512	2.42	0.2491	3.9	40.0
ResSKNet	SSDP	AM-softmax	512	2.33	0.2298	3.4	36.3

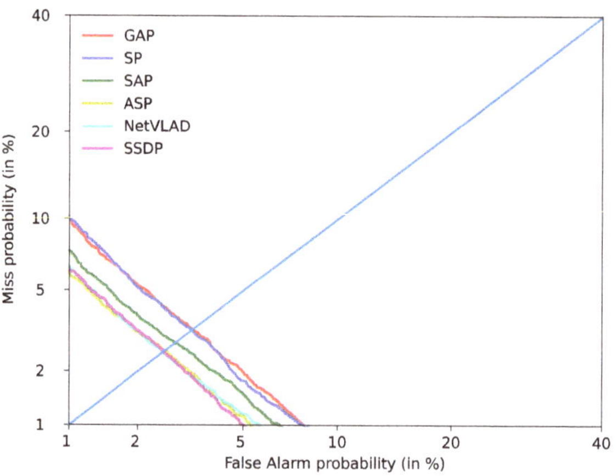

Figure 4. DET curve of aggregation models on Voxceleb1-O.

5.2. Results of the Speaker Recognition Systems on Voxceleb1-O

Next, our proposed ResSKNet-SSDP end-to-end speaker recognition system is compared and evaluated on the Voxceleb1-O test set with these current advanced speaker recognition systems, as shown in Table 6. As some of the methods are not open source, we directly applied the results from the paper. In the previous work, RawNet3 [40] and ECAPA-TDNN [12] were the most advanced methods in speaker recognition systems with

EER/DCF of 2.35%/0.2513 and 2.38%/0.2349. However, our ResSKNet-SSDP end-to-end speaker recognition system outperforms the previous best results, achieving the lowest EER/DCF of 2.33%/0.2298. This proves that the ResSKNet-SSDP end-to-end speaker recognition system has a greater feature extraction ability and is a lightweight speaker recognition system suitable for practical applications.

Table 6. Results of different systems on the Voxceleb1-O test set.

System	Aggregation	Loss	Dims	Voxceleb1-O	
				EER (%)	DCF
ThinResNet-34 [19]	NetVLAD	AM-softmax	512	3.32	0.3391
ThinResNet-34 [19]	GhostVLAD	softmax	512	2.85	-
ResNet-34 [21]	GAP	AM-softmax	512	4.83	0.3809
ResNet-50 [21]	GAP	AM-softmax	512	3.95	0.3471
RawNet2 [41]	SP	AAM-softmax	512	2.48	0.2337
RawNet3 [41]	SP	AAM-softmax	512	2.35	0.2513
CNN+Transformer [40]	GAP	softmax	512	2.56	-
PA-ResNet50 [42]	MRMHA	AAM-softmax	512	3.32	-
ECAPA-TDNN [12]	SP	AM-softmax	128	2.38	0.2349
x-vector [7]	AP	AM-softmax	1500	3.85	0.3515
Res2Net [16]	SAP	AM-softmax	512	2.65	0.2613
ResNeXt [16]	SAP	AM-softmax	512	2.79	0.2817
Y-vector [43]	SP	AM-softmax	512	2.78	0.2694
S-vector [44]	SP	softmax	512	2.76	0.3015
ResSKNet-SSDP	SSDP	AM-softmax	512	2.33	0.2298

To evaluate the performance of the ResSKNet-SSDP end-to-end speaker recognition system more comprehensively, we tested it again in the more extensive and challenging Voxceleb1-E and Voxceleb1-H test sets, as shown in Table 7. In Voxceleb1-E, which uses the entire Voxceleb1 as the test set, the proposed ResSKNet-SSDP end-to-end speaker recognition system still achieves the best results with an EER/DCF of 2.44%/0.2559. In the Voxceleb1-H test set, using the same country and gender, the differences in accent and intonation decreased, and they were more difficult to distinguish. As a result, the EER/DCF increased for all systems. However, our proposed ResSKNet-SSDP end-to-end speaker recognition system still holds the lead with an EER/DCF of 4.10%/0.3502 below other methods. This demonstrates that the ResSKNet-SSDP end-to-end speaker recognition system can obtain more distinguishable features and thus better distinguish between speakers with higher similarity. In the three test sets of Voxceleb, our designed ResSKNet-SSDP also maintains a great advantage, and the DET curve is always below the other systems' curves, as shown in Figure 5.

Table 7. Results of different systems on the Voxceleb1-E and Voxceleb1-H test sets.

System	Aggregation	Loss	Dims	Voxceleb1-E		Voxceleb1-H	
				EER (%)	DCF	EER (%)	DCF
ThinResNet-34 [19]	NetVLAD	AM-softmax	512	3.24	0.3471	5.57	0.4479
ResNet-34 [21]	GAP	AM-softmax	512	4.92	0.4218	7.97	0.5091
ResNet-50 [21]	GAP	AM-softmax	512	4.42	0.3822	6.71	0.4514
RawNet2 [41]	SP	AAM-softmax	512	2.87	0.2701	4.33	0.4091
x-vector [7]	AP	AM-softmax	1500	4.01	0.3868	7.33	0.4893
Res2Net [16]	SAP	AM-softmax	512	2.71	0.2653	4.42	0.3702
ResNeXt [16]	SAP	AM-softmax	512	2.74	0.2993	4.58	0.4036
Y-vector [43]	SP	AM-softmax	512	2.64	0.2701	4.33	0.3775
ResSKNet-SSDP	SSDP	AM-softmax	512	2.44	0.2559	4.10	0.3502

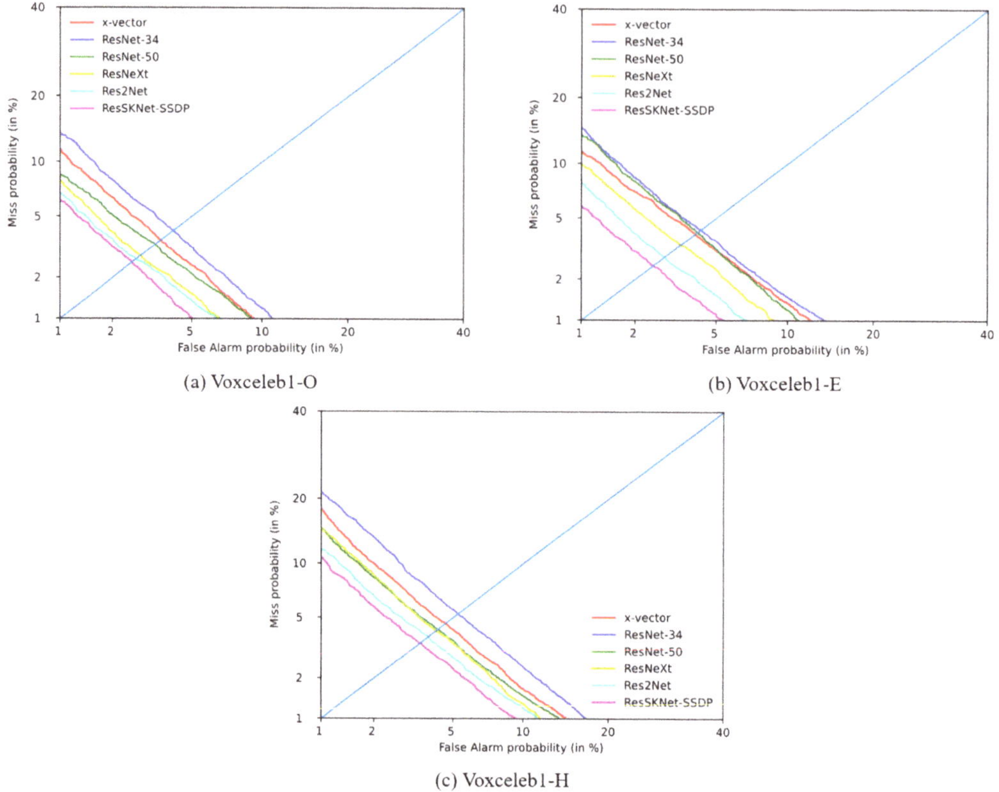

(a) Voxceleb1-O

(b) Voxceleb1-E

(c) Voxceleb1-H

Figure 5. DET curve of different systems on Voxceleb1 test set.

5.3. Results of the Speaker Recognition Systems on CN-Celeb

Table 8 shows the test results of these speaker recognition systems on CN-Celeb. It can be seen that the results on CN-Celeb are worse because the registered and tested discourse is shorter in this dataset. For example, the EER/DCF on CN-Celeb is substantially and obviously increased compared to the x-vector results for Voxceleb1-O. These results clearly show that state-of-the-art speaker recognition systems cannot inherently deal with the complexity introduced by multiple genres. In previous work, EDINet achieved the best identification results on CN-Celeb, with an EER of 12.8%. The ResSKNet-SSDP end-to-end speaker recognition system we have designed achieves slightly better performance than EDINet, with an EER/DCF of 12.81%/0.5051. In addition, we can clearly see that the DET curves of all of the systems have shifted significantly upwards compared to the Voxceleb dataset, but the curve of ResSKNet-SSDP is still lower than the other systems, as shown in Figure 6. The ResSKNet-SSDP end-to-end speaker recognition system achieved an excellent performance on all of the different test sets, proving that it is more efficient, lighter, and more suitable for practical application.

Table 8. Results of different systems on the CN-celeb test set.

Model	Aggregation	Loss	Dims	EER (%)	DCF
PA-ResNet-50 [42]	MRHMA	AM-softmax	256	13.21	0.5409
ResNet-34 [21]	SP	AM-softmax	512	16.52	0.6092
EDINet [45]	GAP	cosine loss	256	12.56	-
ResNet-50 [21]	SAP	AM-softmax	512	15.38	0.5691
ThinResNet-34 [19]	ASP	AM-softmax	512	18.37	0.6097
x-vector [7]	NetVLAD	AM-softmax	512	16.59	0.5864
Res2Net [16]	SAP	AM-softmax	512	13.72	0.5393
ResNeXt [16]	SAP	AM-softmax	512	14.55	0.5645
ResSKNet-SSDP	SSDP	AM-softmax	512	12.28	0.5051

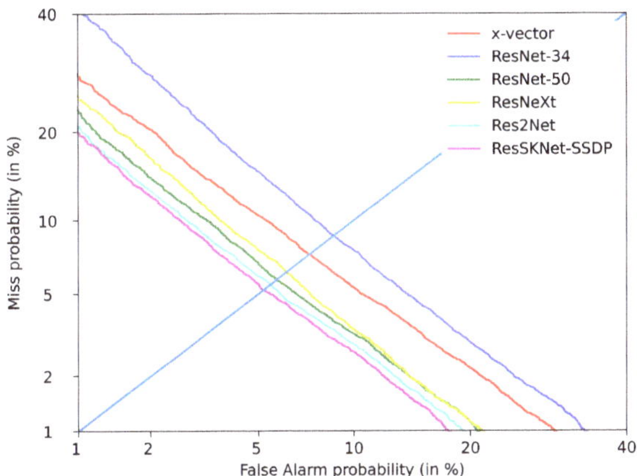

Figure 6. DET curve of systems on CN-celeb.

5.4. Visualization Analysis

We used the visualization method by Kye S M et al. to visualize the effectiveness of the proposed ResSKNet-SSDP end-to-end speaker identification system. We formed the visualization map after the dimensionality reduction in the speaker identity features by the t-SNE [46]. In Voxceleb1-H, fifty speakers were randomly selected to be represented by different colors. Each person randomly selected ten audios, and then ten randomly extracted three-second test segments from each audio. There were a total of 5000 three-second test segments obtained. The visualization maps of x-vector, Res2Net, ResSKNet, and ResSKNet-SSDP are shown in Figure 7.

(a) shows the speaker feature visualization graph of x-vector. It is observed that the feature extraction ability of x-vector is weak. The result is that the visualization graph is also very poorly classified, with many classification errors and collisions occurring. This suggests that the speaker features obtained by x-vector are not discriminative. (b) shows the visualization of Res2Net. It is noticeable that its visualization has been significantly improved compared to (a). It has fewer classification errors and collisions. It verifies the effectiveness and robustness of Res2Net. It also shows that Res2Net has a better feature extraction ability, and the speaker features are more discriminative. However, the intra-class distance of the speaker features is bigger, and the inter-class is smaller, and there are still some classification errors and collisions, as shown in (b) in Figure 7c shows the visualization map of ResSKNet. Comparing the visualization map of Res2Net, we can see that the collision in the visualization map of ResSKNet is significantly improved, and almost no speaker features collide. The inter-class distance increases significantly. It also proves that ResSKNet effectively improves the feature extraction ability by acquiring larger receptive fields and generating more combinations of receptive fields. However, there are still a few cases of classification errors and large intra-class distances in ResSKNet, as shown in (c) in Figure 7d represents the visualization map of ResSKNet-SSDP. It not only has no classification errors, but also has a larger inter-class distance and closer intra-class distance. This indicates that our proposed SSDP is efficient. It effectively improves the feature extraction ability of ResSKNet-SSDP to achieve more distinguishable features and it gives the speaker features from the same speaker a higher similarity.

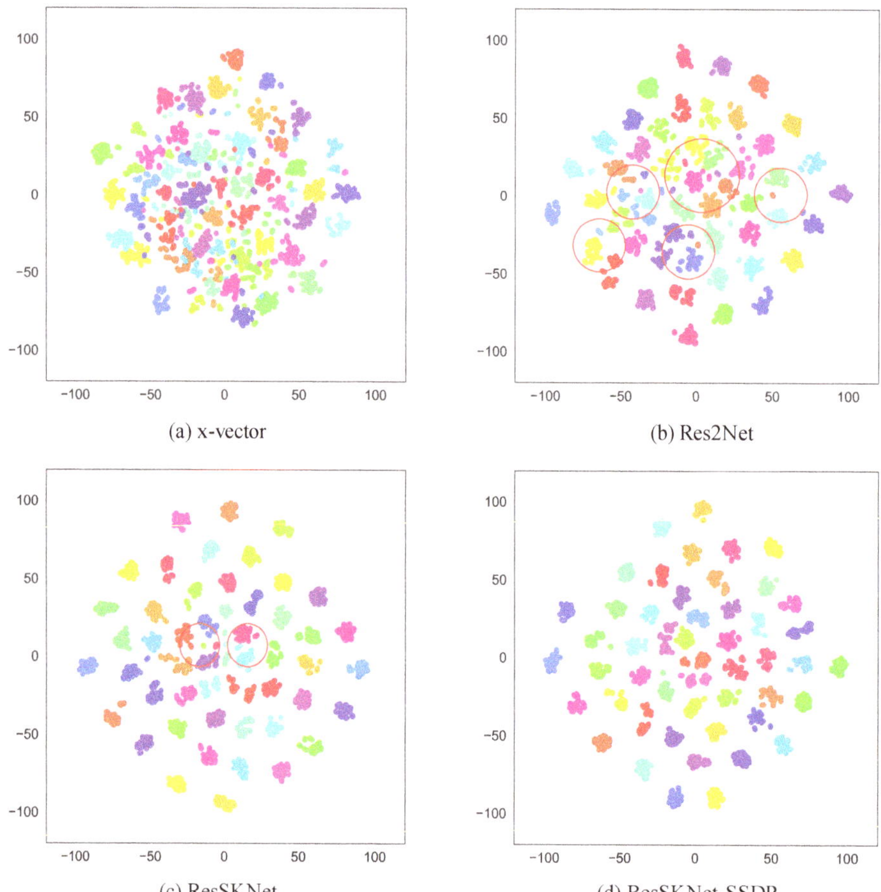

Figure 7. t-SNE visualization results.

6. Conclusions

In this work, we designed a speaker recognition architecture, ResSKNet-SSDP, with an improved feature extraction capability and improved adaptation to the speaker recognition task. The ResSKNet-SSDP network models the long-term contexts more effectively through the ResSKNet network structure. In addition, it also introduces the SSDP aggregation model to capture the short- and long-term changes of the frame-level features, aggregating variable-length frame-level features into fixed-length, more differentiated utterance-level features. We achieved the lowest EER/DCF of 2.33%/0.2298, 2.44%/0.2559, 4.10%/0.3502, and 12.28%/0.5051 on the noisy Voxceleb1 and CN-Celeb test sets, outperforming many of the existing methods and effectively improving the accuracy of the end-to-end speaker recognition system. Our proposed ResSKNet-SSDP end-to-end speaker recognition system has 3.1 M fewer parameters and 31.6 ms less inference time compared to the lightest x-vector, but 35.1% better performance, which indicates that it is a more efficient and lightweight structure. It is also more suitable for practical application.

The work we have conducted has improved the feature extraction ability of the speaker recognition system. When performing speaker recognition, we need to map the utterance-level features into the features space for similarity metrics. Therefore, the essence of speaker recognition is a metric learning problem and how to make the network more effective for metric learning is also a critical issue. The loss function is the key to solving this problem. In future work, we will focus on improving the loss function. In addition, we will also extend the proposed approach to other voice applications, such as language recognition and emotion recognition.

Author Contributions: F.D.: Conceptualization, Software, Writing—review & editing; L.D.: Conceptualization, Methodology, Software, Writing—original draft; P.J.: Writing—review & editing, Resources; G.Z.: Supervision, Writing—review and editing; Q.Y.: Data curation. All authors have read and agreed to the published version of the manuscript.

Funding: This work is supported by National Natural Science Foundation of China [grant number 61972324] and Sichuan Science and Technology Program [grant number 2021YFS0313 and 2021YFG0133].

Institutional Review Board Statement: Not applicable.

Informed Consent Statement: Not applicable.

Data Availability Statement: "Voxcceleb Data set" at https://www.robots.ox.ac.uk/~vgg/data/voxceleb accessed on 14 January 2023. "CN-Celeb Data set" at http://www.openslr.org/82/ accessed on 14 January 2023.

Conflicts of Interest: The authors declare no conflict of interest.

References

1. Ioffe, S. Probabilistic Linear Discriminant Analysis. In Proceedings of the European Conference on Computer Vision 2006, Graz, Austria, 7–13 May 2006; Leonardis, A., Bischof, H., Pinz, A., Eds.; Lecture Notes in Computer Science. Springer: Berlin/Heidelberg, Germany, 2006; Volume 3954. [CrossRef]
2. Dehak, N.; Kenny, P.J.; Dehak, R.; Dumouchel, P.; Ouellet, P. Front-End Factor Analysis for Speaker Verification. *IEEE Trans. Audio Speech Lang. Process.* **2011**, *19*, 788–798. [CrossRef]
3. Cai, W.; Chen, J.; Li, M. Exploring the Encoding Layer and Loss Function in End-to-End Speaker and Language Recognition System. In Proceedings of the Speaker and Language Recognition Workshop (Odyssey 2018), Les Sables d'Olonne, France, 26–29 June 2018; pp. 74–81. [CrossRef]
4. Variani, E.; Lei, X.; McDermott, E.; Moreno, I.L.; Gonzalez-Dominguez, J. Deep Neural Networks for Small Footprint Text-Dependent Speaker Verification. In Proceedings of the 2014 IEEE International Conference on Acoustics, Speech, and Signal Processing, Florence, Italy, 4–9 May 2014; pp. 4052–4056. [CrossRef]
5. He, K.; Zhang, X.; Ren, S.; Sun, J. Deep Residual Learning for Image Recognition. In Proceedings of the 2016 IEEE Conference on Computer Vision and Pattern Recognition (CVPR), Las Vegas, NV, USA, 27–30 June 2016; pp. 770–778. [CrossRef]
6. Rahman Chowdhury, F.R.; Wang, Q.; Moreno, I.L.; Wan, L. Attention-Based Models for Text-Dependent Speaker Verification. In Proceedings of the 2018 IEEE International Conference on Acoustics, Speech, and Signal Processing (ICASSP), Calgary, AB, Canada, 15–20 April 2018; pp. 5359–5363. [CrossRef]
7. Snyder, D.; Garcia-Romero, D.; Sell, G.; Povey, D.; Khudanpur, S. X-Vectors: Robust DNN Embeddings for Speaker Recognition. In Proceedings of the 2018 IEEE International Conference on Acoustics, Speech, and Signal Processing (ICASSP), Calgary, AB, Canada, 15–20 April 2018; pp. 5329–5333. [CrossRef]
8. Kanagasundaram, A.; Sridharan, S.; Sriram, G.; Prachi, S.; Fookes, C. A Study of X-Vector Based Speaker Recognition on Short Utterances. In Proceedings of the 20th Annual Conference of the International Speech Communication Association, Graz, Austria, 15–19 September 2019; pp. 2943–2947. [CrossRef]
9. Snyder, D.; Garcia-Romero, D.; Sell, G.; McCree, A.; Povey, D.; Khudanpur, S. Speaker Recognition for Multi-Speaker Conversations Using X-Vectors. In Proceedings of the 2019 IEEE International Conference on Acoustics, Speech, and Signal Processing (ICASSP), Brighton, UK, 12–17 May 2019; pp. 5796–5800.
10. Povey, D.; Cheng, G.; Wang, Y.; Li, K.; Xu, H.; Yarmohammadi, M.; Khudanpur, S. Semi-Orthogonal Low-Rank Matrix Factorization for Deep Neural Networks. In Proceedings of the Interspeech 2018, Hyderabad, India, 2–6 December 2018; pp. 3743–3747.
11. Yu, Y.-Q.; Li, W.-J. Densely Connected Time Delay Neural Network for Speaker Verification. In Proceedings of the Interspeech 2020, Shanghai, China, 25–29 October 2020; pp. 921–925.

12. Desplanques, B.; Thienpondt, J.; Demuynck, K. ECAPA-TDNN: Emphasized Channel Attention, Propagation and Aggregation in TDNN Based Speaker Verification. In Proceedings of the Interspeech 2020, Shanghai, China, 25–29 October 2020; pp. 3830–3834. [CrossRef]

13. Bian, T.; Chen, F.; Xu, L. Self-attention-based speaker recognition using cluster-range loss. *Neurocomputing* **2019**, *368*, 59–68. [CrossRef]

14. Heo, H.S.; Lee, B.J.; Huh, J.; Chung, J.S. Clova Baseline System for the Voxceleb Speaker Recognition Challenge 2020. *arXiv* **2020**, arXiv:2009.14153.

15. Yao, W.; Chen, S.; Cui, J.; Lou, Y. Multi-Stream Convolutional Neural Network with Frequency Selection for Robust Speaker Verification. *arXiv* **2020**, arXiv:2012.11159.

16. Zhou, T.; Zhao, Y.; Wu, J. ResNeXt and Res2Net Structures for Speaker Verification. In Proceedings of the 2021 IEEE Spoken Language Technology Workshop (SLT), Shenzhen, China, 19–22 January 2021; pp. 301–307. [CrossRef]

17. Hu, J.; Shen, L.; Albanie, S.; Sun, G.; Wu, E. Squeeze-and-Excitation Networks. *IEEE Trans. Pattern Anal. Mach. Intell.* **2020**, *42*, 2011–2023. [CrossRef] [PubMed]

18. Li, X.; Wang, W.; Hu, X.; Yang, J. Selective Kernel Networks. In Proceedings of the 2019 IEEE/CVF Conference on Computer Vision and Pattern Recognition (CVPR), Long Beach, CA, USA, 15–20 June 2019; pp. 510–519.

19. Nagrani, A.; Chung, J.S.; Xie, W.; Zisserman, A. Voxceleb: Large-scale speaker verification in the wild. *Comput. Speech Lang.* **2020**, *60*, 101027, ISSN 0885-2308. [CrossRef]

20. Wang, M.; Feng, D.; Su, T.; Chen, M. Attention-Based Temporal-Frequency Aggregation for Speaker Verification. *Sensors* **2022**, *22*, 2147. [CrossRef] [PubMed]

21. Chung, J.S.; Huh, J.; Mun, S. Delving into VoxCeleb: Environment Invariant Speaker Recognition. In Proceedings of the Odyssey 2020: The Speaker and Language Recognition Workshop, Tokyo, Japan, 2–5 November 2020; pp. 349–356. [CrossRef]

22. Kye, S.M.; Chung, J.S.; Kim, H. Supervised Attention for Speaker Recognition. In Proceedings of the 2021 IEEE Spoken Language Technology Workshop (SLT), Shenzhen, China, 19-22 January 2021; pp. 286–293. [CrossRef]

23. Okabe, K.; Koshinaka, T.; Shinoda, K. Attentive Statistics Pooling for Deep Speaker Embedding. In Proceedings of the Interspeech 2018, Hyderabad, India, 2–6 December 2018; pp. 2252–2256. [CrossRef]

24. Georges, M.; Huang, J.; Bocklet, T. Compact speaker embedding: Lrx-vector. In Proceedings of the Interspeech 2020, Shanghai, China, 25–29 October 2020; pp. 3236–3240. [CrossRef]

25. Razmjouei, P.; Kavousi-Fard, A.; Dabbaghjamanesh, M.; Jin, T.; Su, W. Ultra-Lightweight Mutual Authentication in the Vehicle Based on Smart Contract Blockchain: Case of MITM Attack. *IEEE Sens. J.* **2021**, *21*, 15839–15848. [CrossRef]

26. Sahba, A.; Sahba, R.; Rad, P.; Jamshidi, M. Optimized IoT Based Decision Making For Autonomous Vehicles in Intersections. In Proceedings of the 2019 IEEE 10th Annual Ubiquitous Computing, Electronics & Mobile Communication Conference (UEMCON), New York, NY, USA, 10–12 October 2019; pp. 0203–0206.

27. Kavousi-Fard, A.; Nikkhah, S.; Pourbehzadi, M.; Dabbaghjamanesh, M.; Farughian, A. IoT-based data-driven fault allocation in microgrids using advanced μPMUs. *Ad Hoc Netw.* **2021**, *119*, 102520. [CrossRef]

28. Arandjelović, R.; Gronat, P.; Torii, A.; Pajdla, T.; Sivic, J. NetVLAD: CNN Architecture for Weakly Supervised Place Recognition. *IEEE Trans. Pattern Anal. Mach. Intell.* **2017**, *40*, 1437–1451. [CrossRef] [PubMed]

29. Wang, F.; Cheng, J.; Liu, W.; Liu, H. Additive Margin Softmax for Face Verification. *IEEE Signal Process. Lett.* **2018**, *25*, 926–930. [CrossRef]

30. Wang, Q.; Wu, B.; Zhu, P.; Li, P.; Zuo, W.; Hu, Q. ECA-Net: Efficient Channel Attention for Deep Convolutional Neural Networks. In Proceedings of the 2020 IEEE/CVF Conference on Computer Vision and Pattern Recognition (CVPR), Seattle, WA, USA, 13–19 June 2020; pp. 11531–11539. [CrossRef]

31. Lee, J.; Nam, J. Multi-Level and Multi-Scale Feature Aggregation Using Sample-Level Deep Convolutional Neural Networks for Music Classification. *arXiv* **2017**, arXiv:1706.06810.

32. Gao, Z.; Song, Y.; McLoughlin, I.; Li, P.; Jiang, Y.; Dai, L.-R. Improving Aggregation and Loss Function for Better Embedding Learning in End-to-End Speaker Verification System. In Proceedings of the Interspeech, 2019, Graz, Austria, 15–19 September; 2019; pp. 361–365.

33. Miao, X.; McLoughlin, I.; Wang, W.; Zhang, P. D-MONA: A dilated mixed-order non-local attention network for speaker and language recognition. *Neural Netw.* **2021**, *139*, 201–211. [CrossRef]

34. Li, L.; Liu, R.; Kang, J.; Fan, Y.; Cui, H.; Cai, Y.; Vipperla, R.; Zheng, T.F.; Wang, D. CN-Celeb: Multi-genre speaker recognition. *Speech Commun.* **2022**, *137*, 77–91. [CrossRef]

35. Park, D.S.; Chan, W.; Zhang, Y.; Chiu, C.-C.; Zoph, B.; Cubuk, E.D.; Le, Q.V. Specaugment: A simple data augmentation method for automatic speech recognition. In Proceedings of the Interspeech 2019, Graz, Austria, 15–19 September 2019.

36. Sitaula, C.; He, J.; Priyadarshi, A.; Tracy, M.; Kavehei, O.; Hinder, M.; Withana, A.; McEwan, A.; Marzbanrad, F. Neonatal Bowel Sound Detection Using Convolutional Neural Network and Laplace Hidden Semi-Markov Model. *IEEE ACM Trans. Audio Speech Lang. Process.* **2022**, *30*, 1853–1864. [CrossRef]

37. Burne, L.; Sitaula, C.; Priyadarshi, A.; Tracy, M.; Kavehei, O.; Hinder, M.; Withana, A.; McEwan, A.; Marzbanrad, F. Ensemble Approach on Deep and Handcrafted Features for Neonatal Bowel Sound Detection. *IEEE J. Biomed. Health Inform.* **2022**. [CrossRef] [PubMed]

Sensors **2023**, *23*, 1203

38. Kingma, D.; Ba, J. Adam: A Method for Stochastic Optimization. In Proceedings of the 2nd International Conference on Learning Representations, ICLR 2014, Banff, AB, Canada, 14–16 April 2014.
39. Sadjadi, S.O.; Kheyrkhah, T.; Tong, A.; Greenberg, C.; Reynolds, D.; Singer, E.; Mason, L.; Hernandez-Cordero, J. The 2016 Nist Speaker Recognition Evaluation. In Proceedings of the Interspeech 2017: Conference of the International Speech Communication Association, Stockholm, Sweden, 20–24 August 2017; pp. 1353–1357. [CrossRef]
40. Wang, R.; Ao, J.; Zhou, L.; Liu, S.; Wei, Z.; Ko, T.; Li, Q.; Zhang, Y. Multi-View Self-Attention Based Transformer for Speaker Recognition. *arXiv* **2021**, arXiv:2110.05036.
41. Jung, J.W.; Kim, Y.J.; Heo, H.S.; Lee, B.J.; Kwon, Y.; Chung, J.S. Pushing the Limits of Raw Waveform Speaker Recognition. *arXiv* **2022**, arXiv:2203.08488v2.
42. Wei, Y.; Du, J.; Liu, H.; Wang, Q. CTFALite: Lightweight Channel-specific Temporal and Frequency Attention Mechanism for Enhancing the Speaker Embedding Extractor. In Proceedings of the Interspeech 2022, Incheon, Korea, 18–22 September 2022; pp. 341–345. [CrossRef]
43. Zhu, G.; Jiang, F.; Duan, Z. Y-Vector: Multiscale Waveform Encoder for Speaker Embedding. In Proceedings of the Interspeech 2021, Brno, Czech Republic, 30 August–3 September 2021; pp. 96–100. [CrossRef]
44. Mary, N.J.M.S.; Umesh, S.; Katta, S.V. S-Vectors and TESA: Speaker Embeddings and a Speaker Authenticator Based on Transformer Encoder. *IEEE ACM Trans. Audio Speech Lang. Process.* **2020**, *30*, 404–413. [CrossRef]
45. Li, J.; Liu, W.; Lee, T. EDITnet: A Lightweight Network for Unsupervised Domain Adaptation in Speaker Verification. In Proceedings of the Interspeech 2022, Incheon, Korea, 18–22 September 2022; pp. 3694–3698. [CrossRef]
46. Van der Maaten, L.; Hinton, G. Visualizing Data using t-SNE. *J. Mach. Learn. Res.* **2008**, *9*, 2579–2605.

Article

Broadband Air-Coupled Ultrasound Emitter and Receiver Enable Simultaneous Measurement of Thickness and Speed of Sound in Solids

Klaas Bente [1,2], Janez Rus [3,4], Hubert Mooshofer [5], Mate Gaal [1,*] and Christian Ulrich Grosse [3]

[1] Bundesanstalt für Materialforschung und -prüfung BAM, 8.4 Acoustic and Electromagnetic Methods, Unter den Eichen 87, 12205 Berlin, Germany
[2] Health and Medical University, Olympischer Weg 1, 14471 Potsdam, Germany
[3] Chair of Non-Destructive Testing, Centre for Building Materials, Technical University of Munich, Franz-Langinger-Straße 10, 81245 Munich, Germany
[4] Laboratory of Wave Engineering, Swiss Federal Institute of Technology Lausanne, Station 11, 1015 Lausanne, Switzerland
[5] Siemens AG, Corporate Technology, Otto-Hahn-Ring 6, 81739 München, Germany
* Correspondence: mate.gaal@bam.de

Abstract: Air-coupled ultrasound sensors have advantages over contact ultrasound sensors when a sample should not become contaminated or influenced by the couplant or the measurement has to be a fast and automated inline process. Thereby, air-coupled transducers must emit high-energy pulses due to the low air-to-solid power transmission ratios (10^{-3} to 10^{-8}). Currently used resonant transducers trade bandwidth—a prerequisite for material parameter analysis—against pulse energy. Here we show that a combination of a non-resonant ultrasound emitter and a non-resonant detector enables the generation and detection of pulses that are both high in amplitude (130 dB) and bandwidth (2 µs pulse width). We further show an initial application: the detection of reflections inside of a carbon fiber reinforced plastic plate with thicknesses between 1.7 mm and 10 mm. As the sensors work contact-free, the time of flight and the period of the in-plate reflections are independent parameters. Hence, a variation of ultrasound velocity is distinguishable from a variation of plate thickness and both properties are determined simultaneously. The sensor combination is likely to find numerous industrial applications necessitating high automation capacity and opens possibilities for air-coupled, single-side ultrasonic inspection.

Keywords: thermoacoustic emitter; optical microphone; air-coupled ultrasound; local resonance; thickness measurement; thickness resonance

Citation: Bente, K.; Rus, J.; Mooshofer, H.; Gaal, M.; Grosse, C.U. Broadband Air-Coupled Ultrasound Emitter and Receiver Enable Simultaneous Measurement of Thickness and Speed of Sound in Solids. *Sensors* **2023**, *23*, 1379. https://doi.org/10.3390/s23031379

Academic Editor: Farook Sattar

Received: 23 December 2022
Revised: 19 January 2023
Accepted: 22 January 2023
Published: 26 January 2023

1. Introduction

Most currently used air-coupled ultrasound (ACU) transducers can be classified as either piezoelectric or capacitive in nature [1–5]. However, developments in recent years have resulted in fundamentally new approaches to generate and detect ultrasound in air [6]. Thermoacoustic emitters were proven to provide high amplitude and high-bandwidth acoustic signals with frequencies of up to 1 MHz [7]. At the same time, highly sensitive optical microphones were developed that cover the same frequency range [8–11]. This new generation of emitters and receivers can overcome a former fundamental restriction of air-coupled ultrasound: only transducers with pulse durations above 10 µs provided sufficient amplitude for practical applications. Previously proposed methods to generate and detect pulses shorter than 10 µs resulted in low amplitudes [1,12]. However, high amplitudes are mandatory for air-coupled ultrasound applications due to the low energy transmission coefficients at the involved solid-to-air interfaces.

Here we demonstrate the combined use of a broadband, thermoacoustic emitter and an optical microphone and show an initial application: the simultaneous determination of thickness and sound velocity of a solid material. Carbon fiber-reinforced plastic (CFRP) plates with different thicknesses were used to show the applicability for a currently relevant material [13]. A transmission setup and time-of-flight data processing were implemented.

Previously proposed techniques that measure thickness and speed of sound simultaneously rely on the thickness resonances (TR) of the investigated sample [5]. Such techniques, however, require the determination of the frequency-dependent transmission coefficient, necessitating some advanced data processing and data fitting. More importantly, the correct transducer combination is required for each new sample. The sensor combination presented here is simpler to use, and the transducer pair works for all samples that have a TR frequency below a critical value of 1 MHz.

A plate thickness measurement has been demonstrated previously using a laser pulse to generate the ultrasound and an optical microphone as a detector [14]. Since the ultrasound was generated directly in the specimen, the distance between the source of the ultrasound and the detector was not constant during the scan. It was thus not possible to extract the plate thickness independently from the plate material (and vice versa), as in the case for our method. In [14], the plate thickness and plate curvature were measured, while the corresponding sound velocity in the plate needed to be known or measured at the reference point. This is not necessary for the method introduced in this work.

2. Materials and Methods

2.1. Transducers

The thermoacoustic effect describes the generation of sound from heat. Well-known cases are thunder and spark discharge [15], but also less commonly known cases such as laser-induced breakdown [16] and thermophones [7,17,18] are capable of transforming heat into acoustic waves. Physically best understood are thermophones, which use a thin Ohmic conductor to transform electric energy into heat in the surrounding fluid. When the heat deposition occurs fast enough, a finite volume of air is heated, which is equivalent to an increase in pressure. Thin films on a curved surface are well suited for material testing [7]. Such setups hold the advantages of a thermoacoustically active 2D area that can be used for pressure generation and beam focusing at the same time. Typical materials are indium tin oxide for the thin film and silica glass for the substrate. The transducer used in this study consisted of a 200 nm indium tin oxide film on a borosilicate glass substrate with a curvature of 55 mm and radial electrode placement (see Figure 1). Thermoacoustic transducers generate ultrasound directly in air and do not rely on mechanical or resonant vibrations. This enables the generation of short pulses, necessary for the method proposed in this work.

The receiver was an optical microphone (Eta 450 Ultra, XARION Laser Acoustics GmbH, Vienna, Austria). The working principle is based on a rigid Fabry-Pérot laser interferometer with two miniaturized mirrors, where sound waves in air change the refractive index, alter the optical wavelength and the light transmission of the pair of mirrors. Like the thermoacoustic emitter, the microphone does not rely on resonantly vibrating components to detect ultrasound and hence enables broadband signal detection.

It is the combination of a high-bandwidth transmitter and a high-bandwidth receiver that lets us resolve the in-plate reflections in time even for millimeter thin composite plates, as discussed later in this article. The measured signal of the setup with and without a specimen is shown in Figure 2.

2.2. Calculation of Thickness and Speed of Sound

High-power microsecond pulses in CFRP plates enable exact determination of the pulse arrival time t_{ToF} and the simultaneous measurement of the in-plate reflection period t_{TR}, as shown in Figure 2. This enables the determination of plate thickness and longitudinal ultrasound velocity in the investigated plate from a single measurement.

We show this by expanding on the time-of-flight calculations from [19], in which only one of the two parameters could be determined independently. In addition, note that reference [19] reports a speed of sound measurements with uncertainties mostly in the range of 20–50% due to long pulse widths.

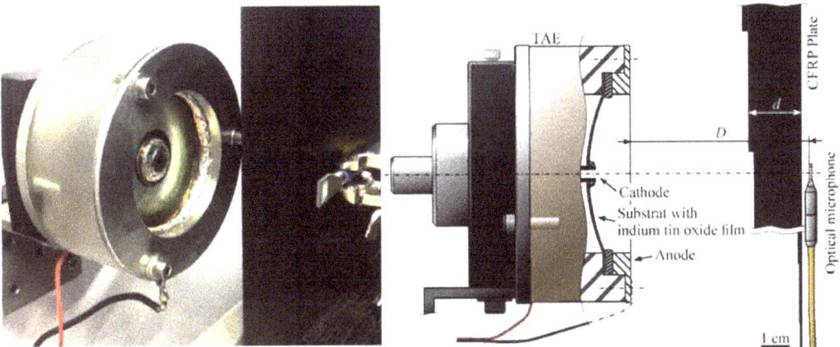

Figure 1. Image and schematic of the air-coupled ultrasound transmission setup. Sending and receiving transducers were positioned on both sides of the CFRP step wedge. The emitter consists of a thin indium tin oxide film in between two electric poles. The receiver was an optical microphone. The distance between transmitter and receiver D and the thickness of the plate could be determined experimentally.

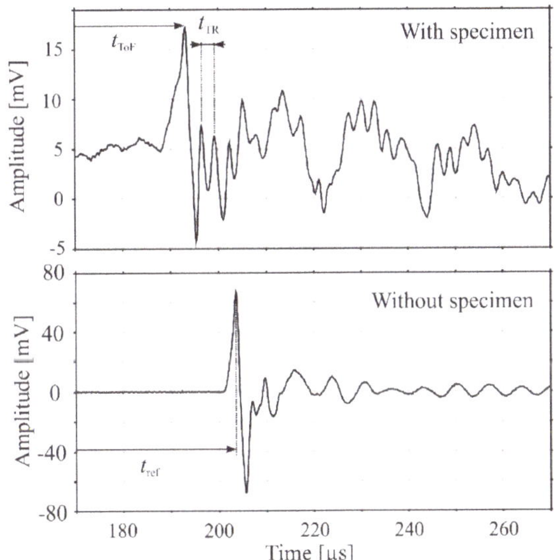

Figure 2. Representative signals for the proposed technique. The upper and lower graphs show a measured signal with and without a specimen, respectively. The times t_{TR} and t_{ToF} are independent parameters, which enable the discrimination between the influence of the plate thickness and the material properties. An initial measurement of the time of flight between the transducers (t_{ref}) enables the determination of absolute values for both parameters simultaneously. The observable signal oscillations after the pulse detection feature components other than only the in-plate reflections and are further analyzed in the signal measurement and processing section.

In the following, t_{ToF} is the time the pulse travels through air and sample, D is the corresponding distance between ultrasound source and detector, d is the material thickness, and v_A and v_M are the speed of sound in air and in the analyzed material, respectively. With these definitions, we can express t_{TR} and t_{ToF} as

$$t_{TR} = \frac{2d}{v_M} \qquad (1)$$

and

$$t_{ToF} = \frac{d}{v_M} + \frac{D - d}{v_A}. \qquad (2)$$

These two equations can be solved for the plate thickness d and the longitudinal ultrasound velocity in the plate material v_M. Setting up two transducers with a distance D is typically error prone with uncertainties in the order of magnitude of 1 mm. This leads to a similar error in the calculation of d. Therefore, it is best practice to replace the measurement of D by a measurement of the time of flight of a pulse between the two transducers with no sample t_{ref}, meaning $D = t_{ref} \cdot v_A$. These considerations lead to

$$d = \left(t_{ref} - t_{ToF} + \frac{t_{TR}}{2} \right) \cdot v_A \qquad (3)$$

and

$$v_M = \frac{2d}{t_{TR}}. \qquad (4)$$

Both parameters, d and v_M, can be estimated from the independent parameters t_{ToF} and t_{TR}, obtained from a single ultrasonic signal. The only prerequisites for this obtainment are the detectability of the in-plate reflections and the constant distance between the transducers.

2.3. Experimental Setup

The air-coupled ultrasound transmission setup can be divided into four parts: the sending part, the receiving part, the manipulator, and the piloting computer. The sending part consisted of the USPC 4000 Airtech system (Hillger NDT GmbH, Braunschweig, Germany), a voltage divider, an Agilent 33500B (Keysight Technologies, Santa Rosa, CA, USA) arbitrary waveform generator (AWG), an in-house power amplifier and the already described thermoacoustic transducer. The USPC 4000 Airtech system generated electric trigger signals, which were adjusted to 5 V and transmitted to the AWG. The pulse width was adjusted at the AWG and the pulse was transmitted to the in-house power amplifier and transmitted to the thermoacoustic transducer.

The power amplifier was designed such that pulses in the order of magnitude of several 100 ns to several 100 µs and 30 kW could be generated for a thermoacoustic load with a resistance of 8 Ω.

The receiving part consisted of the optical microphone, its data processing unit, and an analog-to-digital converter, which was part of the USPC 4000 Airtech system. The manipulator was a FlatScan 1000 (Hillger NDT GmbH, Braunschweig, Germany), controlled by the USPC 4000 Airtech system. Hillgus software (Hillger NDT GmbH, Braunschweig, Germany) was used to synchronize the sent and received signals and to control the manipulator position.

The transducers were facing one another, and their distance was varied until a maximum amplitude could be measured on the receiving side. This distance was 54 mm for all thicknesses. The pulse width of the electric pulse was set to 2 µs at an amplitude of 375 V. Together with the 7.8 Ohm of the transducer, this resulted in an approximately 18 kW electrical sending power. The pulses were transmitted at a 75 Hz repetition rate and the sample was scanned with a spatial resolution of 0.15 mm along the scanning axis. Three adjacent lines with 0.15 mm distance were scanned and the results averaged. This avoided artefacts caused by local sample inhomogeneities. The signal converter of the optical microphone was set to 20 dB amplification and the data acquisition software Hillgus amplified the signal by another 9 dB for all performed scans. The temperature was 23 °C and the relative humidity was 40%.

2.4. Specimen

The step wedge was a CFRP made from an epoxy prepeg of type HexPly(R). The fiber density was 1.78 g/cm^3 and the resin density 1.22 g/cm^3. The fiber orientation was quasi isotropic with 120 layers over a total thickness of 20.2 mm.

The specimen features a surface roughness below the smallest wavelength components of sound in both interface media, parallel front and back surfaces, and a sufficiently low dispersion. These parameters are typical prerequisites for air-coupled ultrasonic testing.

3. Results

3.1. Signal Measurement and Processing

Linear scans were performed using plates with thicknesses ranging from 1 mm to 5.1 mm (Figure 3a) and from 6.1 mm to 20.2 mm (Figure 3b), as indicated above the figures. Figure 3a,b have different time windows since the trigger time delays were not the same for both scans. The delays were chosen such that the first break and the subsequent in-plate reflections are fully captured for all plate thicknesses.

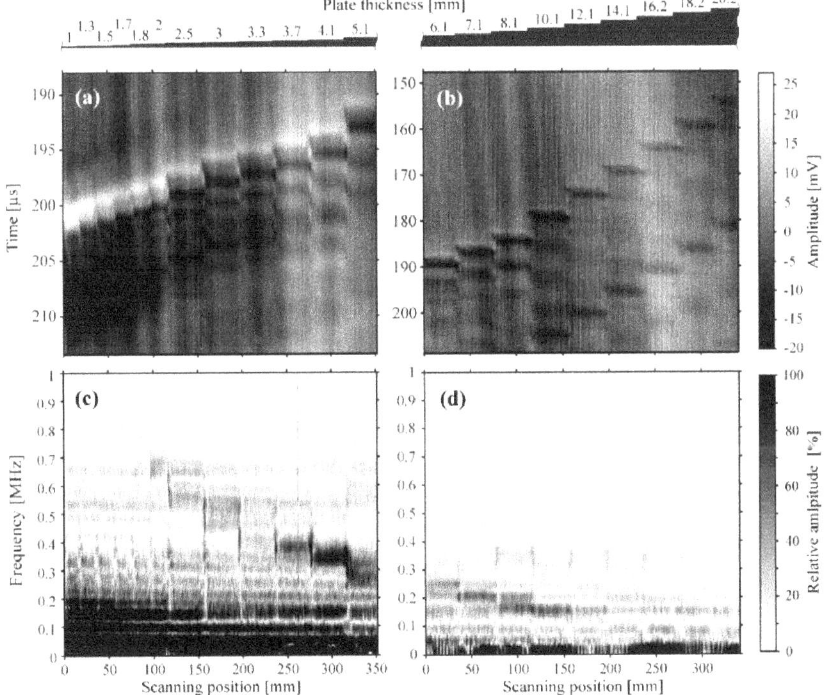

Figure 3. B-scans for thinner (**a**) and thicker (**b**) plate thickness ranges. The signals in frequency domain are shown in S-scans (B-scans in frequency domain) for the same positions of both linear scans in (**c,d**). The TR frequency is inversely proportional to the plate thickness, while the ToF is linearly proportional to the plate thickness.

The obtained signals were converted to frequency domain for all scanning positions. The linear scans of the plate thicknesses are shown in S-(spectrum) scans (Figure 3c,d). The term S-scan is used to designate B-scans in frequency domain. In order to improve the signal-to-noise ratio, only the time window between 10 µs before and 32 µs after the first break of the signal was used. It was converted to frequency domain, after applying a Hamming window. The amplitude of the S-scans is expressed relative to the maximum peak value of the TR frequency of both linear scans.

The goal of the signal processing is to determine t_{ToF} and t_{TR} in an automated process. The value of t_{TR} is determined in frequency space. The in-plate reflections and the TR frequency are distinguishable for the plate thicknesses ranging from 1.7 mm to 10.1 mm. Figure 3c,d clearly show that the TR frequency peak value is proportional to d^{-1}. Although our system is comparatively broadband, it is to some extent possible to observe its characteristic behavior. Each signal was superimposed by two main oscillations with period times of roughly 10 µs and 30 µs (see Figure 2). These signals were likely caused by multiple reflections of the pulse inside the microphone cavity and between the microphone and the plate surface, respectively. Such interferences leads to differences in the quality of the signal for certain plate thicknesses. For example, the detected TR frequencies close to 660 kHz, 540 kHz, 340 kHz, and 150 kHz have higher amplitude than others. However, despite these interferences, the method allows for reliable detection of the TR frequency peak over a broad frequency range.

The value of t_{ToF} was determined in time domain. A strong correlation between the plate thickness and t_{ToF} is shown in Figure 3a,b. The in-air reflections between the specimen and the optical microphone are visible in Figure 3b. They arrive approximately 30 µs after the first break of the US signal. This corresponds to the microphone-to-specimen distance of 4.8 mm.

3.2. Comparison of Measured and Reference Values

The first step of the analysis was the extraction of t_{ToF} and t_{TR} from the signal for all scanning positions. t_{ToF} was reliably obtained by picking the time of the maximum signal levels (shown in Figure 3a) for the scan over the lower plate thickness range (1 mm to 5.1 mm). For greater thicknesses (6.1 mm to 20.2 mm), the first negative peak is more pronounced than the first positive peak (Figure 3b). The arrival times were thus more reliably obtained by picking the time of the minimum signal level. The in-air reflections, which arrived after the first break of the signal and might have larger amplitude, were neglected. The mean value of the arrival time difference between the minimum and the maximum signal values of all the scanning locations was added to the arrival times for the scan of the lower thickness range in order to synchronize the scans of both thickness ranges.

The parameter t_{ToF} was obtainable by this method for all plate thicknesses. This was not the case for the second parameter t_{TR}, which was not measurable for the whole thickness range. As can be seen in Figure 3, the in-plate reflections (in time domain) and TR peak (frequency domain) are visible only for the plate thicknesses between 1.7 mm and 10.1 mm. This is due to the frequency range of our measurement system—up to 0.8 MHz for the applied excitation parameters. The upper limit of the plate thickness is defined by the ultrasound attenuation level in the plate. The condition for the successful measurement is that the in-plate reflections are detectable. t_{TR} was obtained by inverting the TR peak (visible in Figure 3c,d).

In Figure 4, we present the plate thicknesses obtained by inserting t_{ToF} and t_{TR} in Equations (3) and (4). The only parameter in this equation that was not measured directly by our experiment is sound velocity in air which proportionally influences all the measurement points and amounted to 345.7 m/s in our experiment. The thicknesses measured by our ACU method are compared to values obtained by the reference measurement with a micrometer screw (precision higher than 0.01 mm). In Table 1, statistical values are provided for each of the plate thicknesses: mean value of the air-coupled measurement, its standard deviation, number of measurements at each of the plate thicknesses (obtained by the linear scan), and the number of the outliers, which we excluded to calculate the standard deviation. We did not exclude any values in Figure 4.

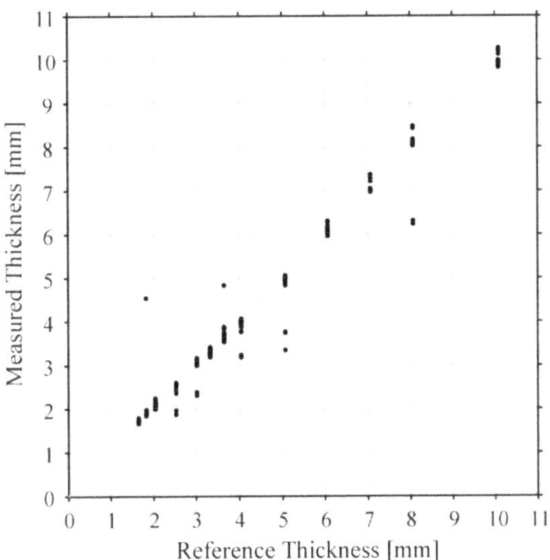

Figure 4. Plate thicknesses obtained by simultaneous measurement of t_{TR} and t_{ToF} using Equation (3). As t_{TR} and t_{ToF} are independent parameters, we can simultaneously estimate longitudinal sound velocity in the sample material (approximately 2800 m/s) using Equation (4), without a reference measurement at a known thickness, which is required for conventional contact ultrasonic measurements. This is only possible because our method is broadband and air coupled.

Table 1. Plate thicknesses determined with the presented ultrasound method with reference values.

Reference [mm]	Air-Coupled— Mean Value [mm]	Air-Coupled—SD [1] [mm]	Number of Measurements	Number of Excluded Outliers
1.65	1.70	0.03	23	0
1.83	1.89	0.03	22	1
2.04	2.11	0.05	22	0
2.53	2.55	0.05	45	2
3.02	3.06	0.04	43	2
3.33	3.32	0.04	44	0
3.65	3.62	0.07	46	1
4.05	3.99	0.06	46	2
5.07	4.96	0.06	38	3
6.07	6.07	0.08	43	0
7.07	7.03	0.07	44	0
8.07	8.12	0.08	45	2
10.90	9.93	0.08	46	0

[1] SD = standard deviation.

The outliers are caused by the false picking of the time of arrival or the TR peak frequency. The two parameters might not refer to the same thickness, especially at the (discrete) transition from one thickness to another. The wave propagation properties could be affected by the thickness transitions, which could alter the TR frequency (Figure 3). The standard deviations of the air-coupled thickness measurements remained below 0.09 mm for the plate thicknesses from 1.65 mm and 10.1 mm.

The longitudinal sound velocity in the plate material was measured for all scanning locations by the same experiment using Equation (4). Its mean value over the whole scan is 2800 m/s with the standard deviation of 160 m/s. A reference measurement of the speed of sound in the utilized plate was performed using a contact ultrasound time of flight measurement with a 2.25 MHz transducer. The technique produced 2896 m/s, which is well within the first standard deviation of the proposed air-coupled technique.

4. Discussion

Our experimental setup enables, for the first time, the resolution of in-plate reflections at plate thicknesses ranging from 1.65 mm to 10 mm without physical contact to the specimen. In contrast to conventional pulse echo methods based on contact ultrasound, the time of flight and the period of in-plate reflections are, in our setup, independent parameters. This opens the possibility to measure the speed of sound and material thickness simultaneously from a single ultrasonic signal. In other words, it is possible to measure plate thickness without actually knowing the ultrasound propagation speed in the sample material. This new paradigm only assumes the prior knowledge of the speed of sound in air and a reference time-of-flight measurement without the specimen. The latter will deliver the distance between the transducers, which should and can (due to the air-coupled setup) be kept constant during the whole experiment.

The experiments were conducted at a CFRP plate due to its relevance in research and industry. The applicability of our technique to a certain object will depend on the object's composition of speed of sound, thickness, and dispersion. Dispersion might influence our technique stronger than other material analysis approaches as it may vary the shape of the in-plate reflections. Applying our technique to a wide range of material parameter combinations will be a subject of future research.

Temperature changes and airflows are likely to occur in industrial applications. The airflows can influence air-coupled ultrasound measurements when the overall distance between the sender and the receiver in air is several centimetres long. Such flows might disrupt the method by altering the measured time of flight in air. It is, however, unlikely that temperature changes will influence our measurement since they occur on different time scales than the ones relevant for our method—200 µs.

We measured signal components up to 0.8 MHz. However, the non-resonant working principles of the utilized emitter and receiver do not limit this frequency range. A limiting factor is the energy per pulse that is proportional to the pulse length and amplitude. The pulse width (e.g., full width at half maximum, τ_{FWHM}), however, determines the lower limit of detectable plate thicknesses d_{limit} via $d_{\mathrm{limit}} = c \cdot \tau_{\mathrm{FWHM}}/2$. Higher amplitudes will be required for future applications that require even shorter pulse lengths. With further developments in the efficiency of thermoacoustic emitters and the sensitivity of optical receivers, sub-microsecond pulses are likely to be provided for future applications.

The proposed technique is one initial possibility for the utilization of short pulses and non-resonant transducers and more applications are likely to follow. As our method is able to distinguish the in-plate reflections, it provides the potential to allow single-side air-coupled ultrasonic inspection and to resolve the direct reflection from the back-wall echo. Localized material properties of the specimen or the inspected features could be characterized by the analysis of the local resonances in the ultrasonic frequency range, as it has been demonstrated using laser generated ultrasound [15,20]. The broadband thermoacoustic emitter carries high potential to replace laser pulse ultrasound excitation [21], which is more expensive and evokes special safety-related concerns. As both transducers also function in liquids, expanding the scope of the technique to immersion testing is likely to succeed in future research.

Author Contributions: Conceptualization, H.M., M.G., C.U.G., K.B. and J.R.; methodology, K.B., J.R. and H.M.; software, J.R.; validation, J.R. and H.M.; formal analysis, J.R.; investigation, K.B., J.R. and H.M.; resources, M.G. and C.U.G.; data curation, J.R.; writing—original draft preparation, K.B. and J.R.; writing—review and editing, H.M., M.G., C.U.G., K.B. and J.R.; visualization, J.R.; supervision, M.G. and C.U.G.; project administration, M.G. and C.U.G. All authors have read and agreed to the published version of the manuscript.

Funding: This research received no external funding.

Institutional Review Board Statement: Not applicable.

Informed Consent Statement: Not applicable.

Data Availability Statement: Not applicable.

Acknowledgments: Klaas Bente would like to thank Peter J. Hanley for his support.

Conflicts of Interest: The authors declare no conflict of interest.

References

1. Wright, W.; Hutchins, D. Air-coupled ultrasonic testing of metals using broadband pulses in through-transmission. *Ultrasonics* **1998**, *37*, 19–22. [CrossRef]
2. Gan, T.; Hutchins, D.; Billson, D.; Schindel, D. The use of broadband acoustic transducers and pulse-compression techniques for air-coupled ultrasonic imaging. *Ultrasonics* **2001**, *39*, 181–194. [CrossRef] [PubMed]
3. Waag, G.; Hoff, L.; Norli, P. Air-coupled ultrasonic through-transmission thickness measurements of steel plates. *Ultrasonics* **2015**, *56*, 332–339. [CrossRef] [PubMed]
4. Vössing, K.J.; Gaal, M.; Niederleithinger, E. Air-coupled ferroelectret ultrasonic transducers for nondestructive testing of wood-based materials. *Wood Sci. Technol.* **2018**, *52*, 1527–1538. [CrossRef]
5. Álvarez-Arenas, T.E.G. Simultaneous determination of the ultrasound velocity and the thickness of solid plates from the analysis of thickness resonances using air-coupled ultrasound. *Ultrasonics* **2010**, *50*, 104–109. [CrossRef] [PubMed]
6. Gaal, M.; Kotschate, D.; Bente, K. Advances in air-coupled ultrasonic transducers for non-destructive testing. *Proc. Meet. Acoust. Acoust. Soc. Am.* **2019**, *38*, 030003. [CrossRef]
7. Daschewski, M.; Boehm, R.; Prager, J.; Kreutzbruck, M.; Harrer, A. Physics of thermo-acoustic sound generation. *J. Appl. Phys.* **2013**, *114*, 114903. [CrossRef]
8. Fischer, B. Optical microphone hears ultrasound. *Nat. Photon.* **2016**, *10*, 356–358. [CrossRef]
9. Preisser, S.; Rohringer, W.; Liu, M.; Kollmann, C.; Zotter, S.; Fischer, B.; Drexler, W. All-optical highly sensitive akinetic sensor for ultrasound detection and photoacoustic imaging. *Biomed. Opt. Express* **2016**, *7*, 4171–4186. [CrossRef] [PubMed]
10. Rus, J.; Fischer, B.; Grosse, C.U. Photoacoustic inspection of CFRP using an optical microphone. In Proceedings of the SPIE, Optical Measurement Systems for Industrial Inspection XI 2019, Munich, Germany, 24–27 June 2019; p. 1105622.
11. Rus, D.J.; Kulla, J.-C.; Grosse, C.U. Air-Coupled Ultrasonic Inspection of Fiber-Reinforced Plates Using an Optical Microphone. *e-J. Nondestruct. Test.* **2019**.
12. Schindel, D.; Hutchins, D.; Zou, L.; Sayer, M. The design and characterization of micromachined air-coupled capacitance transducers. *IEEE Trans. Ultrason. Ferroelectr. Freq. Control* **1995**, *42*, 42–50. [CrossRef]
13. Rus, J.; Gustschin, A.; Mooshofer, H.; Grager, J.-C.; Bente, K.; Gaal, M.; Pfeiffer, F.; Grosse, C.U. Qualitative comparison of non-destructive methods for inspection of carbon fiber-reinforced polymer laminates. *J. Compos. Mater.* **2020**, *54*, 4325–4337. [CrossRef]
14. Rus, J.; Grosse, C.U. Thickness measurement via local ultrasonic resonance spectroscopy. *Ultrasonics* **2020**, *109*, 106261. [CrossRef] [PubMed]
15. Kotschate, D.; Gaal, M.; Kersten, H. Acoustic emission by self-organising effects of micro-hollow cathode discharges. *Appl. Phys. Lett.* **2018**, *112*, 154102. [CrossRef]
16. Qin, Q.; Attenborough, K. Characteristics and application of laser-generated acoustic shock waves in air. *Appl. Acoust.* **2004**, *65*, 325–340. [CrossRef]
17. Wente, E.C. The Thermophone. *Phys. Rev.* **1922**, *19*, 333. [CrossRef]
18. La Torraca, P.; Bobinger, M.; Pavan, P.; Becerer, M.; Zhao, S.; Koebel, M.; Cattani, L.; Lugli, P.; Larcher, L. High Efficiency Thermoacoustic Loudspeaker Made with a Silica Aerogel Substrate. *Adv. Mater. Technol.* **2018**, *3*, 1800139. [CrossRef]
19. Schindel, D.; Hutchins, D. Through-thickness characterization of solids by wideband air-coupled ultrasound. *Ultrasonics* **1995**, *33*, 11–17. [CrossRef]
20. Janez, R.; Grosse, C.U. Local Ultrasonic Resonance Spectroscopy: A Demonstration on Plate Inspection. *J. Nondestruct. Eval.* **2020**, *39*, 1–12.
21. Rus, J. Contact-Free Non-Destructive Inspection by Broadband Ultrasound. Ph.D. Thesis, Technische Universität München, München, Germany, 2021.

Article

Pipeline Leakage Detection Based on Secondary Phase Transform Cross-Correlation

Hetao Liang [1], Yan Gao [2,*], Haibin Li [1], Siyuan Huang [1], Minghui Chen [1] and Baomin Wang [3]

[1]   Huaneng Hunan Yueyang Power Generation Co., Ltd., Yueyang 414002, China
[2]   Institute of Acoustics, Chinese Academy of Sciences, Beijing 100190, China
[3]   Huaneng Clean Energy Research Institute, China Huaneng Group Co., Ltd., Beijing 102209, China
*   Correspondence: gaoyan@mail.ioa.ac.cn; Tel.: +86-10-8254-7794

Abstract: Leaks from pipes and valves are a reputational issue in industry. Maintenance of pipeline integrity is becoming a growing challenge due to the serious socioeconomic consequences. This paper presents a secondary phase transform (PHAT) cross-correlation method to improve the performance of the acoustic methods based on cross-correlation for pipeline leakage detection. Acoustic emission signals generated by pipe leakage are first captured by the sensors at different locations, and are subsequently analyzed using the cross-correlation curve to determine whether leakage is occurring. When leakage occurs, time delay estimation (TDE) is further carried out by peak search in the cross-correlation curve between the two sensor signals. In the analysis, the proposed method calculates the secondary cross-correlation function before the PHAT operation. A sinc interpolation method is then introduced for automatic searching the peak value of the cross-correlation curve. Numerical simulations and experimental results confirm the improved performance of the proposed method for noise suppression and accurate TDE compared to the basic cross-correlation method, which may be beneficial in engineering applications.

Keywords: acoustic emission; leak detection; cross-correlation; phase transform; time delay estimation

Citation: Liang, H.; Gao, Y.; Li, H.; Huang, S.; Chen, M.; Wang, B. Pipeline Leakage Detection Based on Secondary Phase Transform Cross-Correlation. *Sensors* **2023**, *23*, 1572. https://doi.org/10.3390/s23031572

Academic Editors: Farook Sattar, Niladri Bihari Puhan and Reza Fazel-Rezai

Received: 21 December 2022
Revised: 28 January 2023
Accepted: 30 January 2023
Published: 1 February 2023

1. Introduction

Maintaining the reliability and integrity of pipeline networks is becoming an increasingly critical challenge in industry. A large number of pipes and valves exist in power plants for the transport of liquids and gases, and delays in detecting and repairing the damaged pipe sections can result in significant financial loss and hazardous situations. In long-term service, the pipeline networks are constantly affected by high temperature, high pressure, corrosion and damage, all of which makes them susceptible to leakage. Leakage detection is crucially important for a company when making asset management decisions and maintaining the resilience of the pipeline system. In recent decades, much attention has been paid to targeting leakage effectively and efficiently in academic and industrial communities [1–3].

Traditionally, leak detection surveys are conducted manually, for example, by listening to the sound of leaks and applying soap bubbles to suspicious locations. These methods have low inspection efficiency. In recent years, rapid progress in leakage detection methods have been made, including the mass balance method, flow model method, infrared thermography, optical fiber method and acoustic method [4–10]. These methods have their own advantages and disadvantages in terms of detection accuracy, efficiency, installation mode and cost.

For the mass balance method, monitoring the quantity of the fluid flowing through the inlet and outlet of the pipeline may be undertaken to determine whether leakage is occurring [11]. The principle of this method is relatively straightforward at the expense of its efficacy and reliability, because the change caused by a suspected leak can only be reflected after a period of time, especially for small flow leakage. Furthermore, more variables and parameters, such as fluid viscosity, pipe resistance, etc. need to be taken into account in the flow model. As such, a real-time flow model of the pipeline can be developed, which yields more reliable results compared to the mass balance method. Infrared thermography is mainly suitable and successful for leakage detection in pipe networks with obvious temperature characteristics, such as heating pipes [12]. Due to high temperature resolution, it has proven to be sensitive to small temperature differences and hence effective for detecting small leakages, which greatly improves the applicability of the method. More recently, optical fiber sensing technology has been developed for the detection of the temperature changes, vibration and strain caused by leakage in the pipe system [13]. To date, the optical fiber method has shown great promise due to its detection accuracy and efficacy. However, there are certain difficulties in installation for the pipelines buried under the soil or with wrapping materials.

Acoustic methods are commonly used for leakage detection due to the reliability, low cost and easy installation, including the traditional listening method, the negative pressure wave method and the acoustic emission method [14–16]. In principle, when leakage occurs, acoustic signals are generated in the vicinity of the leak, which are often seen as acoustic emission signals. The acoustic energy is transmitted in the pipe system in a variety of modes in broadband frequencies, which can be measured by acoustic and vibration sensors. Correspondingly, the leakage can be detected and located by analyzing the transmitted signals that carry the information of leak sources [17].

Acoustic emission based on cross-correlation is robust and widely used for pipeline leakage detection. The basic cross-correlation (BCC) process suffers from essential drawbacks, i.e., although this method is generally effective for white noise, the detection performance suffers to some degree for other-colored noise. To overcome this problem, an appropriate windowing or frequency weight function is introduced in the generalized cross-correlation (GCC) methods to pre-whiten the measured signals before cross-correlation. Nevertheless, the performance of the GCC methods varies in leak detection surveys due to the uncertainties in the practical engineering environments. Of particular interest is the phase transform (PHAT) cross-correlation, which uses only the phase information directly related to the difference between the arrival times of the leak noise at the sensors locations [18,19]. This method has proven to be effective for leak detection in a high signal-to-noise (SNR) environment. Otherwise, the detection accuracy suffers, in that the effects of background noise are neglected in the pre-whitening process. In recent years, some new time delay estimation (TDE) algorithms have been proposed. Ji et al. proposed a PHAT-β GCC algorithm by changing the exponential regulator β of the pre-whitening weight function in the 3D localization of transformer patrol robot [20]. Cui et al. proposed a variable step normalized LMS adaptive filter for leak localization in water-filled plastic pipes [21]. The algorithm transforms the TDE into the parameter estimation of the filter with advantages of variable step iterative learning. Alternative signal processing processes have been attempted to improve the leak detection accuracy [22–26], such as wavelet analysis, empirical model decomposition, neural network and deep learning methods. Whilst the advantages of these processes are immediately apparent, they also bring additional problems, such as large number of detection samples, long training time and complex calculated model.

In this paper, a secondary phase transform (PHAT) cross-correlation method is proposed to accentuate the information directly related to leakage. The method can suppress effectively the interfering effects of background noise by calculating the secondary cross-correlation function before the PHAT operation. Additionally, a sinc interpolation method is introduced for automatic searching the peak value of the secondary PHAT cross-correlation function. This can further improve the TDE accuracy since the error caused in the selection of the maximum value is significantly reduced. The remainder of this paper is as follows. In Section 2, the background of pipeline leakage detection is briefly discussed, followed by a detailed description of the proposed secondary phase transform (PHAT) cross-correlation. In Section 3 a numerical model is described, and some results are presented to show the detectability of the method for pipeline leakage detection. In Section 4 experimental work is carried out to verify the effectiveness of the proposed method in comparison with the BCC. Finally, conclusions are drawn in Section 5.

2. Methodology

2.1. Background of Pipeline Leak Detection

Figure 1 depicts the schematic diagram of the process of pipeline leak detection. The pipeline leakage will cause the fluid to overflow under the action of pressure and produce acoustic signals, which propagate along the pipeline upward and downstream. Acoustic/vibration sensors are generally installed at the pipe fittings, for example values and fire hydrants, to capture the transmitted leak signals.

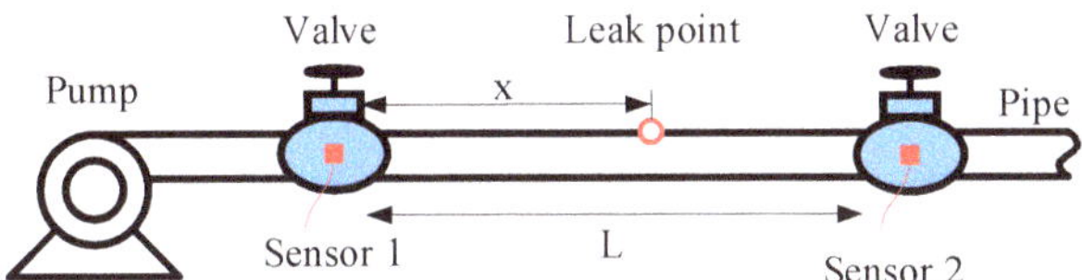

Figure 1. Schematic diagram of the process of pipeline leak detection.

Assume that the acoustic signals $x_1(t)$ and $x_2(t)$ received by sensor 1 and sensor 2 are expressed as follows:

$$\begin{aligned} x_1(t) &= s(t) + n_1(t) \\ x_2(t) &= s(t-D) + n_2(t) \end{aligned} \tag{1}$$

where $s(t)$ and $s(t-D)$ represent the propagating leakage signals that travel along the pipe and are captured by sensors 1 and 2; the time delay D denotes the difference in the arrival times of the corresponding leakage signals at the sensors' locations; and $n_1(t)$ and $n_2(t)$ denote the background noise at two sensor locations.

The BCC between the acoustic signals, $x_1(t)$ and $x_2(t)$, can be obtained by

$$\begin{aligned} R_{12}(\tau) &= E[x_1(t)x_2(t-\tau)] \\ &= E\{[s(t)+n_1(t)][s(t-D-\tau)+n_2(t-\tau)\} \\ &= R_{ss}(\tau-D) + R_{sn_2}(\tau) + R_{sn_1}(\tau-D) + R_{n_1n_2}(\tau) \end{aligned} \tag{2}$$

where $R_{12}(\ )$ denotes the BCC algorithm; $R_{ss}(\ )$ is the BCC function between leakage signals $s(t)$ and $s(t-D)$; $R_{sn1}(\ )$ and $R_{sn2}(\ )$ are the BCC functions between the leakage signal $s(t)$ and background noise signals $n_1(t)$ and $n_2(t)$ respectively; and $R_{n1n2}(\ )$ is the BCC function between background noise $n_1(t)$ and $n_2(t)$. Generally, for random white noise, leakage signal and background noise signals are independent and uncorrelated with each other. In this case, $R_{sn1}(\ )$, $R_{sn2}(\ )$ and $R_{n1n2}(\ )$ in Equation (1) are equal to 0.

The above analysis suggests that if there is no leakage in the pipeline, the acoustic signals received by the two sensors are background noise, resulting in the cross-correlation result being close to zero. In contrast, when leakage occurs in the pipeline, the acoustic signals received by the two sensors include the additional transmitted leakage signals, which leads to the correlation result being non-trivial. This is further demonstrated in the simulations in the next section.

Applicable leak detection process involves two phases. When pipeline leakage is initially identified, background noise may produce signals with characteristics similar to leakage signals. Thus, it is critical to take accurate measurements to localize the exact location of the leak. Equation (2) can be further simplified as

$$R_{12}(\tau) = E[s(t)s(t-D-\tau)] = R_{ss}(\tau-D) \tag{3}$$

It is apparent that when $\tau - D = 0$, the BCC function $R_{12}(\tau)$ achieves the maximum value. In this case, the time delay between the two leakage signals $\tau = D$. The time delay D is subsequently obtained by automatic searching the cross-correlation function $R_{12}(\tau)$ for its peak value by using the sinc-interpolation. This can avoid the error caused by manually selecting the maximum value.

$$D = \mathrm{argmax}\left[\sum_{\tau=1}^{T} R_{12}(\tau)\mathrm{Sinc}(t - \tau)\right] \tag{4}$$

where argmax denotes the argument corresponding to the maximum value of the function; T is the signal length and Sinc denotes the sinc-interpolation.

Referring to Figure 1, given that the distance between sensor 1 and sensor 2 is L, the propagation wavespeed is v, and the time delay is D, the relative location from sensor 1 to the leak source, x can be obtained by

$$x = \frac{L - vD}{2} \tag{5}$$

In the BCC method, the time delay is estimated by searching the peak of the cross-correlation function. This method has low computational complexity and is easy to be programmed. Accurate TDE can be achieved provided that leakage signals and background noise are uncorrelated with each other. However, when the uncorrelated assumption is violated in actual leak detection surveys, the TDE obtained by the BCC method will be in significant error due to the smearing effects on the main peak or even false peaks.

2.2. Secondary PHAT Cross-Correlation

In order to suppress the interfering effects of background noise, this paper proposes a secondary PHAT generalized cross-correlation method. The process of the proposed algorithm is shown in Figure 2, with detailed description as follows:

Step 1: calculate the auto-correlation function $R_{11}(\tau)$ by performing auto-correlation operation on sensor signal $x_1(t)$

$$R_{11}(\tau) = E[x_1(t)x_1(t - \tau)] = E\{[s(t)+n_1(t)][s(t - \tau)+n_1(t - \tau)]\} \tag{6}$$

Step 2: calculate the cross-correlation function $R_{12}(\tau)$ by performing the cross-correlation operation on sensor signals $x_1(t)$ and $x_2(t)$

$$R_{12}(\tau) = E[x_1(t)x_2(t - \tau)] = E\{[s(t)+n_1(t)][s(t - D - \tau)+n_2(t - \tau)]\} \tag{7}$$

Step 3: calculate the secondary cross-correlation function $R_{RR}(\tau)$ by performing cross-correlation operation on the above auto-correlation function $R_{11}(\tau)$ and cross-correlation function $R_{12}(\tau)$

$$R_{RR}(\tau) = E[R_{11}(t)R_{12}(t - \tau)] = E\{[R_{ss}(t) + R_{sn_1}(t) + R_{n_1s}(t) + R_{n_1n_1}(t)] \tag{8}$$

Step 4: the phase transform operation is used to weight the cross-power spectral density of $R_{RR}(\tau)$ in the frequency domain. Finally, the secondary PHAT cross-correlation function, $R_{S\text{-}PHAT}(\tau)$, is obtained by performing the inverse Fourier transform

$$R_{S\text{-}PHAT}(\tau) = \int_{-\infty}^{\infty} G_{RR}(\omega)\varphi_{12}(\omega)e^{-j\omega\tau}d\omega \tag{9}$$

$$\varphi_{12}(\omega) = \frac{1}{|G_{RR}(\omega)|} \tag{10}$$

where $\varphi_{12}(\omega)$ is the weighting function of the secondary PHAT cross-correlation in the frequency domain.

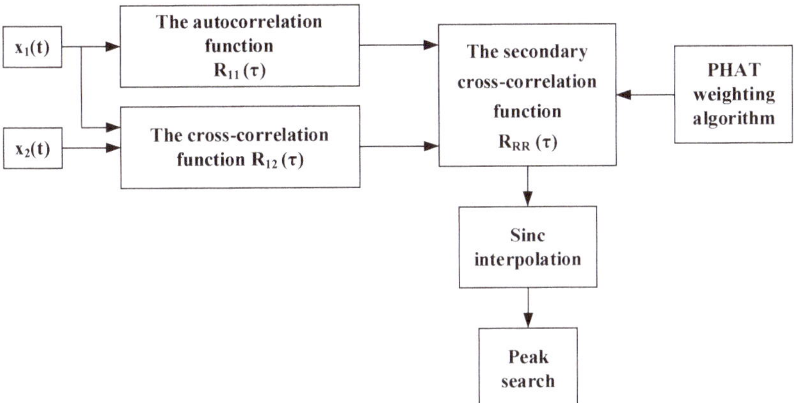

Figure 2. Schematic of the implementation of the secondary PHAT cross-correlation.

3. Simulations

Simulation results are presented for comparing the performance of the proposed secondary PHAT cross-correlation and the BCC method for leak identification and localization.

3.1. Leak Identification

3.1.1. Pipe without Leakage

In the simulation model, white noise is used to simulate the signals received by the sensors when there is no leakage. Figure 3 plots the sensor signals in the time and frequency domains. The BCC function between the two sensor signals is calculated and shown in Figure 4. As can be seen from Figure 4, the BCC result value is very small. As anticipated in Section 2, it confirms that no leakage is occurring in the pipeline.

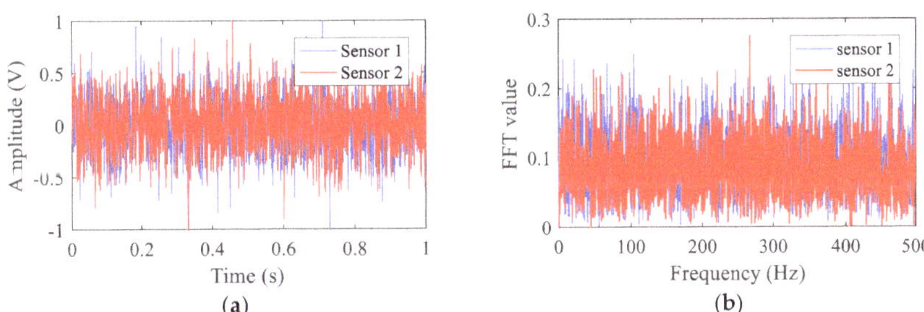

Figure 3. Simulated signals in the pipe without leakage: (**a**) in the time domain; (**b**) in the frequency domain.

3.1.2. Pipe with Leakage

When leakage occurs, the signals received by sensor 1 and sensor 2 are from the same leakage point with different arrival times. For simplicity, in the analysis, the sensor signal 1 is set to be the delayed signal of sensor 2 with 10 sampling points. The time domain and frequency domain characteristics of two sensor signals 1 and 2 are shown in Figure 5a,b, respectively. The corresponding BCC function is plotted in Figure 6. As expected, a distinct peak is found in the cross-correlation result. Further check of the BCC results for the pipes with and without leakage shows the effectiveness of the cross-correlation method for pipe leakage identification.

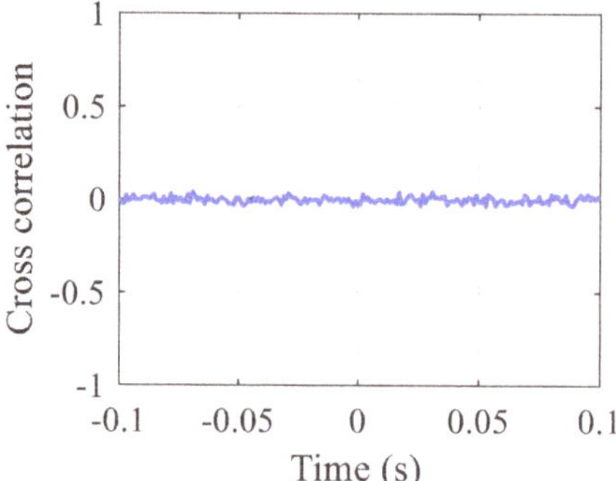

Figure 4. The BCC result of the simulated sensor signals for the pipe without leakage.

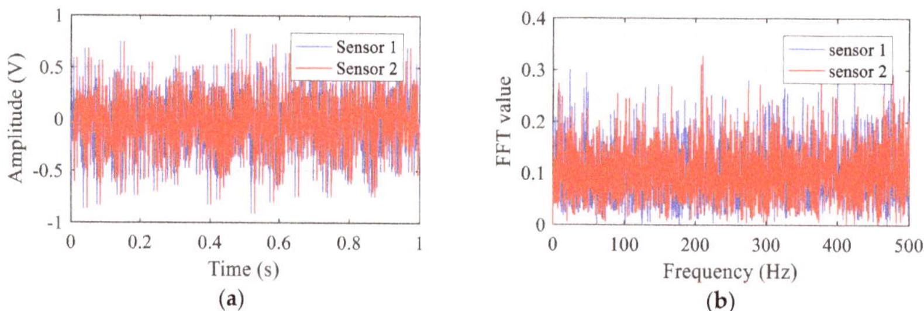

(a)

(b)

Figure 5. Simulated signals in the pipe with leakage: (**a**) in the time domain; (**b**) in the frequency domain.

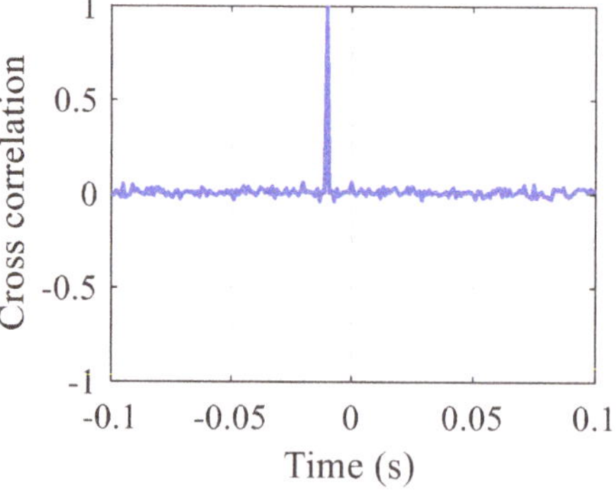

Figure 6. The BCC result of the simulated sensor signals for the pipe with leakage.

3.2. Leak Localization

Evaluation of the performance of TDE using the BCC and the secondary PHAT cross-correlation methods are performed under a high SNR (SNR = 10 dB) and a low SNR (SNR = −10 dB). Figure 7 plots the TDE curves of the two methods. The oscillatory behavior of the BCC result is obviously shown in Figure 7a,c, in particular, in the case of low SNR. As shown in Figure 7c, the fluctuations in the curve lead to a series of local maxima. However, these anomalous peaks are not generated by the leakage source; they are caused by noise interference instead. If these pseudo-peaks are mistakenly selected, the error in TDE will be significant. In contrast, the proposed secondary PHAT cross-correlation method leads to more reliable results for both high and low SNRs with single pronounced peaks corresponding to the actual time delay with some fluctuations in the magnitude being less distinctive than expected. This is due to the reason that the proposed method has the ability to pre-whiten the autocorrelation and cross-correlation, thus effectively reducing the interference effects of spurious peaks.

In order to better demonstrate the real peak corresponding to the actual time delay, Figure 8 plots the local results marked by the red boxes in Figure 7. For the BCC, the actual time delay is more discernible in Figure 8a, whereas it cannot be observed readily in Figure 7c. The simulation results in this section confirm that the proposed secondary PHAT cross-correlation method offers potential improvement for TDE over the BCC method.

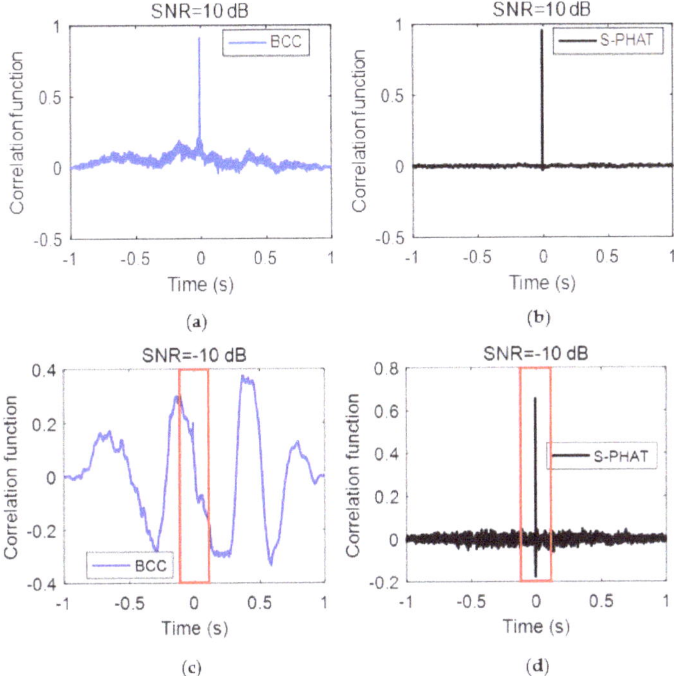

Figure 7. Comparison of the TDE curves of the simulated sensor signals: (**a**) the BCC method under SNR = 10 dB; (**b**) the secondary PHAT cross-correlation method under SNR = 10 dB; (**c**) the BCC method under SNR = −10 dB; and (**d**) the secondary PHAT cross-correlation method under SNR = −10 dB.

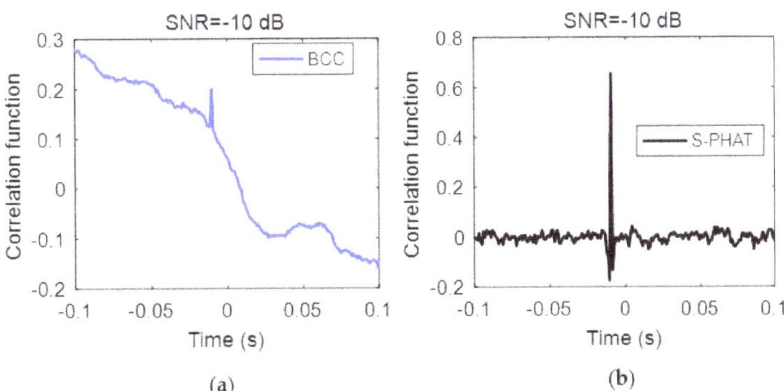

Figure 8. Local results marked by red box in Figure 7: (**a**) the BCC method under SNR = −10 dB; and (**b**) the secondary PHAT cross-correlation method under SNR = −10 dB.

4. Experiments

4.1. Experimental Setup

In order to evaluate the performance of the proposed method for pipeline leakage detection, experimental results from a PE water pipe rig are discussed. Tests were carried out at the leak detection facility built in the laboratory, as shown in Figure 9. The test pipe section had a length of 6 m and the wall thickness of 5 mm. The inlet and outlet of the pipe were, respectively, equipped with regulating valves. With reference to the figure, there was a simulated leak nozzle in the middle of the pipe section. Two hydrophones were used to capture the acoustic signals generated by the leak. The sensor model was a B&K 8103 hydrophone and the effective frequency range was 3 Hz–80 kHz. The distances between sensors 1 and 2 and the leakage was 0.9 m and 0.1 m, respectively. The leak signals were captured by using B&K PUSLE 3050 multi-channel signal acquisition instruments with a sampling rate of 8192 Hz. Similar to the analysis in the simulations, measurements were made in the cases of leak and no leak and in the pipe section.

The experimental procedures are briefly addressed as follows:

Step 1: Block the leak hole in the test pipe section to simulate the absence of leakage;

Step 2: Close the outlet valve and open the inlet valve of the test pipe section. Pressurize the test pipe by filling it with water. When the pressure reaches 1 bar and is stable, the sensors collect the acoustic signals in the pipe without leakage;

Step 3: Close the inlet valve and open the outlet valve of the test pipe section. Depressurize the test section. Then replace the nozzle with the leakage hole of 1 mm in diameter to simulate the pipeline leakage;

Step 4: Close the outlet valve and open the inlet valve of the test pipe section. Pressurize the test pipe by filling it with water. When the pressure reaches 1 bar and is stable, the sensors collect the leakage signals in the pipe.

4.2. Results and Discussions

When no leakage occurs, the signals measured by sensors 1 and 2 are plotted in Figure 10a,b in the time and frequency domains, respectively. In this circumstance, the signals include mainly ambient and system noise. Similar trends are demonstrated in Figure 10 for two sensor signals in both the time and frequency domains, indicating background noise dominates in the measured data. The noise signals have amplitudes of about 2.5 V, which is mainly concentrated at low frequencies below 50 Hz.

Figure 9. Experimental setup of the PE water pipe.

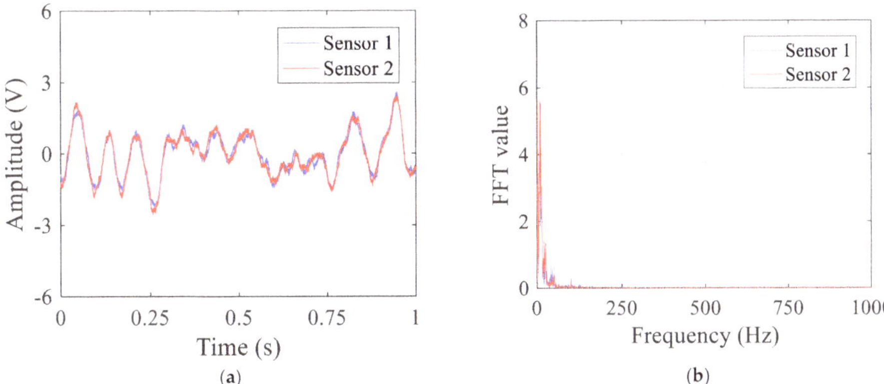

Figure 10. Test signals in the pipe without leakage: (**a**) in the time domain; and (**b**) in the frequency domain.

Next the cross-correlation methods are adopted to determine whether there exists leakage in the pipe section. Comparison of the BCC and the proposed secondary PHAT cross-correlation results for TDE curves is shown in Figure 11. Clearly there are a series of local peaks in the BCC results, as plotted in Figure 11a. In addition, these peak values are considerably larger, indicating that pipe leakage is likely to occur. This is, however, misleading, due to the fact that the BCC result is easily corrupted by the strong inference of background noise. In comparison, the TDE curve given by the secondary PHAT cross-correlation leads to very small result values in magnitude, which is close to 0 for the entire time-history. This confirms that the proposed method is marginally affected by background noise, and thereby can be adopted to monitor the pipe conditions more convincingly.

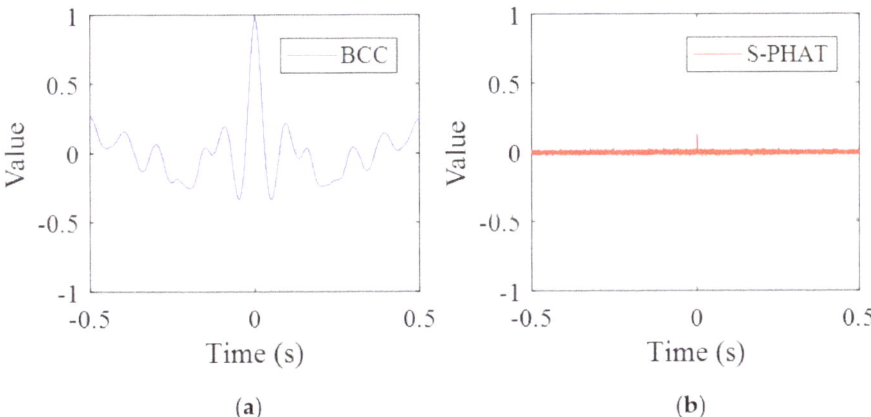

Figure 11. The TDE curves for the test pipe without leakage: (**a**) the BCC method; and (**b**) the secondary PHAT cross-correlation method.

In the case of pipe leakage, the sensor signals are plotted in Figure 12. Compared to the signals without leakage plotted in Figure 10, the sensor signals have very different characteristics in both the time and frequency domains. Besides, it can be seen from Figure 12 that the signal at sensor 2 has larger amplitude compared to that at sensor 1 due to the closer distance relative to the leak source, and hence less attenuation. As shown in Figure 12b, most of the frequency components of the sensor signals are mainly concentrated in the frequency range up to 600 Hz. It must be noted that, similar to the case of no leakage, at lower frequencies below 50 Hz, the measured signals are dominated by background noise.

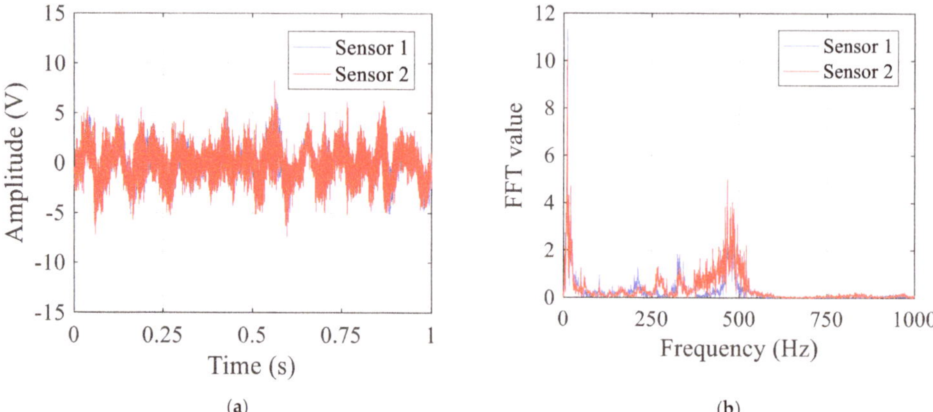

Figure 12. Test signals in the pipe with leakage: (**a**) in the time domain; and (**b**) in the frequency domain.

The cross-correlation methods are now applied to sensor signals for leak localization. Comparison of the TDE curves of the BCC and the proposed secondary PHAT cross-correlation are shown in Figure 13. Both methods are effective for leakage identification, since the main peak values of the BCC and the proposed methods are significantly large, confirming the presence of the pipe leakage.

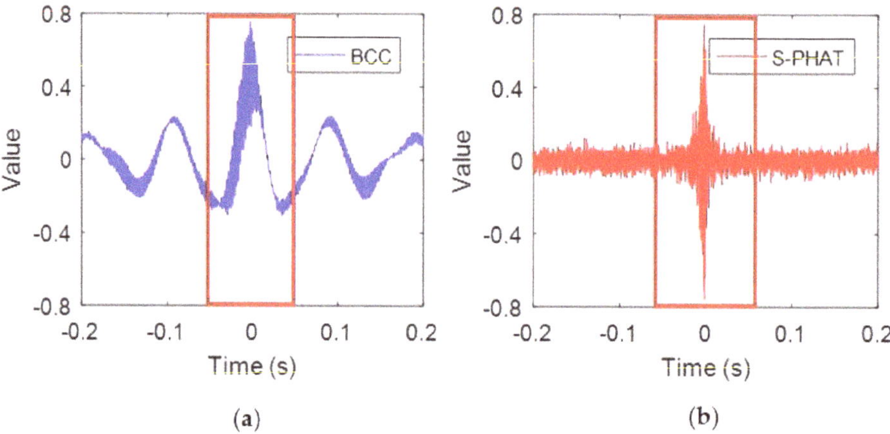

Figure 13. The TDE curves for the test pipe with leakage: (**a**) the BCC method; and (**b**) the secondary PHAT cross-correlation method.

As demonstrated in the simulations, several spurious peaks are found in the BCC results for test data as shown in Figure 13a. To better compare the TDE results, Figure 14 plots the local results marked by the red boxes around the main peak in [−0.05 s, 0.05 s] in Figure 13. The local result plotted in Figure 14a shows that the main peak has some fluctuations with close amplitudes. Again, they are caused by the inference of background noises at the test site. If the anomalous peaks are mistakenly selected as the peak corresponding to the time delay, an appreciable error will be given in the leak detection results. By comparison, the proposed secondary PHAT cross-correlation outperforms the BCC in terms of leak identification and localization. As stated above, the proposed methods via the pre-whitening process can effectively suppress the interference of background noise, hence producing more prominent peak corresponding to the actual time delay.

As can be found from the experimental setup, the distance difference between the two sensors is 0.8 m. In the measurements, the propagation wavespeed obtained is determined to be 310 m/s. As a result, the actual time delay is calculated to be 2.7 ms. It can be seen from Figure 14 that the BCC has several peaks around the actual time delay, while the proposed method leads to only one distinct peak. Their corresponding time delay results and calculation error are listed in Table 1. Compared to the BCC, the proposed secondary PHAT cross-correlation method is capable of suppressing other additional peaks unrelated to the time delay information.

Table 1. Leak localization results.

Method	Peak	Real Time (ms)	Time Delay Estimation	Error (%)	Mean Error (%)
	Peak 1		2.69	3.46	
	Peak 2		2.56	1.54	
BCC	Peak 3	2.6	2.44	6.15	21.54
	Peak 4		2.32	10.77	
	Peak 5		0.37	85.77	
S-PHAT	Peak 1		2.44	6.15	6.15

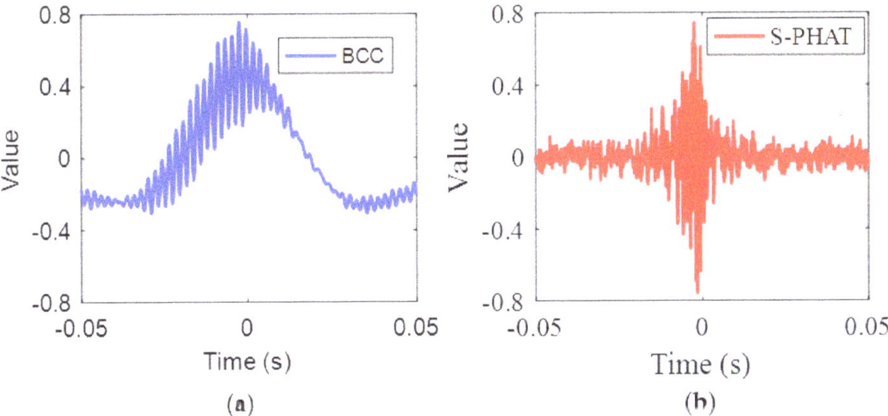

Figure 14. Local results marked by a red box in Figure 13: (**a**) the BCC method; and (**b**) the secondary PHAT cross-correlation method.

5. Conclusions

In order to suppress the undesirable effects of background noise on pipeline leakage detection, a secondary PHAT cross-correlation has been proposed. The paper introduces the principle of the BCC and secondary PHAT cross-correlation methods for the process of pipeline leakage detection. Simulations have been conducted to compare the performance of the proposed method with the BCC method for leak identification and localization in the pipe, with and without leakage. It has been found from the cross-correlation curve that when there is no leakage, the correlation result is close to zero. However, when leakage occurs, the correlation curve will have an obvious peak value corresponding to the time delay resulting from the pipe leakage. The simulation results have shown that the proposed method has sharper peaks than the traditional BCC method in a low-SNR environment. Experiments were carried out on the pipeline test rig made in the laboratory. Test results have shown that the BCC method has multiple peaks, while the proposed secondary PHAT method has only one peak. Comparing the leakage detection error of different peaks, it can be seen that the proposed secondary PHAT method outperforms the traditional

BCC method in terms of both the number of peaks and the average error. The findings suggest that the secondary PHAT cross-correlation method has been proposed as a potential solution to pipeline leakage detection in complex background environments, for example, in power plants or other industrial concerns.

Author Contributions: H.L. (Hetao Liang) carried out investigation and wrote the article; Y.G. contributed to supervising the work and revising the article; H.L. (Haibin Li) and M.C. conducted the experiments; S.H. and B.W. coordinated the work. All authors have read and agreed to the published version of the manuscript.

Funding: This research was funded by the Science and Technology Project of Huaneng Group Headquarters (No. HNKJ21-HF197).

Acknowledgments: The authors gratefully acknowledge the support of the Science and Technology Project of Huaneng Group Headquarters (No. HNKJ21-HF197).

Conflicts of Interest: The authors declare no conflict of interest.

References

1. Meribout, M. Gas Leak-Detection and Measurement Systems: Prospects and Future Trends. *IEEE Trans. Instrum. Meas.* **2021**, *70*, 4505813. [CrossRef]
2. Cui, X.; Yan, Y.; Guo, M.; Han, X.; Hu, Y. Localization of CO_2 Leakage from a Circular Hole on a Flat-Surface Structure Using a Circular Acoustic Emission Sensor Array. *Sensors* **2016**, *16*, 1951. [CrossRef] [PubMed]
3. Gao, F.; Lin, J.; Ge, Y.; Lu, S.; Zhang, Y. A Mechanism and Method of Leak Detection for Pressure Vessel: Whether, When, and How. *IEEE Trans. Instrum. Meas.* **2020**, *69*, 6004–6015. [CrossRef]
4. Ostapkowicz, P.; Bratek, A. Accuracy and Uncertainty of Gradient Based Leak Localization Procedure for Liquid Transmission Pipelines. *Sensors* **2021**, *21*, 5080. [CrossRef] [PubMed]
5. Li, J.; Zheng, Q.; Qian, Z.; Yang, X. A novel location algorithm for pipeline leakage based on the attenuation of negative pressure wave. *Process. Saf. Environ. Prot.* **2019**, *123*, 309–316. [CrossRef]
6. Sekhavati, J.; Hashemabadi, S.H.; Soroush, M. Computational methods for pipeline leakage detection and localization: A review and comparative study. *J. Loss Prev. Process. Ind.* **2022**, *77*, 104771. [CrossRef]
7. Abdulshaheed, A.; Mustapha, F.; Ghavamian, A. A pressure-based method for monitoring leaks in a pipe distribution system: A Review. *Renew. Sustain. Energy Rev.* **2017**, *69*, 902–911. [CrossRef]
8. Ahmad, S.; Ahmad, Z.; Kim, C.-H.; Kim, J.-M. A Method for Pipeline Leak Detection Based on Acoustic Imaging and Deep Learning. *Sensors* **2022**, *22*, 1562. [CrossRef]
9. Lu, H.; Iseley, T.; Behbahani, S.; Fu, L. Leakage detection techniques for oil and gas pipelines: State-of-the-art. *Tunn. Undergr. Space Technol.* **2020**, *98*, 103249. [CrossRef]
10. Puust, R.; Kapelan, Z.; Savic, D.A.; Koppel, T. A review of methods for leakage management in pipe networks. *Urban Water J.* **2010**, *7*, 25–45. [CrossRef]
11. Yang, S.; Talbot, R.W.; Frish, M.B.; Golston, L.M.; Aubut, N.F.; Zondlo, M.A.; Gretencord, C.; McSpiritt, J. Natural Gas Fugitive Leak Detection Using an Unmanned Aerial Vehicle: Measurement System Description and Mass Balance Approach. *Atmosphere* **2018**, *9*, 383. [CrossRef]
12. Yahia, M.; Gawai, R.; Ali, T.; Mortula, M.; Albasha, L.; Landolsi, T. Non-Destructive Water Leak Detection Using Multitemporal Infrared Thermography. *IEEE Access* **2021**, *9*, 72556–72567. [CrossRef]
13. Stajanca, P.; Chruscicki, S.; Homann, T.; Seifert, S.; Schmidt, D.; Habib, A. Detection of Leak-Induced Pipeline Vibrations Using Fiber—Optic Distributed Acoustic Sensing. *Sensors* **2018**, *18*, 2841. [CrossRef] [PubMed]
14. Hu, Z.; Salman, T.; Tarek, Z. A comprehensive review of acoustic based leak localization method in pressurized pipelines. *Mech. Syst. Signal Process.* **2021**, *161*, 107994. [CrossRef]
15. Niri, E.D.; Farhidzadeh, A.; Salamone, S. Nonlinear Kalman Filtering for acoustic emission source localization in anisotropic panels. *Ultrasonics* **2014**, *54*, 486–501. [CrossRef]
16. Murvay, P.-S.; Silea, I. A survey on gas leak detection and localization techniques. *J. Loss Prev. Process. Ind.* **2012**, *25*, 966–973. [CrossRef]
17. Cui, X.; Yan, Y.; Ma, Y.; Ma, L.; Han, X. Localization of CO_2 leakage from transportation pipelines through low frequency acoustic emission detection. *Sens. Actuators A Phys.* **2016**, *237*, 107–118. [CrossRef]
18. Ma, Y.; Gao, Y.; Cui, X.; Brennan, M.J.; Almeida, F.C.; Yang, J. Adaptive Phase Transform Method for Pipeline Leakage Detection. *Sensors* **2019**, *19*, 310. [CrossRef]
19. Gao, Y.; Liu, Y.; Ma, Y.; Cheng, X.; Yang, J. Application of the differentiation process into the correlation-based leak detection in urban pipeline networks. *Mech. Syst. Signal Process.* **2018**, *112*, 251–264. [CrossRef]
20. Ji, H.; Cui, X.; Gao, Y.; Ge, X. 3-D Ultrasonic Localization of Transformer Patrol Robot Based on EMD and PHAT-β Algorithms. *IEEE Trans. Instrum. Meas.* **2021**, *70*, 9004810. [CrossRef]

21. Cui, X.; Gao, Y.; Ma, Y.; Liu, Y.; Jin, B. Variable Step Normalized LMS Adaptive Filter for Leak Localization in Water-Filled Plastic Pipes. *IEEE Trans. Instrum. Meas.* **2022**, *71*, 9600511. [CrossRef]
22. Kong, Q.; Jiang, G.; Liu, Y.; Sun, J. Location of the Leakage From a Simulated Water-Cooling Wall Tube Based on Acoustic Method and an Artificial Neural Network. *IEEE Trans. Instrum. Meas.* **2021**, *70*, 4502618. [CrossRef]
23. Jin, Y.; Shan, C.; Wu, Y.; Xia, Y.; Zhang, Y.; Zeng, L. Fault Diagnosis of Hydraulic Seal Wear and Internal Leakage Using Wavelets and Wavelet Neural Network. *IEEE Trans. Instrum. Meas.* **2019**, *68*, 1026–1034. [CrossRef]
24. Guo, C.; Wen, Y.; Li, P.; Wen, J. Adaptive noise cancellation based on EMD in water-supply pipeline leak detection. *Measurement* **2016**, *79*, 188–197. [CrossRef]
25. Han, X.; Cao, W.; Cui, X.; Gao, Y.; Liu, F. Plastic Pipeline Leak Localization Based on Wavelet Packet Decomposition and Higher Order Cumulants. *IEEE Trans. Instrum. Meas.* **2022**, *71*, 3520911. [CrossRef]
26. Meng, Q.; Lang, X.; Lin, M.; Cai, Z.; Zheng, H.; Song, H.; Liu, W. Leak Localization of Gas Pipeline Based on the Combination of EEMD and Cross-Spectrum Analysis. *IEEE Trans. Instrum. Meas.* **2021**, *71*, 9501209. [CrossRef]

Article

Modal Decomposition of Acoustic Emissions from Pencil-Lead Breaks in an Isotropic Thin Plate

Xinyue Yao [1,*], Benjamin Steven Vien [1], Nik Rajic [2], Cedric Rosalie [2], L. R. Francis Rose [2], Chris Davies [1] and Wing Kong Chiu [1]

[1] Department of Mechanical and Aerospace Engineering, Monash University, Wellington Rd., Clayton, VIC 3800, Australia

[2] Defence Science and Technology Group, 506 Lorimer St., Fishermans Bend, Port Melbourne, VIC 3207, Australia

* Correspondence: xinyue.yao@monash.edu

Abstract: Acoustic emission (AE) testing and Lamb wave inspection techniques have been widely used in non-destructive testing and structural health monitoring. For thin plates, the AEs arising from structural defect development (e.g., fatigue crack propagation) propagate as Lamb waves, and Lamb wave modes can be used to provide important information about the growth and localisation of defects. However, few sensors can be used to achieve the in situ wavenumber–frequency modal decomposition of AEs. This study explores the ability of a new multi-element piezoelectric sensor array to decompose AEs excited by pencil lead breaks (PLBs) on a thin isotropic plate. In this study, AEs were generated by out-of-plane (transverse) and in-plane (longitudinal) PLBs applied at the edge of the plate, and waveforms were recorded by both the new sensor array and a commercial AE sensor. Finite element analysis (FEA) simulations of PLBs were also conducted and the results were compared with the experimental results. To identify the wave modes present, the longitudinal and transverse PLB test results recorded by the new sensor array at five different plate locations were compared with FEA simulations using the same arrangement. Two-dimensional fast Fourier Transforms were then applied to the AE wavefields. It was found that the AE modal composition was dependent on the orientation of the PLB direction. The results suggest that this new sensor array can be used to identify the AE wave modes excited by PLBs in both in-plane and out-of-plane directions.

Keywords: acoustic emission; wave propagation; Lamb wave; modal decomposition; structural health monitoring; spectrum analysis; 2D FFT; pencil lead break test

Citation: Yao, X.; Vien, B.S.; Rajic, N.; Rosalie, C.; Rose, L.R.F.; Davies, C.; Chiu, W.K. Modal Decomposition of Acoustic Emissions from Pencil-Lead Breaks in an Isotropic Thin Plate. *Sensors* **2023**, *23*, 1988. https://doi.org/10.3390/s23041988

Academic Editors: Farook Sattar and Filippo Ubertini

Received: 4 January 2023
Revised: 6 February 2023
Accepted: 8 February 2023
Published: 10 February 2023

1. Introduction

There has been a significant amount of research on the development of structural health monitoring techniques to detect fatigue crack growth [1–3] in aircraft components. The uncontrolled propagation of an internal fatigue crack in a structural component under loading can lead to catastrophic failure, so the early detection of fatigue cracks is desirable. Many non-destructive testing (NDT) methods have been applied to crack detection in metallic structures, including ultrasonics, eddy current, and X-ray computed tomography [4–7].

Acoustic emission (AE) is one of the few passive NDT techniques available for detecting, characterising, quantifying damages, and predicting system failure [8–12]. AE is defined as the rapid release of transient elastic waves from localised sources in solids [13]. Such emissions occur in metal, rock, cement, and composites [9,14–17]. Research has been undertaken on AEs generated during static tensile tests and fatigue tests of metallic material [8,10,18–21].

Generally, the analysis of AE signals mainly focuses on two aspects: hit-related signatures and waveform features. Hit-related signatures include parameters such as rise time, amplitude, and count rate, and waveform features include dominant frequencies, wave modes, and power spectral entropy [22]. In a study of hit-related features, Roberts and Talebzadeh conducted an experiment on compact steel specimens [10], finding that it was possible to use the linear correlation between the AE count rate and the crack propagation rate to predict the remaining life of a structure. A similar linear relationship has also been indicated in other studies [23,24]. As a result, the count rate can be used as an empirical tool for predicting the fatigue life remaining in specimens. In addition, many researchers have used cluster analysis, a data analysis method for classification of AE hits, to identify different types of AE sources [17,25,26].

A number of researchers have suggested that research should focus not only on the AE hit-related features, but also on the AE waveform features [20,27–29]. To capture AE waveforms, the most popular types of AE sensors used are bonded piezoelectric sensors and commercial AE sensors. Because the acoustic environment in most structural applications is noisy, it is customary to exclude from consideration emissions that do not exceed a predetermined signal threshold [30]. While excluding low-amplitude AE signals can lead to information loss, it is a simple and effective noise-mitigation technique and is widely used in practical implementations of AE testing.

In investigations of AE waveform features, researchers have focused on the dominant frequency contribution in the waveforms. In a study in which hits were synchronised with fatigue loading and clustered into nine groups according to the time domain signal and frequency spectrum of waveforms, Bhuiyan et al. found that these groups of hits had distinct peak frequencies for particular loads; the authors associated these with different sources [31]. Other researchers have also reported a correspondence between dominant frequencies and different AE sources [25,32–34], suggesting that the frequency composition of AE waveforms may be relatable to the type of source. However, other studies have also shown that the frequency spectrum of AE waveforms can vary depending on the type of sensor used [31]. As a result, it is important to consider the dynamic response characteristics of a sensor when undertaking such studies [35].

Due to the possible variability in the monitored frequencies, wave mode identification is potentially a better option for providing information about AE sources. The identification and analysis of contributing modes in an AE is a practice referred to as Modal Acoustic Emission (MAE), which previous studies have shown to be an effective method of AE source localisation and identification [36,37]. Hamstad [38] showed that in-plane pencil lead break (PLB) tests excite a distinct ratio of the flexural to the extensional mode when applied at the near top surface and near mid-plane locations in a plate. In Maslouhi [39], during an increase in the fatigue life percentage from 0% to 100%, the wavelet transform (WT) coefficient of a flexural wave mode increased first and then decreased, while that of an extensional wave mode kept decreasing.

Wave modes can be identified using a range of established methods, including time–frequency plots. Using the frequency–time domain WT method, experimentally obtained frequency–time domain contours are compared to contours determined theoretically from Lamb wave dispersion relations [29,38,39]. However, the frequency–time domain method only works when the WT coefficients of emitted wave modes are well-separated in time because, when the source–sensor distance is very short, the WT coefficients for different wave modes are close to each other and thus become hard to distinguish. Commercially available AE sensors only record waveforms at a single location, which limits the ability of these sensors to be used for modal separation if the source–sensor distance is relatively small.

A potentially more useful way of differentiating wave modes is to conduct a wavenumber–frequency decomposition by performing a two-dimensional fast Fourier transform (2D FFT). However, to determine the wavenumber of contributing wave modes, the AE must be resolved spatially. The Linear Array for Modal Decomposition and Analysis (LAMDA) is a new thin-film piezoelectric sensor array made of polyvinylidene fluoride (PVDF). This sensor array consists of 16 separate equidistant sensing elements, enabling the recording of 16 waveforms simultaneously, thus allowing the wavenumber–frequency decomposition of an AE [40,41]. LAMDA has been used to identify the emitted wave modes of ball-drop impacts [40], but the capabilities of LAMDA in identifying wave modes generated by PLB are yet to be investigated. PLB testing is a well-established method for generating AE sources [38] that are broadly representative of actual AE sources in components under stress and, therefore, are significantly more representative than a ball-drop impact. Consequently, an experimental assessment of the performance of LAMDA in the modal decomposition of PLB signals is considered an important preliminary step toward the eventual implementation of this sensor for the AE monitoring of components under loading.

Using the LAMDA sensor, a commercial AE sensor, and finite element analysis (FEA), this study investigates the capabilities of the LAMDA sensor in decomposing AE wave modes generated by PLBs applied to a thin metal plate. Two different orientations of the PLBs were used to create AE sources, specifically along the longitudinal and the transverse direction of the test plate. In the present work, a novel implementation of the conventional PLB is used, where a reference hole is drilled in the thickness direction of the plate into which the lead is inserted, in order to control the location and orientation of the PLB. These test configurations give rise to multiple Lamb wave modes in the test plate, which are identified through modal decomposition using LAMDA sensors.

2. Methods

2.1. PLB Experiments

In this experimental study, PLB tests were conducted on the edge of a 400 mm × 400 mm × 0.6 mm aluminium alloy 5005 (Al5005) plate (refer to Figure 1a). PLB tests are usually conducted on the surface of the testing plate, but in practice, it is difficult to conduct PLB tests with a high degree of repeatability [42]. To ensure the PLBs were applied at precisely the same location and in the required direction, a 0.5 mm diameter and 0.5 mm depth reference hole was drilled on the surface of the testing plate (see Figure 1b). A hard-black pencil lead of 0.5 mm diameter was used to generate PLB sources. The pencil lead was inserted into the hole (details in Figure 1b) and broken in the out-of-plane (transverse) and in-plane (longitudinal) directions, as shown in Figure 1a.

Two sensing systems were used to record the AE signals. The first system was a Physical Acoustics Corp micro-SHM system with a 20 kHz–1000 kHz analog filter. This system operates via a threshold trigger and has a sampling rate of 10 MHz. The trigger threshold was set to 40 dB to avoid triggering during the insertion of the pencil lead into the drilled hole. The commercial AE sensor used with the system was a PKWDI sensor from Physical Acoustics Corp. This sensor can only record the response at a single point, which, in this study, was a position 70 mm radially away from the centroid of the source, labelled Location 1 in Figure 1c.

The second system was the LAMDA [40,41], which consists of 16 5 mm × 1 mm piezoelectric elements arranged in a linear array with a 1.27 mm pitch, mounted on a thin flexible polymer carrier. The total length of this sensor is ~20 mm. Signal recordings from LAMDA were acquired using a 16-channel Acousto Ultrasonic Structural health monitoring Array Module⁺ [43] with a 43 μs pre-trigger time, a 0.02 μs sampling rate, and a 50 kHz–5 MHz bandwidth. The output from LAMDA consisted of 16 voltage waveforms from the 16 sensing elements in the array. Further details on the architecture of the LAMDA and the hardware can be found in [40,41,43]. LAMDA sensors were bonded to the plate along the radial direction from the source, as illustrated in Figure 1c, so that the wavenumber–frequency plots obtained from the PLB source could be compared with the theoretical curves obtained from DISPERSE without adjustment for the angle of incidence [40]. Five LAMDA sensors were positioned 70 mm radially from the source and arranged in orientations of 0, 45, 90, 135, and 180 degrees with respect to the plate edge, denoted as locations 1–5, respectively, as shown in Figure 1c. Waveforms at locations 1 to 5 were recorded for longitudinal direction PLB tests, while only waveforms at locations 1, 2, and 3 were recorded for the transverse PLB. The reason for this is that for the transverse PLB test, the system is symmetric, so waveforms at locations 4 and 5 should theoretically be the same as those at locations 1 and 2. The PLB tests were repeated 10 times at each location, and the wavenumber–frequency plots shown in Section 3.2 represent an average taken over the 10 PLB tests.

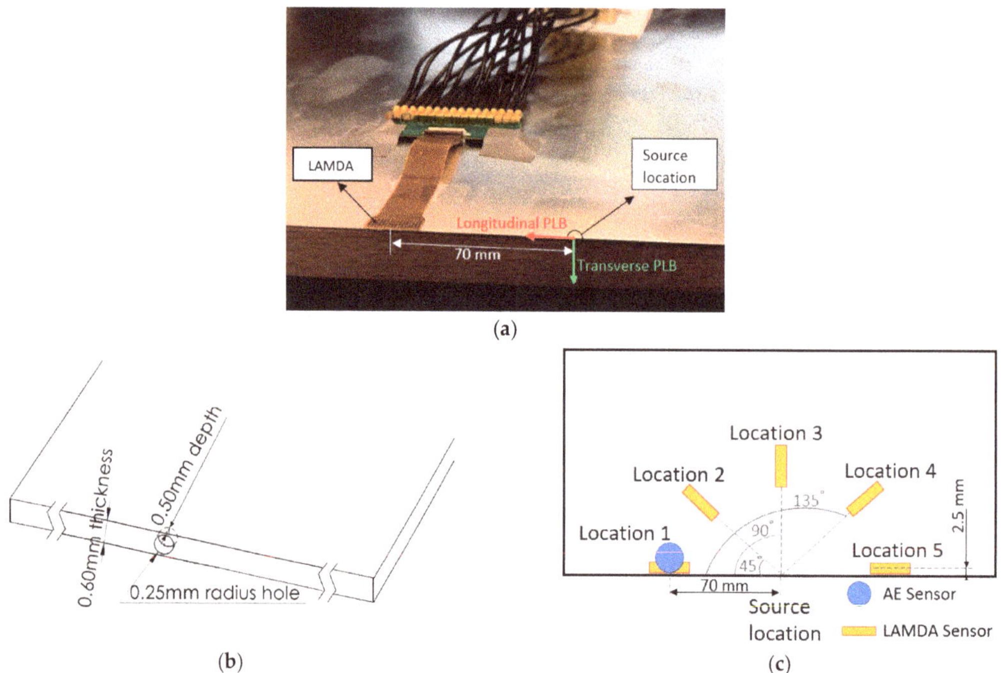

Figure 1. (**a**) PLB test set-up indicating lead-break directions relative to LAMDA orientation; (**b**) schematic showing PLB hole dimensions; (**c**) sensor locations relative to source.

2.2. FEA Analysis

The arrangement shown in Figure 2 was also modelled using the ANSYS 19.2 Explicit Dynamic FEA package. The aim was to compare measurements of the Lamb wave modes generated in the considered plate using PLBs acting in the longitudinal and transverse directions with corresponding numerical predictions.

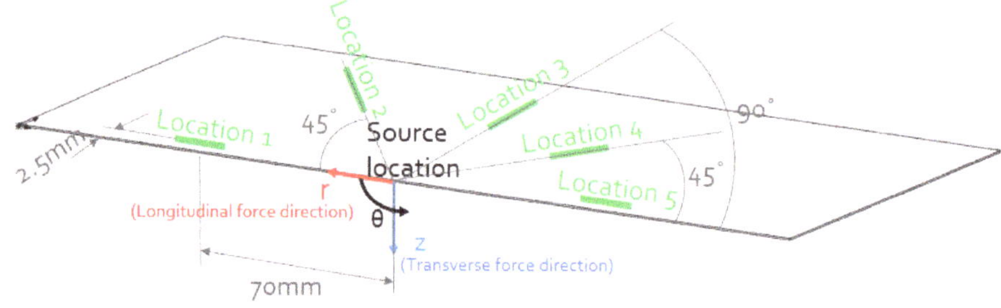

Figure 2. FEA model set up: source position, detecting position, and input force directions.

The plate considered in this simulation had side dimensions of 280 mm × 140 mm × 0.6 mm, and the detection locations were identical to those used in the corresponding experiment described in the previous section. Cylindrical coordinates were used in the velocity probe when extracting waveform information. The origin of the cylindrical coordinate system was at the source location. The out-of-plane direction is defined as the z-axis, the angular direction is defined as the θ-axis, and the radial direction as the r-axis, as shown in Figure 2. The material properties used in this simulation are consistent with those of the plate used in the experiments and for determining the Lamb wave dispersion curves using the DISPERSE software tool [44]; this is, specifically, Young's modulus of 68 GPa and Poisson's ratio of 0.33. The size of the hexahedral element was 0.2 mm in both length and width directions, and 0.075 mm in the plate thickness direction. The 0.2 mm length dimension corresponds to ~10 nodes per wavelength for the shortest wavelength of interest (the A0 mode), which is sufficient [45] to accurately decompose all Lamb wave modes below the maximum frequency of interest of 1 MHz. In the plate thickness direction, the 0.075 mm element dimension was determined to be sufficient for an accurate representation of the mode profiles [46].

A linear ramp forcing function was used to simulate the PLB input force [42]. As the dimensions of the hole were small (~20%), relative to the shortest wavelength, the influence of the hole was neglected, and the force was applied directly as a point force at the mid-plane location of the plate. Transverse and longitudinal 0.5 μs linear ramp nodal input force functions were applied and investigated separately (refer to Figure 2).

The time step used in the simulation was 0.05 μs, which corresponds to 20 time steps per cycle at the maximum considered frequency of 1 MHz, satisfying the ANSYS general requirement for accurate results [47]. At each recording location shown in Figure 2, 16 measurement probes were created at locations corresponding to the centre of each of the LAMDA sensing elements.

2.3. Theoretical Dispersion Curves

The DISPERSE software tool [44] was used to obtain theoretical dispersion curves (wavenumber–frequency characteristics) for S0, SH0 and A0 modes for the aluminium plate in the frequency range of 0–1000 kHz. These theoretical dispersion curves are shown in Figure 3a. Figure 3b shows the corresponding frequency–time plots, determined from the group velocity dispersion curves and the known source to sensor distance.

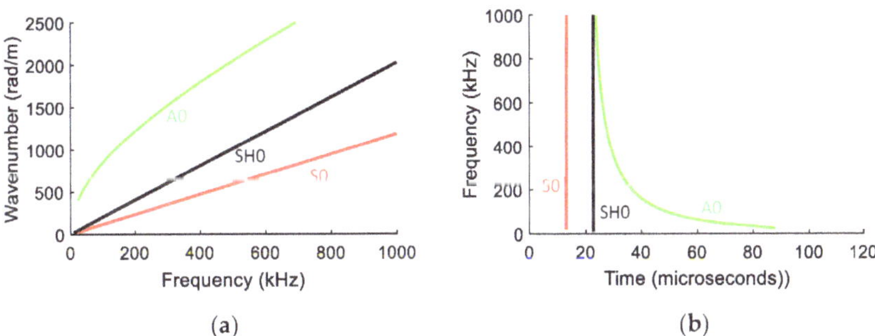

Figure 3. (**a**) Theoretical wavenumber–frequency dispersion curve for 0.6 mm thick aluminium plate. (**b**) Corresponding frequency vs. arrival time 70 mm from the source.

2.4. Validation of Method

To verify the mechanical properties of the aluminium plate, laser vibrometry was used to obtain the Lamb wave dispersion curves experimentally. In addition, in order to explore the impact of the LAMDA sensor on these dispersion curves, laser vibrometry was undertaken on the plate with and without a LAMDA attached. The corresponding dispersion curves were obtained by applying a 2D FFT to the laser vibrometer measurements. A three-dimensional laser vibrometer was used to obtain measurements of the in-plane and out-of-plane displacement components of Lamb waves in the plate. A piezoceramic disc element was used to selectively excite the A0 mode, and a transducer with an angle wedge was used to selectively excite the in-plane S0 mode [48]. To improve the signal-to-noise ratio of these measurements, retroreflective film was used over the scan paths, labelled 1–3 in Figure 4a: route 1 was 80 mm in length, route 2 was 20 mm in length and adjacent to the LAMDA sensor, and route 3 was 20 mm in length and was over the LAMDA sensor.

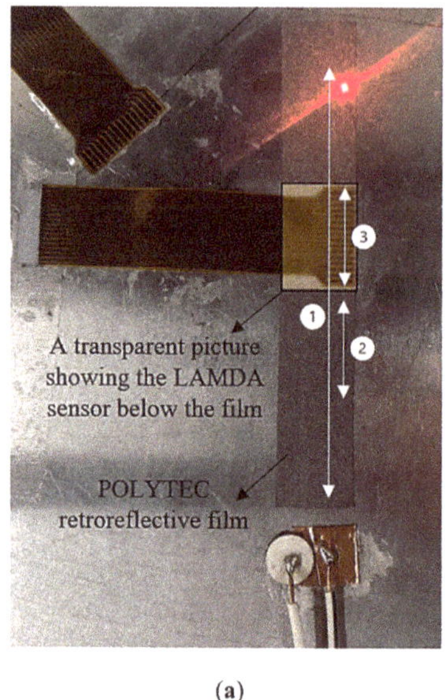

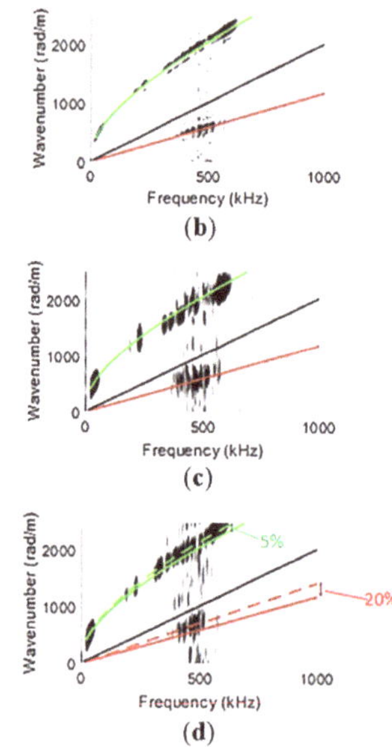

Figure 4. (**a**) Laser scanning routes labelled 1–3. Al5005 plate dispersion curve wavenumber versus frequency (**b**) of route 1 (80 mm); (**c**) route 2 (20 mm); and (**d**) route 3 (20 mm over LAMDA sensor).

As can be seen in Figure 4, there is generally good agreement between the theoretically and experimentally obtained dispersion curves for routes 1 and 2. However, for route 3, there is a small discrepancy of ~5% for the A0 mode, and a slightly larger discrepancy of ~20% for the S0 mode, indicating that the LAMDA has an influence on Lamb wave propagation in the plate. Given that the structure and elastic properties of LAMDA are known, it should be possible to compensate for this influence, but this was not required in the present study as the objective is simply to identify constituent modes.

3. Results and Discussion

3.1. Wavelet Transform Results

In Section 3.1, for the purpose of analysing the time–frequency signals, in order to obtain a direct comparison with the waveform acquired from the commercial AE sensor, the LAMDA, and the FEA at the same distance, the waveform acquired from the 8th (middle) sensing element in LAMDA and the FEA probe at location 1 were compared with those acquired from the Commercial AE sensor. The waveforms collected for the transverse PLB using the Commercial AE sensor, the eighth element of LAMDA, and the eighth probe of FEA are shown in Figure 5(a1), Figure 5(a2) and Figure 5(a3), respectively. Time–frequency plots and the corresponding theoretical curves (Figure 3b) are superimposed in Figure 5(b1–b3). The results of the longitudinal PLB tests are presented in Figure 6 with the same arrangement.

For reference, for the FEA generated plots, zero-time corresponds to the time when the force was applied; in addition, signal contributions arriving after 70 µs in Figure 5(b3) and after 30 µs in Figure 6(b3) correspond to reflections from plate edges, and can thus be ignored.

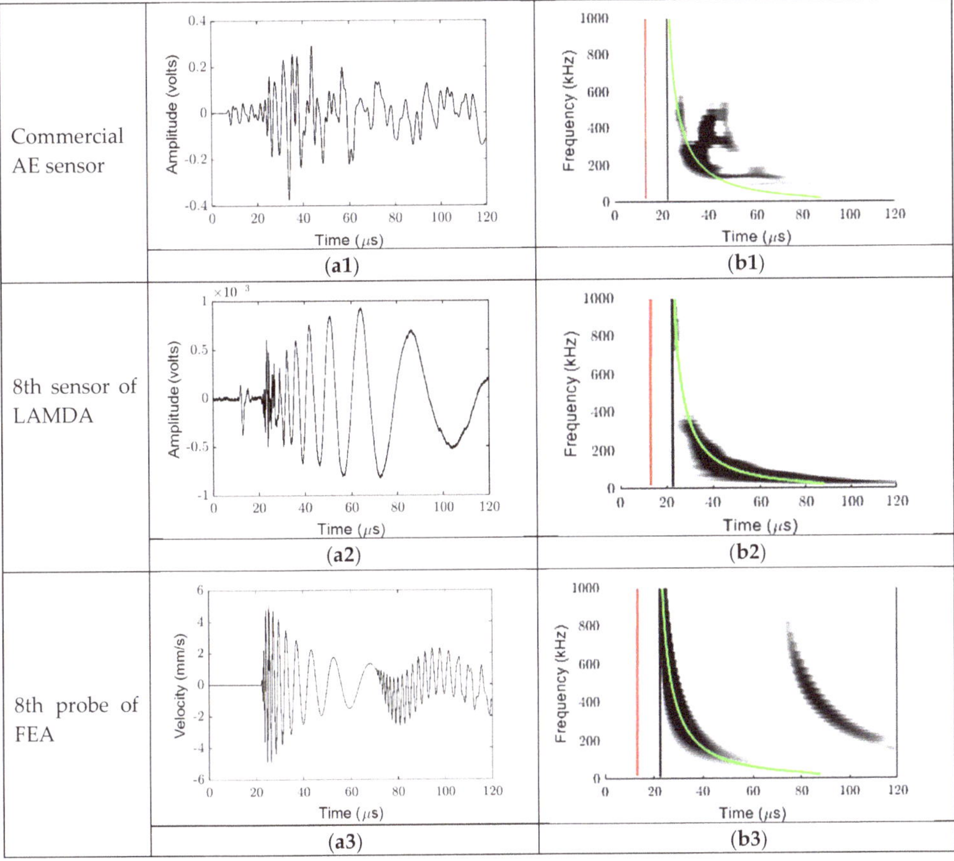

Figure 5. Transverse PLB results. Top row shows waveform (**left**) and corresponding time–frequency plot (**right**) for commercial AE sensor. Middle row as for top row but for LAMDA sensor. Bottom row as for top row but for FEA prediction corresponding to out-of-plane velocity component.

Since measurements from the conventional Commercial AE sensors and LAMDA could not be synchronised with the applied force, the waveform recordings needed to be appropriately time-shifted so that the wave arrival times were consistent with the theoretical group velocities shown in Figure 3b. The required time-shift was determined by ensuring that the experimental frequency–time contour was aligned with the theoretically determined contour. In the case of Figure 5, this contour corresponded to the A0 mode. This process was not as straightforward to apply for the longitudinal PLB results, particularly for the Commercial AE sensor data as it produced a contour markedly different to that of the theoretical result.

From Figure 5, it is clear that the A0 mode is dominant for the transverse PLB orientation, with little evidence of an S0 contribution; for the longitudinal PLB orientation results shown in Figure 6, both A0 and S0 are present in the experimentally obtained measurements; in the simulation results, it is unclear whether the second arrival corresponds to an SH0 mode or an edge wave, which is discussed further in Section 3.2.

The results in Figures 5 and 6 show that the ratio of the S0/A0 modes is dependent on the orientation of the PLB. From the transverse PLB results in Figure 5(b1–b3), the WT coefficient of the S0 mode was extremely small compared to that of the A0 mode. On the other hand, the results obtained with a longitudinal PLB show that the WT coefficient of the S0 mode was about the same as that of the A0 mode, see Figure 6(b1–b3). Therefore, the longitudinal (in-plane) PLB results lead to a higher ratio of the S0/A0 mode compared to the transverse (out-of-plane) PLB. As PLBs on the edge of a plate have been reported to be more similar to the buried AE sources [49], the difference in the modal ratio between the two PLB orientations suggests that LAMDA could potentially be used to distinguish different internal AE sources.

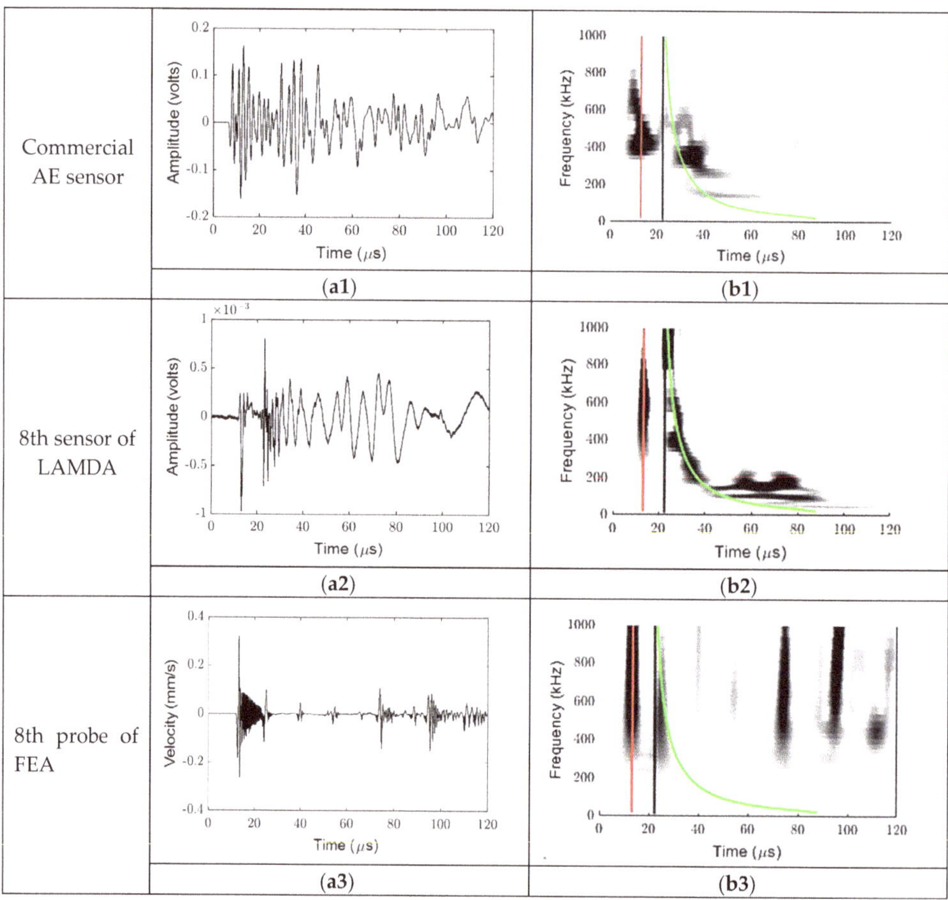

Figure 6. Longitudinal PLB results. Top row shows waveform (**left**) and corresponding time–frequency plot (**right**) for commercial AE sensor. Middle row as for top row but for LAMDA sensor. Bottom row as for top row but for FEA prediction corresponding to radial velocity component.

3.2. 2D FFT Results

The waveforms obtained from FEA were time-gated to exclude reflected wave contributions before applying a 2D FFT. The length of this time gate was determined from the known Lamb waves group velocities. The wavenumber–frequency plots corresponding to the transverse PLB orientation were normalised with respect to the maximum out-of-plane component amplitude at location 1, while those corresponding to the longitudinal PLB orientation were normalised with respect to the maximum radial component amplitude at location 1. After the normalisation, the wavenumber–frequency plots that corresponded to transverse PLBs from the FEA for the angular component contained no information, as the energy was extremely low, and thus was not shown. Similarly, the wavenumber–frequency plots that corresponded to longitudinal PLBs from the FEA for the out-of-plane component contained no information, and thus were not shown. After this, the results from the FEA were compared with experimental results from LAMDA at all detecting locations.

The experimental results and the FEA results that correspond to the transverse PLB orientation for sensing locations 1 to 3 are shown in Figure 7. Figure 7(a1–a3) are wavenumber–frequency plots obtained from FEA waveforms for the radial component (r-axis); Figure 7(b1–b3) are wavenumber–frequency plots obtained from FEA for the out-of-plane component (z-axis); and Figure 7(c1–c3) are wavenumber–frequency plots obtained from waveforms recorded by LAMDA. When the PLB is applied transversely, the geometry and loading of the plate are symmetric about the centreline, so waveforms at locations 4 and 5 should theoretically be the same as at locations 1 and 2, respectively. Thus, only results from locations 1 to 3 are shown here.

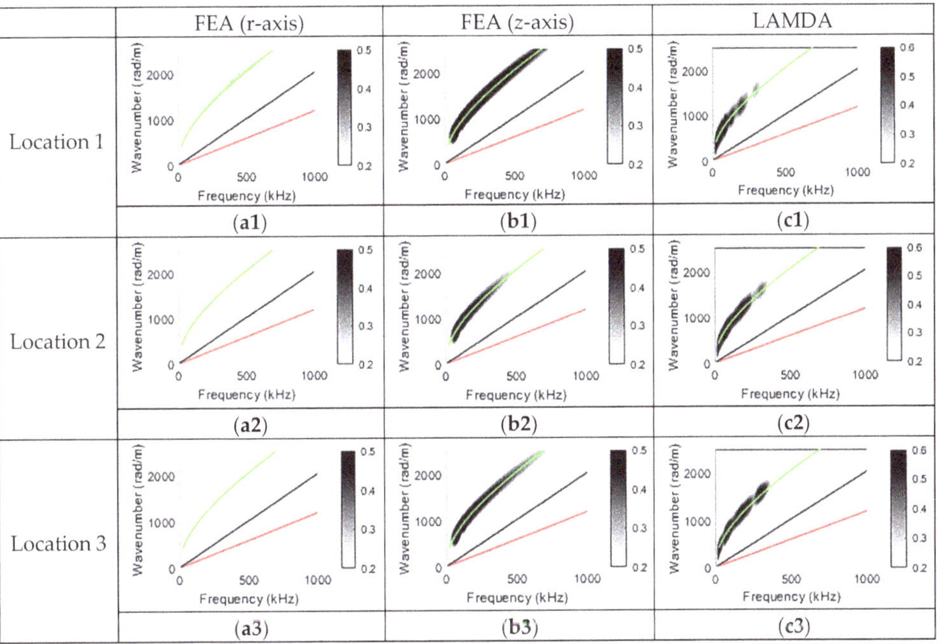

Figure 7. Transverse PLB results. Top row shows wavenumber–frequency plots corresponding to the radial component obtained from FEA, the out-of-plane component obtained from FEA, and from the LAMDA sensor at location 1. Middle row as for top row but for location 2. Bottom row as for top row but for location 3.

Based on the observation of Figure 7, it can be seen that this forcing configuration leads to the generation of dominantly asymmetric wave modes. The wavenumber–frequency plots that correspond to PLB tests using LAMDA sensors were dominated by A0 mode contributions at all three locations, and the wavenumber–frequency plots from FEA, generated using the out-of-plane component, were dominated by a strong A0 mode contribution, and a very weak A0 mode at location 1, which corresponded to the radial component. The dominance of the A0 mode is consistent with findings from Hamstad's study [38], where a monopole out-of-plane source near the mid-plane gave rise to the A0 mode only. Additionally, the wavenumber–frequency plots corresponding to the out-of-plane component in the FEA waveform had a slightly higher energy at location 1 than at locations 2 and 3. A possible reason for this could be the presence of an asymmetric edge wave [50]; however, this was not confirmed. The weak A0 mode present in the radial component results shown in Figure 7(a1) is assumed to correspond to the leakage of the asymmetric edge wave.

Figure 8(a1–a5) are wavenumber–frequency plots of waveforms from FEA that correspond to the radial component (r-axis); Figure 8(b1–b5) are wavenumber–frequency plots of waveforms from FEA that correspond to the angular component (θ-axis); and Figure 8(c1–c5) are wavenumber–frequency plots of waveforms recorded by LAMDA sensors.

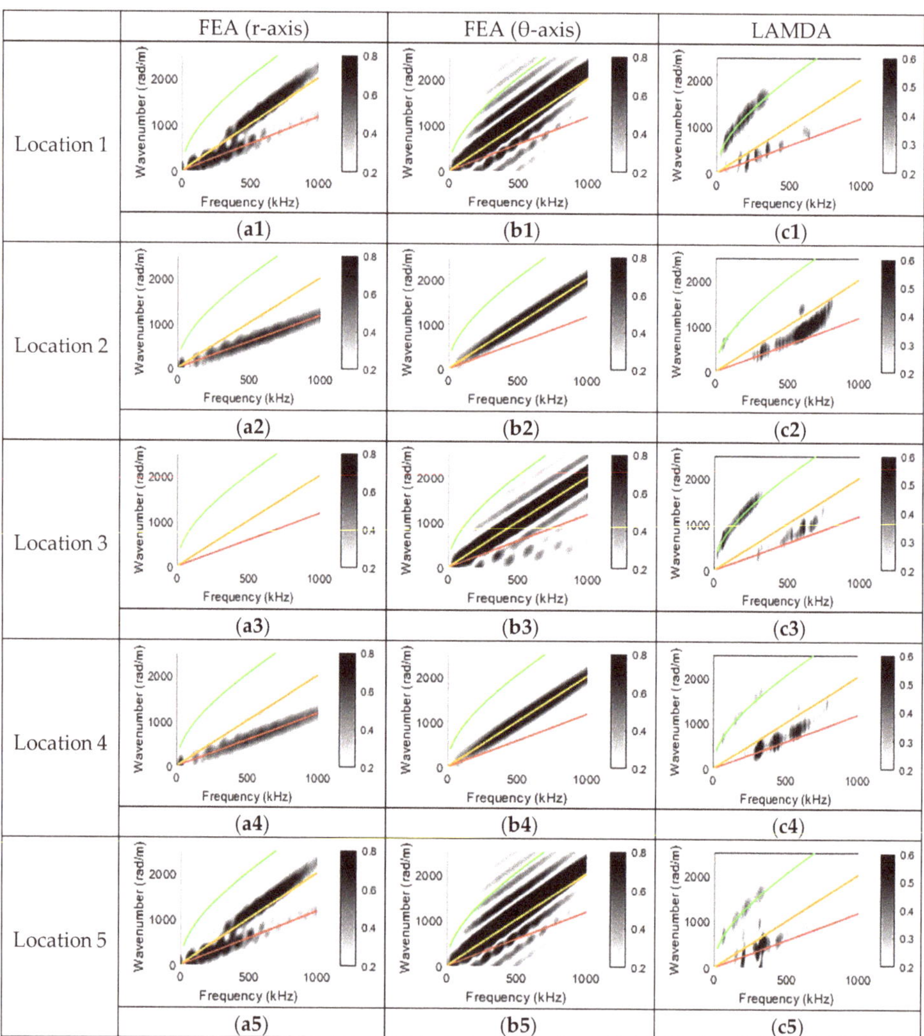

Figure 8. Longitudinal PLB results. Top row shows wavenumber–frequency plots corresponding to the radial component obtained from FEA, the angular component obtained from FEA, and from the LAMDA sensor at location 1. Second row as for top row but for location 2. Third row as for top row but for location 3. Fourth row as for top row but for location 4. Bottom row as for top row but for location 5.

As can be seen in Figure 8(a1–a5), the wavenumber–frequency plots from FEA that correspond to the radial component showed S0 mode content at detecting locations 1, 2, 4, and 5, with almost no energy at location 3. Locations 2 and 4 contained stronger S0 mode contributions above 500 kHz, which was consistent with the corresponding wavenumber–frequency plots obtained from waveforms recorded by LAMDA sensors.

In Figure 8(b1–b5), the wavenumber–frequency plots from FEA waveforms that correspond to the angular component (θ-axis) showed that the SH0 mode appeared to be dominant at locations 2, 3, and 4. However, it was observed that the wave mode contributions at locations 1 and 5 did not align perfectly with the known SH0 mode wavenumber–frequency curve. Moreover, the energy of the suspected SH0 mode at locations 1 and 5 was higher compared to locations 2 and 4. The reason for this was that the symmetrical Rayleigh-like edge wave propagating near the edge of the plate was also captured at locations 1 and 5 [51]. This edge wave is also evident in the radial displacement component at locations 1 and 5, shown in Figure 8(a1,a5). This wave mode could also be confirmed to be an edge wave rather than the SH0 mode from Figure 6(b3), as the velocity of the symmetrical edge wave was expected to be around 90% of the SH0 mode [52].

Locations 1 and 5 and locations 2 and 4 have the same empirical Lamb wave dispersion curves, corresponding to both radial and angular components of FEA and LAMDA sensors. This demonstrates that the longitudinal PLB tests generate a symmetric wave-scattering pattern with respect to the plate centreline.

For the longitudinal PLB tests, both A0 and S0 mode contributions were observed in the wavenumber–frequency plots obtained from the LAMDA sensors at all five locations. Comparing the wavenumber–frequency plots from FEA predictions and the PLB tests, it is observed that the S0 mode observed in Figure 8(c1–c5) was most likely from the radial component of the collected waveform. Furthermore, an A0 mode contribution is observed in the LAMDA wavenumber–frequency plot that is not present in the corresponding FEA predictions. This is because, when conducting the longitudinal PLB test manually, it was almost impossible to keep the force strictly vertical to the plate. As a result, the longitudinal PLB is likely to also have contained a weak transverse component, leading to the excitation of A0. Observing the wavenumber–frequency plots from LAMDA in Figure 7(c1–c3) and Figure 8(c1–c5), it is apparent that a longitudinal PLB results in an additional S0 modal contribution compared to a transverse PLB. For a longitudinal PLB, the S0 modal contribution above 500 kHz is more significant at locations 2 and 4 than at locations 1 and 5.

The presented results indicate that the orientation of a PLB affects the modal composition of the resulting AE. Such modal contributions can be difficult to distinguish using time–frequency analysis only, i.e., synchronising the time–frequency plots with the theoretical group velocity curves may not be sufficient when there are multiple interfering wave modes. By contrast, the wavenumber–frequency decomposition implemented using LAMDA allows contributing wave modes to be determined regardless of time interference. Consequently, LAMDA could provide a useful basis for the modal decomposition of Lamb wave packets generated by other AE sources. Such decomposition may aid in AE source localisation and characterisation.

4. Conclusions

This study has demonstrated the ability of a novel multi-sensor array to achieve the modal decomposition of the acoustic emission from PLBs when different orientations of the acoustic source are applied. It was shown that a transverse PLB produces a dominant A0 mode, while a longitudinal PLB produces a combination of A0 and S0 modes. These results are consistent with FEA simulations and previously reported AE studies. The current work includes the investigation of only zero-order wave modes and the application of the LAMDA sensors on isotropic plates. Ultimately, this capability will be used to study the wave mode composition of the AEs generated during fatigue crack propagation in coupons. Future work will investigate the higher-order wave modes that are excited by acoustic sources arising from damage initiation and development. Future work will also include the use of LAMDA to identify the wave modes generated by the different damage mechanisms in composite panels.

Author Contributions: Conceptualization, X.Y., W.K.C. and N.R.; methodology, X.Y., B.S.V. and W.K.C.; software, X.Y.; validation, B.S.V., C.R., N.R. and W.K.C.; formal analysis, X.Y.; data curation, X.Y.; writing—original draft preparation, X.Y.; writing—review and editing, B.S.V., N.R., W.K.C., C.R., L.R.F.R. and C.D.; supervision, W.K.C. All authors have read and agreed to the published version of the manuscript.

Funding: This research received no external funding.

Institutional Review Board Statement: Not applicable.

Informed Consent Statement: Not applicable.

Data Availability Statement: Not applicable.

Acknowledgments: The authors wish to thank Joel Smithard for his assistance with installing LAMDA sensors.

Conflicts of Interest: The authors declare no conflict of interest.

References

1. Jones, J. Enhancing the Accuracy of Advanced High Temperature Mechanical Testing through Thermography. *Appl. Sci.* **2018**, *8*, 380. [CrossRef]
2. Lee, C.K.; Scholey, J.J.; Wilcox, P.D.; Wisnom, M.R.; Friswell, M.I.; Drinkwater, B.W. Acoustic emission during fatigue crack growth in aluminium plates. In Proceedings of the 9th European Conference on NDT (ECNDT 2006), Berlin, Germany, 25–29 September 2006; pp. 25–29.
3. Joseph, R.; Giurgiutiu, V. Analytical and Experimental Study of Fatigue-Crack-Growth AE Signals in Thin Sheet Metals. *Sensors* **2020**, *20*, 5835. [CrossRef]
4. Everton, S.; Dickens, P.; Tuck, C.; Dutton, B. Evaluation of laser ultrasonic testing for inspection of metal additive manufacturing. *Proc. SPIE* **2015**, *9353*, 935316. [CrossRef]
5. Mazurek, M.; Austin, R. Nondestructive Inspection of Additive Manufactured Parts in the Aerospace Industry. *DSIAC J.* **2016**, *3*, 13–22.
6. Thompson, A.; Maskery, I.; Leach, R.K. X-ray computed tomography for additive manufacturing: A review. *Meas. Sci. Technol.* **2016**, *27*, 072001. [CrossRef]
7. Jiao, S.; Cheng, L.; Li, X.; Li, P.; Ding, H. Monitoring fatigue cracks of a metal structure using an eddy current sensor. *EURASIP J. Wirel. Commun. Netw.* **2016**, *2016*, 118. [CrossRef]
8. Lindley, T.; Palmer, I.; Richards, C. Acoustic emission monitoring of fatigue crack growth. *Mater. Sci. Eng.* **1978**, *32*, 1–15. [CrossRef]
9. Aggelis, D.; Kordatos, E.; Matikas, T. Acoustic emission for fatigue damage characterization in metal plates. *Mech. Res. Commun.* **2011**, *38*, 106–110. [CrossRef]
10. Roberts, T.; Talebzadeh, M. Acoustic emission monitoring of fatigue crack propagation. *J. Constr. Steel Res.* **2003**, *59*, 695–712. [CrossRef]
11. Das, A.K.; Leung, C.K. A fundamental method for prediction of failure of strain hardening cementitious composites without prior information. *Cem. Concr. Compos.* **2020**, *114*, 103745. [CrossRef]
12. Das, A.K.; Leung, C.K. Fast Tomography: A greedy, heuristic, mesh size–independent methodology for local velocity reconstruction for AE waves in distance decaying environment in semi real-time. *Struct. Health Monit.* **2022**, *21*, 1555–1573. [CrossRef]
13. Prosser, W.H. The propagation characteristics of the plate modes of acoustic emission waves in thin aluminum plates and thin graphite/epoxy composite plates and tubes. *J. Acoust. Soc. Am.* **1992**, *92*, 3441–3442. [CrossRef]
14. Eaton, M.; Pullin, R.; Holford, K. Acoustic emission source location in composite materials using Delta T Mapping. *Compos. Part A Appl. Sci. Manuf.* **2012**, *43*, 856–863. [CrossRef]
15. Fortin, J.; Stanchits, S.; Dresen, G.; Gueguen, Y. Acoustic Emissions Monitoring during Inelastic Deformation of Porous Sandstone: Comparison of Three Modes of Deformation. *Pure Appl. Geophys.* **2009**, *166*, 823–841. [CrossRef]
16. Al-Jumaili, S.K.; Eaton, M.J.; Holford, K.M.; Pearson, M.R.; Crivelli, D.; Pullin, R. Characterisation of fatigue damage in composites using an Acoustic Emission Parameter Correction Technique. *Compos. Part B Eng.* **2018**, *151*, 237–244. [CrossRef]
17. Zhou, W.; Zhao, W.-Z.; Zhang, Y.-N.; Ding, Z.-J. Cluster analysis of acoustic emission signals and deformation measurement for delaminated glass fiber epoxy composites. *Compos. Struct.* **2018**, *195*, 349–358. [CrossRef]
18. Cousland, S.; Scala, C. Acoustic emission during the plastic deformation of aluminium alloys 2024 and 2124. *Mater. Sci. Eng.* **1983**, *57*, 23–29. [CrossRef]
19. Scala, C.; Cousland, S. Acoustic emission during fatigue crack propagation in the aluminium alloys 2024 and 2124. *Mater. Sci. Eng.* **1983**, *61*, 211–218. [CrossRef]
20. Bhuiyan, Y.; Lin, B.; Giurgiutiu, V. Acoustic emission sensor effect and waveform evolution during fatigue crack growth in thin metallic plate. *J. Intell. Mater. Syst. Struct.* **2017**, *29*, 1275–1284. [CrossRef]

21. Yao, X.; Vien, B.S.; Davies, C.; Chiu, W.K. Acoustic Emission Source Characterisation during Fatigue Crack Growth in Al 2024-T3 Specimens. *Sensors* **2022**, *22*, 8796. [CrossRef]

22. Das, A.K.; Leung, C.K. Power spectral entropy of acoustic emission signal as a new damage indicator to identify the operating regime of strain hardening cementitious composites. *Cem. Concr. Compos.* **2019**, *104*, 103409. [CrossRef]

23. Chang, H.; Han, E.; Wang, J.Q.; Ke, W. Acoustic emission study of corrosion fatigue crack propagation mechanism for LY12CZ and 7075-T6 aluminum alloys. *J. Mater. Sci.* **2005**, *40*, 5669–5674. [CrossRef]

24. Hyun, J.-S.; Song, G.-W.; Kim, B.-S.; Park, S.-M. The evaluation of fatigue crack propagation by acoustic emission. In Proceedings of the 26th European Conference on Acoustic Emission Testing, Berlin, Germany, 15–17 September 2004.

25. Bi, H.; Li, Z.; Hu, D.; Toku-Gyamerah, I.; Cheng, Y. Cluster analysis of acoustic emission signals in pitting corrosion of low carbon steel. *Mater. Werkst.* **2015**, *46*, 736–746. [CrossRef]

26. Sause, M.; Gribov, A.; Unwin, A.; Horn, S. Pattern recognition approach to identify natural clusters of acoustic emission signals. *Pattern Recognit. Lett.* **2012**, *33*, 17–23. [CrossRef]

27. Geng, R. Modern acoustic emission technique and its application in aviation industry. *Ultrasonics* **2006**, *44* (Suppl. S1), e1025–e1029. [CrossRef]

28. Wirtz, S.F.; Söffker, D. Application of Shape-based Similarity Measures to Classification of Acoustic Emission Waveforms. *Struct. Health Monit.* 2017. [CrossRef]

29. Kaphle, M.; Tan, A.C.C.; Thambiratnam, D.P.; Chan, T.H.T. Identification of acoustic emission wave modes for accurate source location in plate-like structures. *Struct. Control. Health Monit.* **2012**, *19*, 187–198. [CrossRef]

30. Unnorsson, R. Hit Detection and Determination in AE Bursts. In *Acoustic Emission—Research and Applications*; IntechOpen: London, UK, 2013.

31. Bhuiyan, M.Y.; Giurgiutiu, V. The signatures of acoustic emission waveforms from fatigue crack advancing in thin metallic plates. *Smart Mater. Struct.* **2018**, *27*, 015019. [CrossRef]

32. Wisner, B.; Mazur, K.; Perumal, V.; Baxevanakis, K.; An, L.; Feng, G.; Kontsos, A. Acoustic emission signal processing framework to identify fracture in aluminum alloys. *Eng. Fract. Mech.* **2019**, *210*, 367–380. [CrossRef]

33. Pomponi, E.; Vinogradov, A. A real-time approach to acoustic emission clustering. *Mech. Syst. Signal Process.* **2013**, *40*, 791–804. [CrossRef]

34. Saha, I.; Sagar, R.V. Classification of the acoustic emissions generated during the tensile fracture process in steel fibre reinforced concrete using a waveform-based clustering method. *Constr. Build. Mater.* **2021**, *294*, 123541. [CrossRef]

35. Sause, M.; Hamstad, M. Acoustic Emission Analysis. In *Comprehensive Composite Materials II*; Elsevier: Amsterdam, The Netherlands, 2018; pp. 291–326.

36. Hassan, F.; Bin Mahmood, A.K.; Yahya, N.; Saboor, A.; Abbas, M.Z.; Khan, Z.; Rimsan, M. State-of-the-Art Review on the Acoustic Emission Source Localization Techniques. *IEEE Access* **2021**, *9*, 101246–101266. [CrossRef]

37. Yu, F.; Wu, Q.; Okabe, Y.; Kobayashi, S.; Saito, K. The identification of damage types in carbon fiber–reinforced plastic cross-ply laminates using a novel fiber-optic acoustic emission sensor. *Struct. Health Monit.* **2016**, *15*, 93–103. [CrossRef]

38. Hamstad, M. Acoustic emission signals generated by monopole (pencil lead break) versus dipole sources: Finite element modeling and experiments. *J. Acoust. Emiss.* **2007**, *25*, 92–106.

39. Maslouhi, A. Fatigue crack growth monitoring in aluminum using acoustic emission and acousto-ultrasonic methods. *Struct. Control. Health Monit.* **2011**, *18*, 790–806. [CrossRef]

40. Rajic, N.; Rosalie, C.; Van Der Velden, S.; Rose, L.F.; Smithard, J.; Chiu, W.K. A novel high density piezoelectric sensing capability for in situ modal decomposition of acoustic emissions. In Proceedings of the 9th European Workshop on Structural Health Monitoring, Manchester, UK, 10–13 July 2018.

41. Rajic, N.; Rosalie, C.; Vien, B.S.; Van Der Velden, S.; Rose, L.F.; Smithard, J.; Chiu, W.K. In situ wavenumber–frequency modal decomposition of acoustic emissions. *Struct. Health Monit.* **2020**, *19*, 2033–2050. [CrossRef]

42. Sause, M.G. Investigation of pencil-lead breaks as acoustic emission sources. *J. Acoust. Emiss.* **2011**, *29*, 184–196.

43. Smithard, J.; Rajic, N.; Van Der Velden, S.; Norman, P.; Rosalie, C.; Galea, S.; Mei, H.; Lin, B.; Giurgiutiu, V. An Advanced Multi-Sensor Acousto-Ultrasonic Structural Health Monitoring System: Development and Aerospace Demonstration. *Materials* **2017**, *10*, 832. [CrossRef]

44. Brian Pavlakovic, M.L. *Disperse User's Manual*; Non-Destructive Testing Laboratory, Department of Mechanical Engineering, Imperial College: London, UK, 2001.

45. Alleyne, D.; Cawley, P. A two-dimensional Fourier transform method for the measurement of propagating multimode signals. *J. Acoust. Soc. Am.* **1991**, *89*, 1159–1168. [CrossRef]

46. Nadarajah, N. Novel quantitative sizing of delamination in hard-to-inspect locations. In *Mechanical and Aerospace Engineering*; Monash University: Clayton, VIC, Australia, 2017; p. 256.

47. ANSYS. *User's Manual for Revision 5.0*; Swanson Analysis Systems: Canonsburg, PA, USA, 1994.

48. Vien, B.S.; Chiu, W.K.; Rose, L.R.F. Experimental Investigation of Second-Harmonic Lamb Wave Generation in Additively Manufactured Aluminum. *J. Nondestruct. Eval. Diagn. Progn. Eng. Syst.* **2018**, *1*, 14. [CrossRef]

49. Hamstad, M.A. Comparison of wavelet transform and Choi-Williams distribution to determine group velocities for different acoustic emission sensors. *J. Acoust. Emiss.* **2008**, *26*, 40–59.

50. Vien, B.S.; Chiu, W.; Rose, L. An experimental study on the scattering of edge guided waves by a small edge crack in an isotropic plate. In Proceedings of the 8th European Workshop on Structural Health Monitoring (EWSHM 2016), Bilbao, Spain, 5–8 July 2016.
51. Vien, B.; Nadarajah, N.; Rose, F.; Chiu, W.K. Scattering of the Symmetrical Edge-guided Wave by a Small Edge Crack in an Isotropic Plate. *Struct. Health Monit.* 2015. [CrossRef]
52. Vien, B.S. *Lamb Wave Approach to Identify Hidden Cracks in Hard-to-Inspect Areas of Metallic Structures*; Monash University: Clayton, VIC, Australia, 2016.

Article

Using Hybrid HMM/DNN Embedding Extractor Models in Computational Paralinguistic Tasks

Mercedes Vetráb [1,*] and Gábor Gosztolya [1,2]

[1] Institute of Informatics, University of Szeged, H-6720 Szeged, Hungary; ggabor@inf.u-szeged.hu
[2] ELKH-SZTE Research Group on Artificial Intelligence, H-6720 Szeged, Hungary
* Correspondence: vetrabm@inf.u-szeged.hu

Abstract: The field of computational paralinguistics emerged from automatic speech processing, and it covers a wide range of tasks involving different phenomena present in human speech. It focuses on the non-verbal content of human speech, including tasks such as spoken emotion recognition, conflict intensity estimation and sleepiness detection from speech, showing straightforward application possibilities for remote monitoring with acoustic sensors. The two main technical issues present in computational paralinguistics are (1) handling varying-length utterances with traditional classifiers and (2) training models on relatively small corpora. In this study, we present a method that combines automatic speech recognition and paralinguistic approaches, which is able to handle both of these technical issues. That is, we trained a HMM/DNN hybrid acoustic model on a general ASR corpus, which was then used as a source of embeddings employed as features for several paralinguistic tasks. To convert the local embeddings into utterance-level features, we experimented with five different aggregation methods, namely mean, standard deviation, skewness, kurtosis and the ratio of non-zero activations. Our results show that the proposed feature extraction technique consistently outperforms the widely used x-vector method used as the baseline, independently of the actual paralinguistic task investigated. Furthermore, the aggregation techniques could be combined effectively as well, leading to further improvements depending on the task and the layer of the neural network serving as the source of the local embeddings. Overall, based on our experimental results, the proposed method can be considered as a competitive and resource-efficient approach for a wide range of computational paralinguistic tasks.

Keywords: hidden Markov model; deep neural network; embedding; hybrid acoustic model; computational paralinguistics

Citation: Vetráb, M.; Gosztolya, G. Using Hybrid HMM/DNN Embedding Extractor Models in Computational Paralinguistic Tasks. *Sensors* **2023**, *23*, 5208. https://doi.org/10.3390/s23115208

Academic Editors: Farook Sattar, Niladri Bihari Puhan and Reza Fazel-Rezai

Received: 25 April 2023
Revised: 21 May 2023
Accepted: 23 May 2023
Published: 30 May 2023

1. Introduction

Historically, the main research topic of automatic speech processing has been automatic speech recognition (ASR). In ASR, we have to automatically create a transcription for audio (e.g., recording or utterance). From the 1990s to the present, several other topics have received more attention, such as speaker recognition and diarisation ("who's speaking when") [1], speech compression [2], cognitive load measurement [3,4], detecting Parkinson's [5–7] or Alzheimer's [8–10] disease, identifying Multiple Sclerosis symptoms [11] or assessing the level of depression [12]. Besides these tasks, a complete subfield has arisen, concerning phenomena present in human speech, containing tasks such as age and gender recognition [13], emotion recognition [14,15], identifying laughter events [16], estimating the degree of sleepiness [17] or conflict intensity [18] and detecting whether the speaker is intoxicated [19]. These subtopics belong to the field of computational paralinguistic, which has recently started to receive more interest. In this field, instead of generating transcriptions, we seek to identify other phenomena present in a speech signal, focusing on the non-verbal content of human speech. It refers to the non-verbal aspects of human communication such as tone of voice and other vocal cues. These cues play an essential role in understanding human communication and can significantly impact the meaning and interpretation of spoken language. Paralinguistic features are often extracted from audio data in machine learning applications, using techniques such as speech analysis and audio signal processing. With the various acoustic sensors becoming increasingly cheaper (and, in parallel, more and more widespread), they could allow the remote monitoring of speaker traits and states. It can opening up a wide range of potential applications such as warning when a vehicle driver is too tired or sleepy. A major boost for paralinguistic was the Interspeech Computational Paralinguistic Challenge (ComParE, later renamed ACM Multimedia Computational Paralinguistic Challenge), which has been held annually since 2009 [20–22]. It has led to the development of publicly available datasets and produced a consensus among standard methods, tools and evaluation metrics.

We have to consider slight but really important differences between computational paralinguistic and ASR. One important dissimilarity is the focus of the two area. In automatic speech recognition, we concentrate on the spoken content of the speech signal and try to ignore any other information present (such as the age, gender, native language or the inner feelings of the speaker), as these are considered irrelevant. On the contrary, in computational paralinguistics, we focus on one of the latter speaker states and traits, and disregard the actual words uttered. This dissimilarity of focuses actually leads to a significant technical difference as well.

One technical difference came from the relationship between the length of the input and the output. In ASR the output is the correct phone sequence for a given speech recording and the size of the input utterance is roughly proportional to the length of the output: we expect more words uttered over a longer period of time. Technically, this means that in the traditional speech recognition paradigm the classification step is performed at the local level, handling the audio in small, equal-sized parts called frames. In this case, traditional classifiers such as Gaussian Mixture Models (GMMs [23]), and more recently deep neural networks (DNNs [24]), are used to estimate the local likelihood of the different phones and phone-derived classes from standard frame-level features (such as MFCCs [23]). These local likelihood estimates (the classes of the frames) are then combined over the time axis in the subsequent step (for example, using a hidden Markov model [25]) to obtain the utterance-level output [26] (i.e., a time-aligned sequence of phones). In contrast, in the field of computational paralinguistics, we need to associate different lengths of audio recording inputs with a single label output. Here, the input speech signal is split into larger chunks of continuous speech (such as one sentence), and these chunks are treated as separate units. A given artifact or speaker state (e.g., emotion) is assigned to these chunks ("utterances" or "recordings"). From a machine learning perspective, this means that one such utterance will be one machine learning example. The traditional machine learning models can only process fixed-length inputs, so we need to convert our varying-length recordings into fixed-length feature vectors. We have to calculate a fixed-length feature vector out of these varying-length recording, because, traditional classifiers are unable to handle a concatenation of frame-level attributes of varying lengths as input features. Perhaps the most straightforward solution for that is to take the frames and aggregate the local results instead of combining them. The name of aggregation, is refers to a process rather than a specific mathematical method. The use of this statistic conversion allows us to obtain a fixed-sized utterance-level vector, so the length of the output no longer depends on the length of the input recording.

Another important difference between the two area concerns the size of the databases. For ASR nowadays, hundreds or even thousands of hours' worth of databases are available, meaning tens or hundreds of millions training examples for frame-level phoneme classification. This allows researchers to train DNN models, which are known to be data-greedy. By comparison, computational paralinguistics typically has small corpora, because each task usually requires specific recording protocols and annotations. This drawback means that usually there are just a few hundred (or at most a few thousand) examples for a specific paralinguistic subtopic or class. Therefore, traditional classifier models (which can be well trained on very few data) are used instead of end-to-end neural networks. It is common to employ learning methods for classification such as Support Vector Machines (SVM [27]). Deep neural network machine learning techniques (e.g., fine-tuning) are only rarely used in computational paralinguistics. Standard solutions are still dominant, for example low-level descriptors (e.g., energy, spectral and cepstral (MFCC)) and voicing-related attributes for frame-wise computing; statistic conversion techniques, such as mean and standard deviation for aggregation; and classification methods such as SVMs. In the last decade, there have been few research studies that have applied complex machine learning models and usually the performance is strongly task-dependent.

Based on our previous studies [28,29], we developed a method shown in Figure 1 that combines ASR and paralinguistic approaches. For frame-wise computing, we followed standard ASR principles and we used DNNs to perform a frame-level feature extraction. Afterwards, to aggregate these features, we used more or less traditional computational paralinguistics techniques such as standard deviation and kurtosis. In the end, we employed SVM models to perform the classification task.

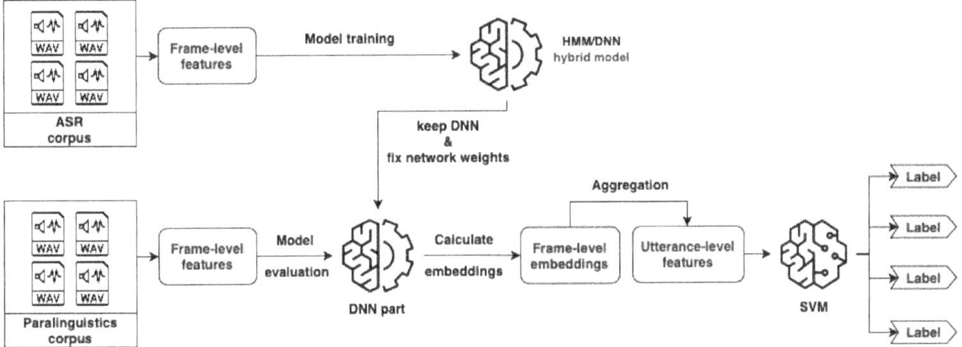

Figure 1. Hybrid HMM/DNN model workflow for paraliguistics task.

2. Processing Paralinguistic Data

Figure 1 shows the complete workflow of our experimental method from the preprocessing of the ASR corpus to the classification of the paralinguistics corpus. As mentioned above, we created a hybrid method that follows ASR principles and computational paralinguistics principles too. In this section, we will describe the workflow in a step-by-step fashion.

2.1. HMM/DNN Hybrid Model

Hidden Markov models (more specifically, HMM/GMMs) used to be the state of the art in automatic speech recognition. They consisted of a local GMM module, being responsible for supplying local (i.e., frame-level) phonetic probability estimates, while the HMM part was responsible for combining these local estimates into utterance-level phone sequences [30]. After deep neural networks were invented, these HMM/GMM models were developed into HMM/DNN [31] hybrid models by replacing the GMM component with a deep neural network, still operating locally (i.e., on the frame level). Soon it became widespread knowledge how to efficiently train and employ HMM/DNN hybrid models. In our study, we seek to employ this knowledge by training such a DNN acoustic model, and using this as the base of our feature extractor for computational paralinguistic tasks.

The HMM/DNN model has two parts. The first part is the deep neural network, while the second part is a hidden Markov model. The outputs of the DNN will be the input of the HMM. The DNN gives frame-level estimations, which will be a posterior probability ($P(c_k \mid x_i)$). The next part, the HMM, expects a class-condition likelihood ($p(x_i \mid c_k)$), so before utilising the output of the DNN in the HMM, we have to transform it. The transformation can be processed with Bayes' theorem. If the posterior estimation is divided by a priori probabilities of phonetic classes ($P(c_k)$); then, we obtain class-condition likelihood value within a scale factor. The a priori probabilities are usually estimated using simple statistical methods. However, the scale factor can be ignored because it has no influence on the subsequent search process.

Nowadays, recurrent neural architectures have become the state of the art in ASR [26]. Applying units such as long short-term memory (LSTM) [32] and Gated Recurrent Units (GRU) [33] as building blocks leads to a better performance. Nevertheless, there are several reasons for employing an HMM/DNN model instead of applying a recurrent neural network. The simple feed-forward DNN structure employed in the HMM/DNN acoustic model makes the training steps easier: it has lower computational complexity and uses less memory. These networks still have a competitive performance [34,35] in the case where training data are scarce.

2.2. DNN Embedding Extraction

To extract embeddings for further classification, first we have to train our hybrid model. We can see the acoustic HMM/DNN model training in the left top corner of Figure 1. Here, we need a larger ASR corpus that has time-aligned phonetic labels. From this corpus, we have to extract frame-level features. The extraction can be handled using different techniques, such as calculating filter banks, deltas, spectrograms or using neural networks. Now, we can use these frame-level features to train our hybrid model for a general language structure. When the training phase is over, we need to make a slight modification to our model to use it for DNN embedding extraction. We have to detach the DNN from the hybrid model and fix its weights. In this case, we are not interested in the original output layer of our DNN, which produces the posterior estimates. Now, we will focus on the previous hidden layers and their activation values, because hidden layers can provide more abstract information. We can see the process of embedding extraction in the left bottom corner of Figure 1. Here, we have to extract frame-level features from a smaller paralinguistics corpus. The length of a frame-level feature has to be the same as a feature from the ASR corpus. The best way to achieve this is to use the same method here as before. Afterwards, we can feed them into the modified deep neural network. The output of the hidden layers will be our embedding features.

2.3. DNN Embedding Aggregation

When we have acoustic frame-level embeddings, we have to convert them into utterance-level features in order to perform a classification. Figure 1 shows the final classification workflow in the bottom right corner. Since databases contain recordings with different lengths, we have a different number of embedded features for each recording. Traditional classifiers handle only fixed-sized inputs for one utterance, so we cannot create utterance-level features with a simple frame-level concatenation. We need to aggregate embeddings and Figure 2 shows the method in more details. This could be performed in a straightforward way by calculating statistical values along their time axis, such as mean, standard deviation or others. The final size of the aggregated vector is independent of the length of the original recording and it only depends on the number of neurons in the last given hidden layer and on the aggregation technique used. In the end, these utterance-level feature vectors can be fed into any traditional classification or regression model, where the output will be a label (class or real number) for each recording.

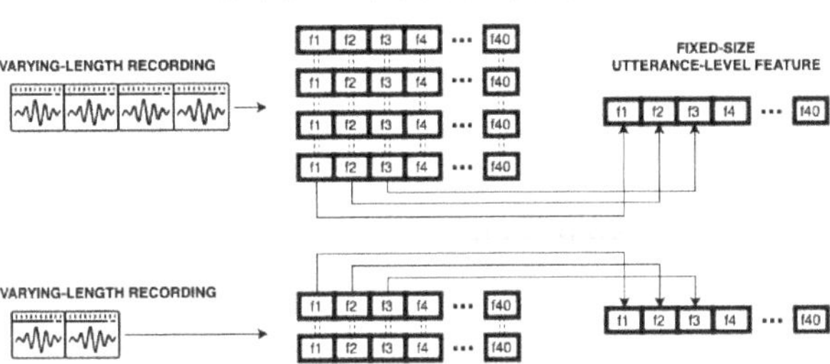

Figure 2. Creating fixed sized feature vectors with statistic conversion.

3. The Databases Used

Next, we will introduce the datasets that we employed in our experiments. Different aspects were taken into account when selecting the databases. On the one hand, we preferred databases that were easily accessible to the research community, and thus the databases used in the ComPare challenge were chosen. The other aspect was that they should be easily comparable, which is why we chose three German language databases so possible language differences would not affect the results. To cover different topics, we used three paralinguistic corpora (AIBO, URTIC and iHEARu-EAT). Although these corpora cover different topics, the recording conditions (such as sampling rate, language and background noise) are quite similar. Table 1 showes a summary from these three paralinguistic database. The fourth database utilized in our experiments (called BEA) was not a paralinguistic one, but it was used for training our hybrid acoustic model.

Table 1. The number of speakers and utterances for the three paralinguistics databases used.

Dataset	Language	No. of Classes	No. of Utterances			No. of Speakers			Total Duration (hh:mm:ss)		
			Train	Dev	Test	Train	Dev	Test	Train	Dev	Test
AIBO	German	5	7578	2381	8257	26	6	25	03:45:18	01:11:46	03:53:44
URTIC	German	2	9505	9596	9551	210	210	210	14:41:34	14:54:47	14:48:18
iHEARu-EAT	German	7	657	287	469	14	6	10	01:20:37	00:31:44	01:00:39

3.1. AIBO

The FAU AIBO Emotion Corpus [36] contains speech taken from 51 native German children. The children were selected from two schools. The database contains 9959 recordings from the Ohm school, and 8257 recordings from the Mont school. The total duration is approximately 9 h. The subjects had to play with a pet robot called AIBO. They were told that AIBO responds to their commands, but it was actually remotely controlled by a human. The Ohm school recordings are commonly used for training (with speaker-wise cross validation). The Mont school recordings were used for the test set. Because of the size of the training set, we were able to define a development set. We kept recordings of 20 children in the training set (7578 utterances) and used recordings of 6 children in the development set (2381 utterances). The original 11 emotional classes were merged to form a 5-class problem. The new classes were constructed from the originals: Anger (angry, irritated, reprimanding), Emphatic, Neutral, Positive (motherese and joyful), and the Rest (helpless, surprised, bored, non-neutral but not belonging to the other categories). This database was also employed in the INTERSPEECH 2009 Emotion Challenge.

3.2. URTIC

The Upper Respiratory Tract Infection Corpus (URTIC) [37] was provided by the Institute of Safety Technology, University of Wuppertal in Germany. It contains native German speech from 630 subjects (248 female, 382 male). The total duration is approximately 45 h.

The recordings have a sampling rate of 44.1 kHz downsampled to 16 kHz. They were split into 28,652 chunks of 3 to 10 s. The participants had to complete different tasks. They had to read short stories (e.g., a well-known story in the field of phonetics "The North Wind and the Sun"), had to produce voice commands (such as stating numbers from 1 to 40) and they also had to narrate spontaneous speech (e.g., say something about their best vacation). The number of tasks varied for each speaker. The database was split speaker-independently into training, development and test sets where each one contained 210 speakers. The training and development sets contained 37 infected participants and 173 participants with no cold. There were two classes, namely cold and no cold. The purpose of the classification was to decide whether the speaker had a cold. This database was also employed in the INTERSPEECH 2017 Computational Paralinguistics Challenge.

3.3. iHEARu-EAT

The iHEARu-EAT corpus [38] was provided by the Munich University of Technology. It contains close-to-native German speech taken from 30 subjects (15 female, 15 male). It was recorded in a quiet, slightly echoing office room. It contains approximately 2.9 h of speech (sampled at 16 kHz). The recordings were segmented into roughly equal parts. The participants had to perform speaking exercises while eating different type of foods. Speakers had to complete different tasks, e.g., read the German version of "The North Wind and the Sun" story, and they had to give a spontaneous narrative about their favourite activity or place. The number of completed activities varied for each speaker because not everyone was willing to eat every type of food offered. The database was split speaker-independently into a training set (20 speakers) and test set (10 speakers). There were seven classes determined by the consistency: apple, nectarine, banana, crisp, biscuit, gummy bear and without any food. The aim of the classification was to recognise what the subject was eating while speaking. These type of foods typically allowed the participants to eat while speaking. This database was also employed in the INTERSPEECH 2015 Computational Paralinguistics Challenge.

3.4. BEA

We used a subset of the BEA Hungarian corpus [39] to pretrain our acoustic model. This was not a specific paralinguistics corpus like the three above, but it is also a speech corpus. It contains only spontaneous speech and it is good for generalising a neural network for speech processing. We applied a subset of this database, which contained the speech of 165 subjects ($\approx$60 h). This subset contained only spontaneous speech with special events such as filled pauses, breathing sounds, laughter, gasps and so on. It had a transcription, where the phonetics set and the special events were also marked.

4. Experimental Setup

4.1. Frame-Level Features

For both the ASR and paralinguistics corpora, the frame-level feature extraction was carried out by 40 Mel-frequency filter banks with the standard values of 25 ms window width and 10 ms frame step. We also extended it with the log-energy value, and calculated the first- and second-order derivatives (i.e., Δ and $\Delta\Delta$ [40]). The final number of features in a frame-level vector was 123.

4.2. HMM/DNN Hybrid Model

We trained the hybrid model with the large BEA corpus. It has a standard feed-forward deep neural network (DNN). During DNN training and evaluation, we used the standard solution of applying a 15-frames-wide sliding window, so the input layer of the network contained $15 \times 123 = 1845$ neurons. The DNN has five hidden layers, where each one contains 1024 ReLU neurons. The final softmax layer of the network had as many neurons as the number of phonetic states, namely 911. For embedding extraction, we used the activation values of the middle five hidden layers (i.e., layer 1, 2, 3, 4 and 5). Each layer generated 1024 sized frame-level feature vector.

4.3. Embedding Aggregation

In the conversion step, we transformed the frame-level embeddings into utterance-level feature vectors by aggregating them with a statistical function along the time axis. The statistical approaches used were the following: arithmetic mean, standard deviation, kurtosis, skewness and "zero ratio". Zero ratio represents how many times out of each embedding an output neuron fired (it means a feature has a non-zero value, as we used ReLU neurons). The used aggregations has the following mathematical formula, if we have N frame-level embeddings in the form $x_1, x_2, \ldots, x_i, \ldots, x_N$:

$$\text{Arithmetic mean: } \overline{x} = \frac{1}{N} \sum_{i=1}^{n} x_i$$

$$\text{Standard deviation: } \sigma = \sqrt{\frac{1}{N} \sum_{i=1}^{N} (x_i - \overline{x})^2}$$

$$\text{Kurtosis} = \frac{1}{N} \sum_{i=1}^{N} \frac{(x_i - \overline{x})^4}{\sigma^4}$$

$$\text{Skewness} = \frac{1}{N} \sum_{i=1}^{N} \frac{(x_i - \overline{x})^3}{\sigma^3}$$

$$\text{Zero ratio} = \frac{1}{N} \sum_{i=1}^{N} y_i, \text{ where } y_i = \begin{cases} 1 & \text{if } x_i > 0, \\ 0 & \text{otherwise} \end{cases}$$

Notice that, having N embeddings with m frames, any of these formulas produce m utterance-level aggregated features.

4.4. Classification

For optimal results, we separated all paralinguistic corpora into training, development and test sets. We determined the optimal parameters of the classifier while training with the training set end evaluating with the development set. After the optimisation, we measured the overall efficiency while training with the combination of training and development sets and evaluating with the test set. During the classification step, our classifier was a Support Vector Machine (i.e., SVM) [41]. We optimized the complexity parameter using 10 powers between 10^{-5} and 10^{0}. In the case of AIBO and URTIC, we always standardised and downsampled the actual training set before feeding it into the SVM. In the case of iHEARu-EAT, we only performed a speaker-wise standardization.

To measure the efficiency, we calculated the Unweighted Average Recall (i.e., UAR) [42] from the posteriors of our SVM model. UAR measures the average recall across all classes without considering class imbalance. To calculate UAR, you compute the recall for each class and then take the average across all classes. Recall, also known as sensitivity or true-positive rate, is the proportion of true-positive instances (correctly identified instances) out of all actual positive instances. UAR is called "unweighted" because it treats each class equally, regardless of class size or prevalence. This makes it suitable for datasets with imbalanced class distributions, where some classes may have significantly fewer instances than others. It provides a balanced view of the overall performance of a classification system, taking into account the performance across all classes equally. In an emotion recognition task with an imbalanced dataset (Happy: 500, Sad: 300, Neutral: 2000), accuracy can be misleading. For instance, a classifier that predicts the majority class (Neutral) for all instances would have high accuracy ($2000/2800 \approx 71.4\%$). However, UAR (Unweighted Average Recall) gives a better evaluation by considering the recall for each class separately. In this case, UAR would indicate poor performance (UAR: $(0 + 0 + 1)/3 \approx 0.333$) as the classifier fails to identify instances of the minority classes (Happy and Sad) while performing well on the majority class.

4.5. Baseline Method

X-vector networks [43] nowadays are receiving more attention in the field of paralinguistics. Previous studies have successfully applied x-vector embeddings in various paralinguistic tasks [5,17]. This feed-forward deep neural network was originally designed for speaker identification, but with a slight modification it can be used for feature extraction as well [43]. Figure 3 shows the structure of our baseline network. It has nine layers, in the following order: five time-delayed frame-level layers, one statistics pooling layer, two segmentation layers and one softmax layer. The statistics pooling layer is responsible for transforming frame-level information into utterance-level information. It aggregates over the output of the fifth frame-level layer and calculates different metrics such as mean or standard deviation. The segmentation layers can capture meaningful information about the speaker, e.g., age and gender. The last layer of the network contains the speaker id.

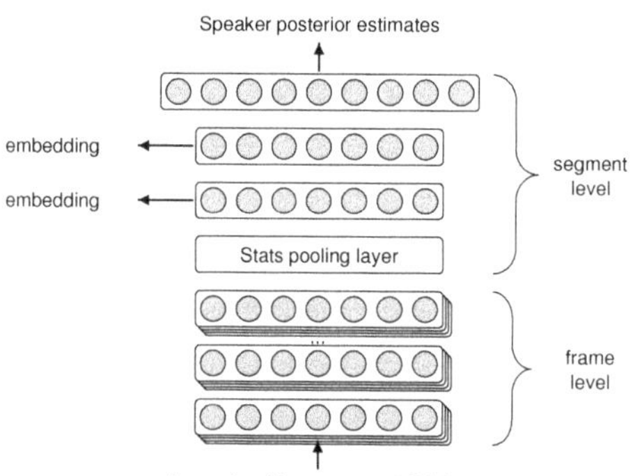

Figure 3. The x-vector neural network structure used as the baseline.

In order to use x-vector DNNs for paralinguistic feature extraction, we trained the model on the BEA corpus and we evaluated separate baselines for each of the AIBO, URTIC and iHEARu-EAT databases. In parallel, we calculated different frame-level features from each database, such as mfcc, fbank and spectrogram values. After training different x-vector models on them, we fixed the weights of the models and removed the last two layers from each. We extracted utterance-level information from the seventh hidden layer as a network embedding and examine traditional classification on them. We also tried out a noise augmentation technique. To find the best baseline, we always carried out a quick search of the frame-level feature sets with and without augmentation for each database separately.

5. Experimental Results

5.1. AIBO

When we used the AIBO database, we had a five-class classification task. Figure 4 shows all the results with different statistic conversion techniques on the development set. The best baseline result was obtained using f-bank features with augmentation and it came to 39.3% (indicated by the grey horizontal line in the figure).

Regarding the layers, we can state that the fourth layer always outperforms the baseline. Moreover, the fourth layer achieved the best performance scores with all the aggregation techniques used. In view of aggregation, there were no significant differences between the robustness of aggregations. Kurtosis and skewness had the worst performance scores overall. In the majority of cases, we cannot beat the baseline with them. The mean and standard deviation performed the best. In most cases, they outperformed the x-vector baseline. The average performance of a conversion technique is represented by a black column. The mean and standard deviation of the fourth and fifth layers had better performance scores than their average layer performance scores and these gave the best results overall.

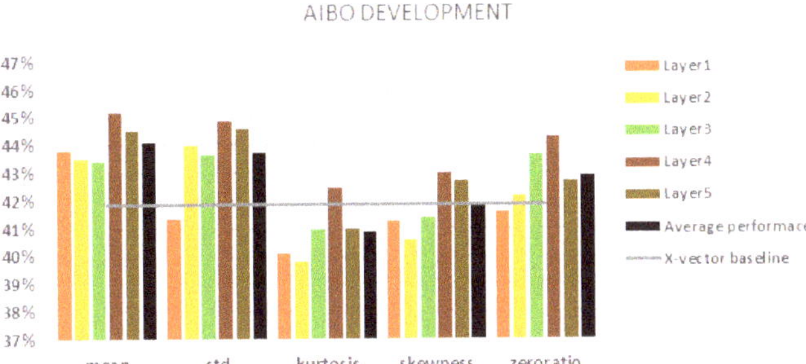

Figure 4. AIBO results on the development set. Extracting embeddings from each of the 5 layers. The black columns denote the average performance of all five layers. The baseline is represented by a grey line.

5.2. URTIC

When we used the URTIC database, we had a two-class classification task. Figure 5 shows all the results obtained with different aggregations with the development set. The best baseline result was obtained using MFCC features with augmentation and it was 66.9% (indicated by the grey horizontal line in the figure).

Here we can state that the third and fourth layers always reach or outperform the average performance of a conversion technique (represented by a black column), but in the majority of cases they cannot beat the x-vector baseline. The standard deviation is a bit more robust than the others, but again there is no significant difference. The kurtosis and skewness statistic conversions again had the worst performance scores. Here, the best results can beat the baseline. One of them is the mean of the third and fourth layers. The other is the zero ratio statistic conversion for the second and fourth layers.

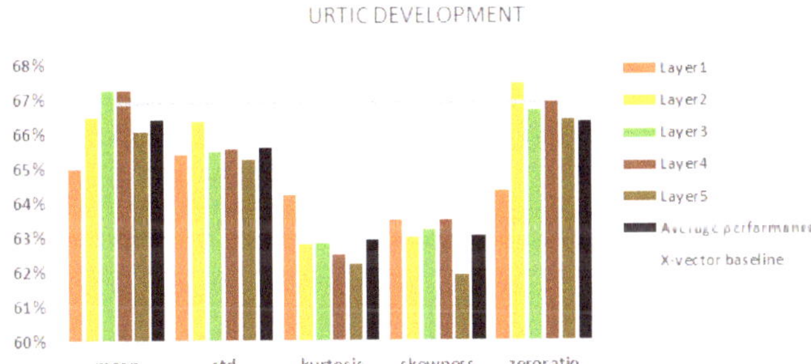

Figure 5. URTIC results on the development set. Embeddings from each of the 5 layers. The black columns denote the average performance of all five layers. Baseline represented by a grey line.

5.3. iHEARu-EAT

With the iHEARu-EAT corpus, we had a seven-class classification task. Figure 6 shows all the results obtained with different aggregations using the development set. The best baseline result was obtained using fbank features with augmentation and it was 58.7% (indicated by the grey horizontal line in the figure).

Here, we can state that all of our embeddings always outperform the baseline. Similar to URTIC, the second and fourth layers perform best and in most cases outperform the local average performance (represented by a black column). The robustness behaviour is similar, but the zero ratio and mean are slightly better. The rest of the aggregations behave just like before. The mean and the standard deviation of the second and fourth layers give the overall best results.

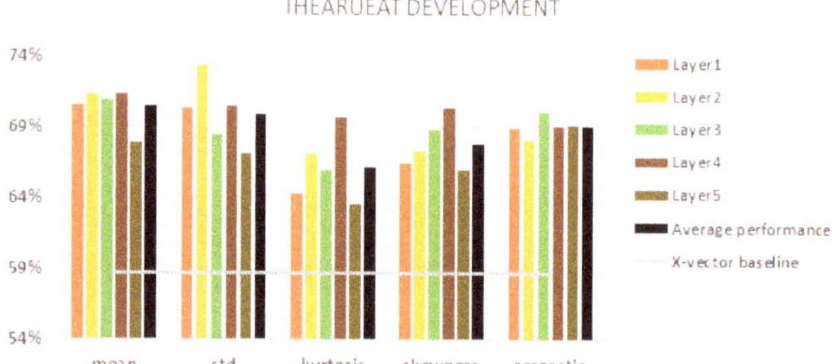

Figure 6. iHEARu-EAT development set results. Embeddings from each of the 5 layers. The black columns denote the average performance of all five layers. Baseline represented by a grey line.

6. Combined Results

A summary of our results from the first series is represented in Table 2. We can see that the HMM/DNN embeddings always outperform the x-vectors. The kurtosis and skewness aggregations perform the worst. The mean, standard deviation and zero ratio techniques behave slightly the same. We also wanted to know the expressive power of the embeddings, so we combined all of the five techniques as well. With this, we improved performance on the development set, but the scores of the test sets dropped. This raises the question of whether there is a specific combination of statistic conversation techniques that gives improvements on the development set while maintaining the ability of generalisation. We tried to determine the best-generalised model that would give better performance scores against future data.

In the second series of experiments, we used sequential forward selection (SFS) to combine multiple aggregated feature vectors. The basic idea behind SFS is to initialise the subset with just the best method, and then iteratively add one more aggregation to the subset based on which combination provides the greatest improvement in performance. To combine a subset of aggregations, we concatenated their utterance-level feature sets. The size of each utterance-level feature vector was as follows: 1024 as one technique, 2048 as a concatenation of two different aggregated vectors, 3072 as a concatenation of three different aggregated vectors, 4096 as a concatenation of four different aggregated vectors and 5120 when we concatenated all the different aggregated vectors.

Table 2. Results of different aggregation techniques with the three different corpora.

	AIBO			URTIC			iHEARu-EAT		
	Layer	**DEV**	**TEST**	**Layer**	**DEV**	**TEST**	**Layer**	**DEV**	**TEST**
mean	4	45.2%	44.0%	4	67.3%	69.3%	2	71.4%	79.0%
std	4	44.8%	44.4%	2	66.4%	68.1%	2	73.3%	74.4%
kurtosis	4	42.5%	40.3%	1	64.2%	60.8%	4	69.7%	69.0%
skewness	4	43.0%	41.2%	1	63.5%	68.3%	4	70.3%	67.3%
zero ratio	4	44.3%	42.1%	2	67.4%	68.8%	3	70.0%	75.5%
all	5	45.5%	44.2%	4	66.0%	65.3%	4	76.6%	74.6%
x-vector	–	41.8%	35.6%	–	66.9%	57.1%	–	58.7%	53.8%

6.1. AIBO

Figure 7 shows the performance scores obtained when we combined the mean statistic conversion technique with all the others. The first chart shows the mean aggregation with layer 4. All of the combinations perform better than their x-vector baseline; however, the mean technique had the best performance scores of 45.2% on the dev set and 44.0% on the test set. The second chart shows the mean statistic conversion technique with layer 5. All of the combinations performed better here as well. The combination of mean, skewness, standard deviation and kurtosis gave the best performance score of 45.3% on the dev set and 44.2% on the test set, but we can obtain almost the same without the kurtosis of 45.1% on the dev set and 43.7% on the test set. We can state that the fifth layer can generalise better if we use the combination of mean+ skewness+ standard deviation+kurtosis techniques, and the fourth layer performs best with only mean statistic conversion. The mean and standard deviation techniques always gave improvements. Instead of the fact that layer 5 had better performance scores than layer 4, we should note that calculating just one aggregation requires less time and memory.

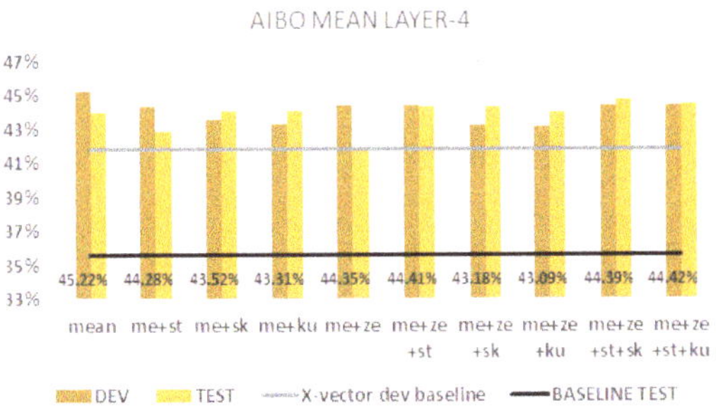
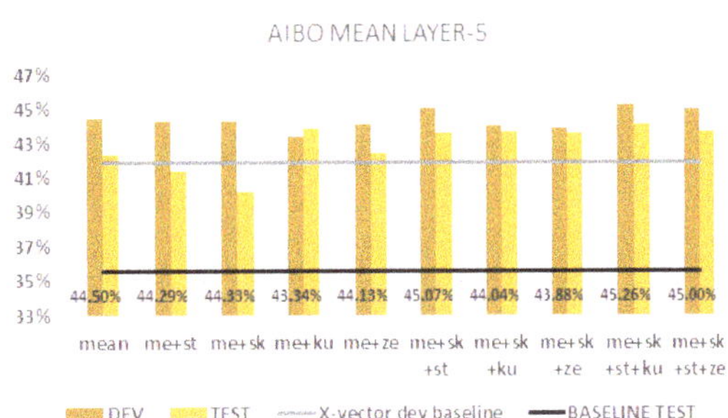

Figure 7. AIBO database classification results with development and test sets. We combine the mean aggregation by SFS. We extract embeddings from layer 4 (first figure) and from layer 5 (second figure).

6.2. URTIC

Figure 8 shows the performance scores obtained when we combined the zero ratio statistic conversion technique with all the others. We can see on the first chart the zero ratio aggregation with layer 2. The best combination (zero ratio+ mean+ standard deviation) had the same performance score (67.4%) on the dev set as the zero-ratio-only option. Note that they have the same performance with the development set, but the combination gives a better performance score (69.6%) on the test set. We can see on the second chart the zero ratio aggregation with layer 4. The first three combinations can outperform the x-vector baseline. If we combine two techniques, the best combination (zero ratio + mean) gives 67.7% with the dev set and 69.5 with the test set. We can state that the fourth layer has the best generalisation if we use the combination of mean and zero ratio techniques. Kurtosis and skewness aggregations always underperform the others. The standard deviation metric can improve the performance with layer 2.

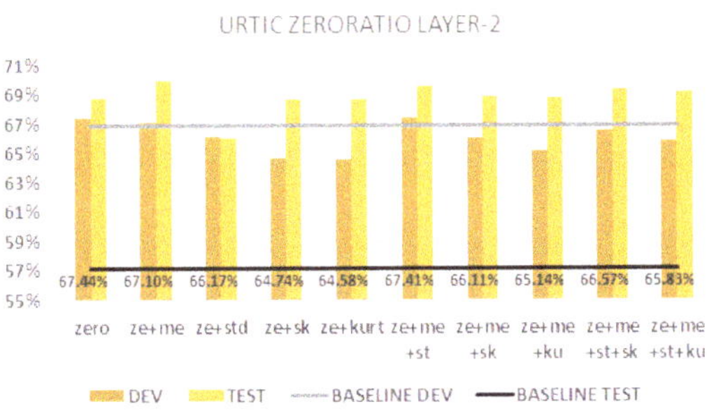
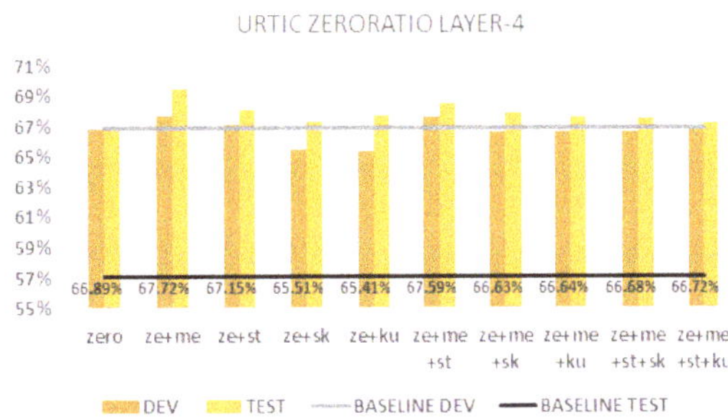

Figure 8. URTIC database classification results with development and test sets. We combine the zero ratio aggregation by SFS. We extract embeddings from layer 2 (first figure) and layer 4 (second figure).

6.3. iHEARu-EAT

Figure 9 shows the performance scores when we combine the standard deviation (i.e., std) statistic conversion technique with all the others. The first chart shows the std aggregation with layer 2. All of the combinations perform better than their x-vector baseline on the development and test sets. In the case of combining four techniques, the zero ratio slightly improves our model and increases its ability to generalise. The best combination is std+kurtosis+zero ratio+mean and it produced a 74.9% performance score on the dev set and 78.3% on the test. The second chart shows the std aggregation with layer 4. All of the combinations perform better here as well. When we combined three techniques (std+skewness+zero ratio), it slightly improved our model and gave a 76.0% performance score on the dev set and 75.0% on the test set. We can state that model trained on features from the second layer can generalise better if we use the combination of mean, zero ratio and skewness techniques. The zero ratio always produces a good improvement.

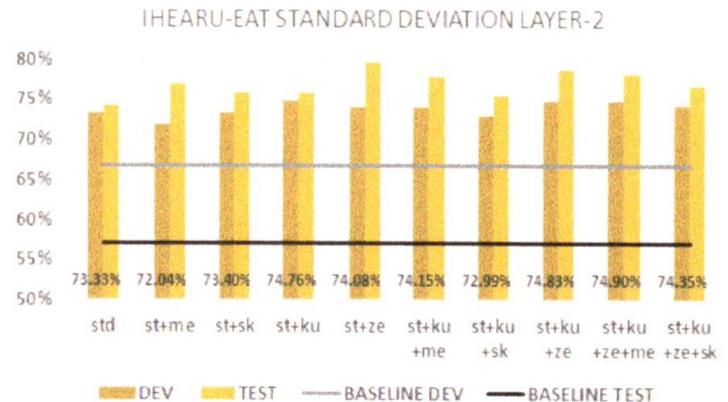

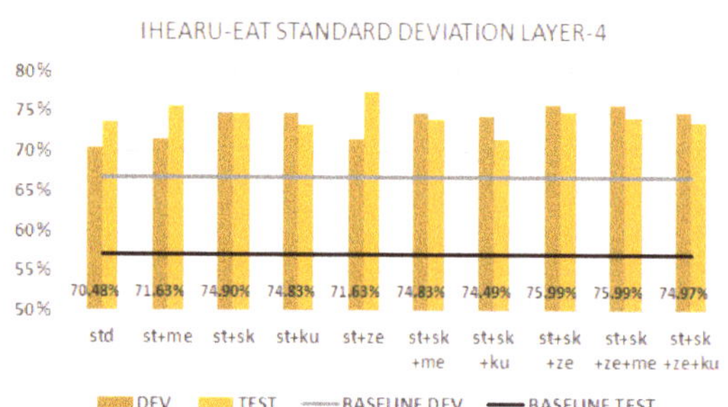

Figure 9. iHEARu-EAT database classification results with development and test sets. We combine the standard deviation aggregation by SFS. We extract embeddings from layer 2 (first figure) and layer 4 (second figure).

A summary of our results from the second series is given in Table 3. We can say that extracting embeddings from the fourth layer always gives the best performance scores. Concatenating aggregations is always a good idea and it helps our model to generalise better, but we should carefully select the techniques used because the best combination may be task-dependent. We should always consider including the mean, standard deviation and/or zero ratio in the combination.

Table 3. The best results obtained by SFS. The base aggregation and layers came from the best corpus-specific aggregations.

AIBO				URTIC				iHEARu-EAT			
Layer	Combination	DEV	TEST	Layer	Combination	DEV	TEST	Layer	Combination	DEV	TEST
4	mean	45.2%	44.0%	2	zero	67.4%	68.8%	2	std	73.3%	74.4%
4	me-ze	44.4%	42.0%	2	ze-me	67.1%	70.0%	2	st-ku	74.8%	75.9%
4	me-ze-st	44.4%	44.3%	2	ze-me-st	67.4%	69.6%	2	st-ku-ze	74.8%	78.9%
4	me-ze-st-ku	44.4%	44.5%	2	ze-me-st-sk	66.6%	69.4%	2	st-ku-ze-me	74.9%	78.3%
5	mean	44.5%	42.3%	4	zero	66.9%	67.1%	4	std	70.5%	73.8%
5	me-sk	44.3%	40.2%	4	ze-me	67.7%	69.5%	4	st-sk	74.9%	74.8%
5	me-sk-st	45.1%	43.7%	4	ze-me-st	67.6%	68.5%	4	st-sk-ze	76.0%	75.0%
5	me-sk-st-ku	45.3%	44.2%	4	ze-me-st-sk	66.8%	67.9%	4	st-sk-ze-me	76.0%	74.3%

7. Discussion

We trained a state of the art hybrid acoustic HMM/DNN model on a large ASR corpus and then used the DNN part to extract frame-level embeddings from smaller paralinguistics corpora. Afterwards, to aggregate these features into utterance levels, we used statistics computational techniques. We chose traditional aggregations, such as standard-deviation and mean, and less traditional ones, such as skewness, kurtosis and zero ratio; these aggregated vectors served as features in our computational paralinguistic classification experiments. In our first experiments, we tested all aggregation techniques individually. Our results indicate that the hybrid acoustic model performed better than the x-vector did. The mean, standard deviation and zero ratio techniques achieve practically the same performance scores.

After obtaining these results, we wanted to improve the expressive power of the embeddings. We chose to investigate the performance of combined aggregation techniques. We tested all the possible combinations of the five techniques. Our results indicate that we were successfully able to extract features from different paralinguistic tasks with our HMM/DNN hybrid acoustic-model-based feature extraction method. Using the second or the fourth layer of the model is always a good choice. As for aggregations, the mean, standard deviation and zero ratio always help improve the performance, but we have to combine these techniques carefully. In the case of kurtosis and skewness aggregations, we can observe varied behaviour in all databases. In the first stage of our research, when we tested each aggregation separately, we could see the trends amongst them. They had the worst performance in terms of each database and each layer. In the second stage of our research, when we tested the combination of aggregations, we could see a similar tendency. When we were deciding which aggregation should be combined next, skewness or kurtosis gave the lowest scores in 13 of 18 cases. Based on these results, it can be stated that skewness and kurtosis aggregation techniques are not able to significantly improve the success rate of paralinguistic task processing.

In the case of aggregation combination, we can see that combining three techniques will always improve our results in any paralinguistic task. On the other hand, the combination of four techniques will behave inconsistently. Although it improves the results on the development set, the results on the test set are often decreasing. This suggests that the generalisation ability of our model is also decreasing. For this reason, when choosing the number of aggregations to combine, it is worth taking into account Occam's razor principle, which states that unnecessarily complex models should not be preferred against simpler ones.

The possible limitations of our approach include the potential language dependency of the extracted embeddings. For further research directions, we see several opportunities. Although our results were competitive even with a Hungarian HMM/DNN hybrid acoustic model for German tasks, and although the x-vector method (used as feature extractor) showed language-independency tendencies before [17], the effect of using an acoustic model trained on the same language should be studied in the future. On the other hand, each of the databases studied here is German-speaking, it could potentially be investigated whether aggregations computed from x-vector embeddings behave similarly on databases of different languages. Additionally, it is unclear how the amount of training material affects the quality of the extracted features. Furthermore, training a DNN is inherently a stochastic procedure due to random weight initialization; therefore, the variance in the classification performance might also prove to be an issue. We plan to investigate these factors in the near future. Another possible direction is whether these aggregations computed from different neural network embeddings follow a similar trend.

Author Contributions: Conceptualization, methodology, supervision, HMM/DNN model training, embedding calculation and writing—review and editing G.G. Aggregation, classification, formal analysis, visualization, writing—original draft preparation, M.V. All authors have read and agreed to the published version of the manuscript.

Funding: This research was supported by the NRDI Office of the Hungarian Ministry of Innovation and Technology (grant no. TKP2021-NVA-09), and within the framework of the Artificial Intelligence National Laboratory Program (RRF-2.3.1-21-2022-00004).

Institutional Review Board Statement: Not applicable.

Informed Consent Statement: Not applicable.

Data Availability Statement: No new data were created or analyzed in this study. Data sharing is not applicable to this article.

Conflicts of Interest: The authors declare no conflict of interest.

References

1. Han, K.J.; Kim, S.; Narayanan, S.S. Strategies to Improve the Robustness of Agglomerative Hierarchical Clustering Under Data Source Variation for Speaker Diarization. *IEEE Trans. Audio, Speech, Lang. Process.* **2008**, *16*, 1590–1601. [CrossRef]
2. Lin, Y.C.; Hsu, Y.T.; Fu, S.W.; Tsao, Y.; Kuo, T.W. IA-NET: Acceleration and Compression of Speech Enhancement Using Integer-Adder Deep Neural Network. In Proceedings of the Interspeech 2019, Graz, Austria, 15–19 September 2019; pp. 1801–1805. [CrossRef]
3. Van Segbroeck, M.; Travadi, R.; Vaz, C.; Kim, J.; Black, M.P.; Potamianos, A.; Narayanan, S.S. Classification of Cognitive Load from Speech Using an i-Vector Framework. In Proceedings of the Fifteenth Annual Conference of the Interspeech 2014, Singapore, 14–18 September 2014; pp. 751–755. [CrossRef]
4. Gosztolya, G.; Grósz, T.; Busa-Fekete, R.; Tóth, L. Detecting the intensity of cognitive and physical load using AdaBoost and Deep Rectifier Neural Networks. In Proceedings of the Fifteenth Annual Conference of the Interspeech 2014, Singapore, 14–18 September 2014; pp. 452–456. [CrossRef]
5. Jeancolas, L.; Mangone, D.P.D.G.; Benkelfat, B.E.; Corvol, J.C.; Vidailhet, M.; Lehéricy, S.; Benali, H. X-Vectors: New Quantitative Biomarkers for Early Parkinson's Disease Detection From Speech. *Front. Neuroinform.* **2021**, *15*, 578369. [CrossRef] [PubMed]
6. Vásquez-Correa, J.; Orozco-Arroyave, J.R.; Nöth, E. Convolutional Neural Network to Model Articulation Impairments in Patients with Parkinson's Disease. In Proceedings of the Interspeech 2017, Stockholm, Sweden, 20–24 August 2017; pp. 314–318. [CrossRef]
7. Kadiri, S.; Kethireddy, R.; Alku, P. Parkinson's Disease Detection from Speech Using Single Frequency Filtering Cepstral Coefficients. In Proceedings of the Interspeech 2020, Shanghai, China, 25–29 October 2020; pp. 4971–4975. [CrossRef]
8. Pappagari, R.; Cho, J.; Joshi, S.; Moro-Velázquez, L.; Żelasko, P.; Villalba, J.; Dehak, N. Automatic Detection and Assessment of Alzheimer Disease Using Speech and Language Technologies in Low-Resource Scenarios. In Proceedings of the Interspeech 2021, Brno, Czech Republic, 30 August–3 September 2021; pp. 3825–3829. [CrossRef]
9. Chen, J.; Ye, J.; Tang, F.; Zhou, J. Automatic Detection of Alzheimer's Disease Using Spontaneous Speech Only. In Proceedings of the Interspeech 2021, Brno, Czech Republic, 30 August–3 September 2021; Volume 2021, pp. 3830–3834. [CrossRef]
10. Pérez-Toro, P.; Klumpp, P.; Hernandez, A.; Arias, T.; Lillo, P.; Slachevsky, A.; García, A.; Schuster, M.; Maier, A.; Nöth, E.; et al. Alzheimer's Detection from English to Spanish Using Acoustic and Linguistic Embeddings. In Proceedings of the Interspeech 2022, Incheon, Republic of Korea, 18–22 September 2022; pp. 2483–2487. [CrossRef]
11. Yamout, B.; Fuleihan, N.; Hajj, T.; Sibai, A.; Sabra, O.; Rifai, H.; Hamdan, A.L. Vocal Symptoms and Acoustic Changes in Relation to the Expanded Disability Status Scale, Duration and Stage of Disease in Patients with Multiple Sclerosis. *Eur. Arch. Otorhinolaryngol* **2009**, *266*, 1759–1765. [CrossRef] [PubMed]
12. Egas-López, J.V.; Kiss, G.; Sztahó, D.; Gosztolya, G. Automatic Assessment of the Degree of Clinical Depression from Speech Using X-Vectors. In Proceedings of the ICASSP 2022—2022 IEEE International Conference on Acoustics, Speech and Signal Processing (ICASSP), Singapore, 23–27 May 2022; pp. 8502–8506. [CrossRef]
13. Přibil, J.; Přibilová, A.; Matoušek, J. GMM-Based Speaker Age and Gender Classification in Czech and Slovak. *J. Electr. Eng.* **2017**, *68*, 3–12. [CrossRef]
14. Mustaqeem; Kwon, S. CLSTM: Deep Feature-Based Speech Emotion Recognition Using the Hierarchical ConvLSTM Network. *Mathematics* **2020**, *8*, 2133. [CrossRef]
15. Zhao, Z.; Bao, Z.; Zhang, Z.; Cummins, N.; Wang, H.; Schuller, B. Attention-Enhanced Connectionist Temporal Classification for Discrete Speech Emotion Recognition. In Proceedings of the Interspeech 2019, Graz, Austria, 15–19 September 2019; pp. 206–210. [CrossRef]
16. Gosztolya, G.; Beke, A.; Neuberger, T. Differentiating laughter types via HMM/DNN and probabilistic sampling. In Proceedings of the Speech and Computer: 21st International Conference, SPECOM 2019, Istanbul, Turkey, 20–25 August 2019; pp. 122–132. [CrossRef]
17. Egas-López, J.V.; Gosztolya, G. Deep Neural Network Embeddings for the Estimation of the Degree of Sleepiness. In Proceedings of the IEEE International Conference on Acoustics, Speech and Signal Processing (ICASSP), Toronto, ON, Canada, 6–11 June 2021; pp. 7288–7292. [CrossRef]
18. Grezes, F.; Richards, J.; Rosenberg, A. Let me finish: Automatic conflict detection using speaker overlap. In Proceedings of the Interspeech 2013, Lyon, France, 25–29 August 2013; pp. 200–204. [CrossRef]

19. Bone, D.; Black, M.P.; Li, M.; Metallinou, A.; Lee, S.; Narayanan, S. Intoxicated speech detection by fusion of speaker normalized hierarchical features and GMM supervectors. In Proceedings of the Twelfth Annual Conference of Interspeech 2011, Lorence, Italy, 28–31 August 2011; pp. 3217–3220 [CrossRef]

20. Schuller, B.; Steidl, S.; Batliner, A. The INTERSPEECH 2009 Emotion Challenge. In Proceedings of the Computational Paralinguistics Challenge (ComParE), Interspeech 2009, Brighton, UK, 6–10 September 2009; pp. 312–315. [CrossRef]

21. Schuller, B.; Steidl, S.; Batliner, A.; Hantke, S.; Hönig, F.; Orozco-Arroyave, J.R.; Nöth, E.; Zhang, Y.; Weninger, F. The INTER-SPEECH 2015 computational paralinguistics challenge: Nativeness, Parkinson's & eating condition. In Proceedings of the Computational Paralinguistics Challenge (ComParE), Interspeech 2015, Dresden, Germany, 6–10 September 2015; pp. 478–482. [CrossRef]

22. Schuller, B.W.; Batliner, A.; Bergler, C.; Mascolo, C.; Han, J.; Lefter, I.; Kaya, H.; Amiriparian, S.; Baird, A.; Stappen, L.; et al. The INTERSPEECH 2021 Computational Paralinguistics Challenge: COVID-19 Cough, COVID-19 Speech, Escalation Primates. In Proceedings of the Computational Paralinguistics Challenge (ComParE), Interspeech 2021, Brno, Czech Republic, 30 August–3 September 2021; pp. 431–435 [CrossRef]

23. Rabiner, L.; Juang, B.H. *Fundamentals of Speech Recognition*; Prentice Hall Signal Processing Series; Pearson College Div: Engelwood Cliffs, NJ, USA, 1993.

24. Rumelhart, D.E.; Hinton, G.E.; Williams, R.J. Learning representations by back-propagating errors. *Nature* **1986**, *323*, 533–536. [CrossRef]

25. Cox, S. *Hidden Markov Models for Automatic Speech Recognition: Theory and Application*; Royal Signals & Radar Establishment: Malvern, UK, 1988.

26. Hinton, G.; Deng, L.; Yu, D.; Dahl, G.E.; Mohamed, A.R.; Jaitly, N.; Senior, A.; Vanhoucke, V.; Nguyen, P.; Sainath, T.N.; et al. Deep Neural Networks for Acoustic Modeling in Speech Recognition: The Shared Views of Four Research Groups. *IEEE Signal Process. Mag.* **2012**, *29*, 82–97. [CrossRef]

27. Boser, B.; Guyon, I.; Vapnik, V. A Training Algorithm for Optimal Margin Classifier. In Proceedings of the Fifth Annual ACM Workshop on Computational Learning Theory, Pittsburgh, PA, USA, 27–29 July 1992; Volume 5, pp 144–152. [CrossRef]

28. Egas-López, J.V.; Balogh, R.; Imre, N.; Hoffmann, I.; Szabó, M.K.; Tóth, L.; Pákáski, M.; Kálmán, J.; Gosztolya, G. Automatic Screening of Mild Cognitive Impairment and Alzheimer's Disease by Means of Posterior-Thresholding Hesitation Representation. *Comput. Speech Lang.* **2022**, *75*, 101377. [CrossRef]

29. Gosztolya, G. Posterior-Thresholding Feature Extraction for Paralinguistic Speech Classification. *Knowl.-Based Syst.* **2019**, *186*, 104943. [CrossRef]

30. Gauvain, J.L.; Lee, C.H. Maximum a posteriori estimation for multivariate Gaussian mixture observations of Markov chains. *IEEE Trans. Speech Audio Process.* **1994**, *2*, 291–298. [CrossRef]

31. Morgan, N.; Bourlard, H. Continuous speech recognition using multilayer perceptrons with hidden Markov models. In Proceedings of the International Conference on Acoustics, Speech, and Signal Processing, Albuquerque, NM, USA, 3–6 April 1990; Volume 1, pp. 413–416. [CrossRef]

32. Hochreiter, S.; Schmidhuber, J. Long Short-term Memory. *Neural Comp.* **1997**, *9*, 1735–1780. [CrossRef] [PubMed]

33. Cho, K.; van Merrienboer, B.; Gulcehre, C.; Bahdanau, D.; Bougares, F.; Schwenk, H.; Bengio, Y. Learning Phrase Representations using RNN Encoder-Decoder for Statistical Machine Translation. In Proceedings of the 2014 Conference on Empirical Methods in Natural Language Processing (EMNLP), Doha, Qata, 26–28 October 2014. [CrossRef]

34. Panzner, M.; Cimiano, P. Comparing Hidden Markov Models and Long Short Term Memory Neural Networks for Learning Action Representations. In Proceedings of the Second International Workshop of Machine Learning, Optimization, and Big Data, MOD 2016, Volterra, Italy, 26–29 August 2016; Volume 10122, pp. 94–105. [CrossRef]

35. Schmitt, M.; Cummins, N.; Schuller, B. Continuous Emotion Recognition in Speech—Do We Need Recurrence? In Proceedings of the Interspeech 2019, Graz, Austria, 15–19 September 2019; pp. 2808–2812. [CrossRef]

36. Steidl, S. *Automatic Classification of Emotion Related User States in Spontaneous Children's Speech*; Logos: Berlin, Germany, 2009.

37. Krajewski, J.; Schieder, S.; Batliner, A. Description of the Upper Respiratory Tract Infection Corpus (URTIC). In Proceedings of the Interspeech 2017, Stockholm, Sweden, 20–24 August 2017.

38. Hantke, S.; Weninger, F.; Kurle, R.; Ringeval, F.; Batliner, A.; Mousa, A.; Schuller, B. I Hear You Eat and Speak: Automatic Recognition of Eating Condition and Food Type, Use-Cases, and Impact on ASR Performance. *PLoS ONE* **2016**, *11*, e0154486. [CrossRef] [PubMed]

39. Neuberger, T.; Gyarmathy, D.; Gráczi, T.E.; Horváth, V.; Gósy, M.; Beke, A. Development of a Large Spontaneous Speech Database of Agglutinative Hungarian Language. In Proceedings of the 17th International Conference, TSD 2014, Brno, Czech Republic, 8–12 September 2014; pp. 424–431. [CrossRef]

40. Deng, L.; Droppo, J.; Acero, A. A Bayesian approach to speech feature enhancement using the dynamic cepstral prior. In Proceedings of the 2002 IEEE International Conference on Acoustics, Speech, and Signal Processing, Orlando, FL, USA, 13–17 May 2002; Volume 1, pp. I-829–I-832. [CrossRef]

41. Chang, C.C.; Lin, C.J. LIBSVM: A library for support vector machines. *ACM Trans. Intell. Syst. Technol.* **2011**, *2*, 27. [CrossRef]

42. Metze, F.; Batliner, A.; Eyben, F.; Polzehl, T.; Schuller, B.; Steidl, S. Emotion Recognition using Imperfect Speech Recognition. In Proceedings of the Interspeech 2010, Chiba, Japan, 26–30 September 2010; pp. 478–481. [CrossRef]
43. Snyder, D.; Garcia-Romero, D.; Sell, G.; Povey, D.; Khudanpur, S. X-Vectors: Robust DNN Embeddings for Speaker Recognition. In Proceedings of the 2018 IEEE international conference on Acoustics, Speech and Signal Processing (ICASSP), Calgary, AB, Canada, 15–20 April 2018; pp. 5329–5333. [CrossRef]

Article

Improving APT Systems' Performance in Air via Impedance Matching and 3D-Printed Clamp

Liu Liu * and Waleed Abdulla *

Department of Electrical, Computer and Software Engineering, The University of Auckland, Auckland 1010, New Zealand
* Correspondence: lliu282@aucklanduni.ac.nz (L.L.); w.abdulla@auckland.ac.nz (W.A.)

Abstract: This paper presents a study on improving the performance of the acoustic piezoelectric transducer system in air, as the low acoustic impedance of air leads to suboptimal system performance. Impedance matching techniques can enhance the acoustic power transfer (APT) system's performance in air. This study integrates an impedance matching circuit into the Mason circuit and investigates the impact of fixed constraints on the piezoelectric transducer's sound pressure and output voltage. Additionally, this paper proposes a novel equilateral triangular peripheral clamp that is entirely 3D-printable and cost-effective. This study analyses the peripheral clamp's impedance and distance characteristics and confirms its effectiveness through consistent experimental and simulation results. The findings of this study can aid researchers and practitioners in various fields that employ APT systems to improve their performance in air.

Keywords: APT system in air; piezoelectric transducer; impedance matching; clamp

1. Introduction

Wireless power transfer (WPT) is an innovative technology representing a crucial advancement in the domain of power transfer [1–3]. This method can potentially revolutionize traditional energy utilization patterns in various applications, such as portable electronic devices, implanted medical devices, integrated circuits, electric vehicles (EVs), and so on. Unlike traditional wires, the WPT technique transmits energy from the power source to the target through the air. Near-field and far-field transmissions constitute this new form of energy. Near-field utilizes the inductive coupling effect of nonradiative electromagnetic fields, which includes both inductive and capacitive mechanisms. Far-field WPT can be achieved using acoustic, optical, and microwave energy carriers.

Acoustic power transfer (APT) is a newer wireless power transfer (WPT) form. Energy transmission is accomplished using sound waves as intermediate energy carriers [4,5]. As shown in Figure 1, piezoelectric transducers transform electrical energy into vibrations and continuously propagate them as pressure waves. A receiver, situated at a specific point along the wave's path, converts the electricity back into sound energy. There are three types of medium in which it can travel: gaseous, fluid, and solid. Compared with other wireless powering techniques, such as inductive power transfer and midrange RF power transmission, acoustic power transfer (APT) has some advantages, such as lower tissue absorption, a shorter wavelength that enables more miniature transducers [6], and an increase in power intensity threshold for safe operation. The main advantage of APT is non-reliance on electromagnetic fields to transmit energy. It can be used in environments with strong electromagnetic fields.

The performance of an acoustic power transfer (APT) system depends on several factors, including the material properties of the piezoelectric transducers [7] and the size and geometry of these transducers. For example, [8] optimized the converter to reduce the system's losses. Another crucial factor that significantly affects the system's performance is impedance matching. While APT technology is commonly used with solid and fluid

Citation: Liu, L.; Abdulla, W. Improving APT Systems' Performance in Air via Impedance Matching and 3D-Printed Clamp. *Sensors* **2023**, *23*, 5347. https://doi.org/10.3390/s23115347

Academic Editors: Farook Sattar, Niladri Bihari Puhan and Reza Fazel-Rezai

Received: 26 April 2023
Revised: 31 May 2023
Accepted: 1 June 2023
Published: 5 June 2023

mediums, its usage in air is rare. This is primarily because the acoustic impedance of piezoelectric transducers differs significantly between air and other mediums, causing a significant mismatch, as illustrated in Table 1. The acoustic impedance of piezoelectric transducers is similar to that of solids and liquids.

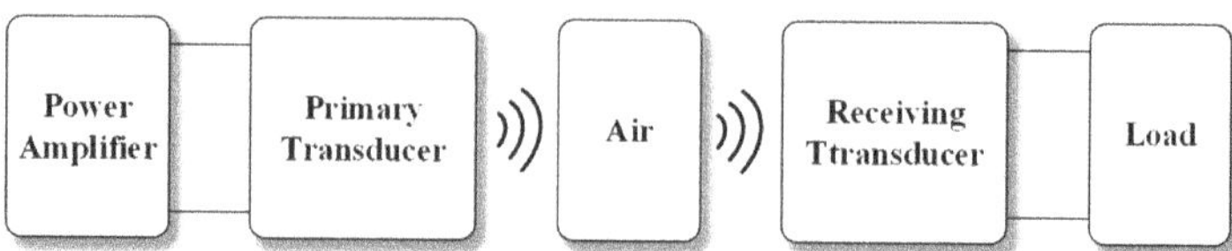

Figure 1. Acoustic power transfer system.

Table 1. Acoustic impendence.

Medium	Acoustic Impendence $Z_0 = \rho c$	Density ρ (kg/m^3)	Sound Speed c (m/s)
Steel	39×10^6	7700	5000
Aluminium	14×10^6	2700	5200
Water	1.44×10^6	1000	1440
Air	410	1.2	340
PZT-5H	36×10^6	7500	4800

In contrast, it is considerably lower in air, resulting in substantial losses when using APT systems in an airborne setting. Therefore, studying the energy transfer of APT in air is of great importance. Due to its minimal impact on biological tissues and the environment, APT can be applied to medical detection and monitoring inside the human body, and it can also offer new technological support for developing smart homes, IoT, and other fields. This paper aims to enhance APT systems' performance in air by implementing impedance matching.

Impedance matching enhances system performance and stability by minimizing reflection and loss during the energy transfer process. Generally, two impedance matching methods are used for APT systems to facilitate energy transfer in air. One traditional approach addresses abrupt impedance changes near the interface of the propagation medium and the piezoelectric transducer by inserting a reflective layer between them. Typically, this layer consists of specialized materials. Ref. [9] proposed a method to quantify the impact of acoustic impedance mismatches on performance and demonstrated that optimizing the thickness of intermediary layers can resolve these impedance issues. The optimization of transducer performance involves proposing fabrication methods and modelling for the matching layers, as described in [10]. Other special materials such as polyether sulfone and polystyrene foam can also be placed between the transducer and the propagation medium to adjust the impedance [11–18]. The ideal intermediate matching layer in an APT system should ideally have an impedance equal to the geometric mean of the propagation medium and the transducer. However, materials with extremely low impedance, which are required for this purpose, are typically not available in pure form or as a single phase. Moreover, using multiple layers of material matching layers increases attenuation.

Alternatively, instead of traditional matching techniques, an acoustic meta-surface can be employed. Acoustic meta-surfaces consist of ultrathin 2D building blocks that allow for manipulation of sound phase. These meta-surfaces find applications in acoustic lenses [19], sound diffusers [20], acoustic holograms [21], as well as impedance-matched surfaces and wavefront engineering. For instance, a simple meta-surface composed of thin membranes and tiny air cavities provides an efficient impedance matching surface for water-to-air communication [22]. Ref. [23] proposed non-local waterborne acoustic meta-surfaces (WAM) with highly non-local features for efficient underwater acoustic control, taking into account fluid–solid interactions (FSIs) to manipulate underwater sound with higher efficiency. However, this method increases the volume of the system.

Finding an impedance matching method that satisfies both low attenuation and no increase in volume is challenging. To overcome this problem, a new impedance matching method for APT systems in air has been developed. The study found that the fixed position of the piezoelectric transducer plays a crucial role in impedance matching, although there are limited studies on this aspect. Typically, the back of the piezoelectric transducer is fixed. To develop a contact model, a comprehensive impedance analysis is performed on systems with structural connections and contact interfaces, using the principle of equivalence between mechanical and electrical characteristics. In this paper, the piezoelectric transducer is fixed on the back [24]. Ref. [25] proposed a self-detached APT system, which can be easily installed and removed in harsh environments. However, it is still fixed on the back of the piezoelectric transducer. However, different positions of the fixed piezoelectric transducer result in different vibration patterns, which can impact system performance. Therefore, studying the fixed position of the piezoelectric transducer is particularly important. In this paper, the effect of impedance matching on the efficiency of the APT system is examined using a one-dimensional circuit model, demonstrating its effectiveness. The influence of the piezoelectric transducer's clamp position on vibration patterns and sound pressure level is simulated using COMSOL. Additionally, a novel detachable peripheral clamp is designed using 3D printing, significantly reducing the cost. The output voltage of the APT system is compared using both the novel peripheral clamp and the traditional clamp, revealing an increase of nearly 200 mV with the new peripheral clamp, highlighting its effectiveness.

2. Matching Theory of Acoustic Impedance

Various modern technologies employ PTs for signal transmission, making PT a critical area of research. However, there are many theories for studying PT, including classical theory [26], Rayleigh's theory [27], Bishop's theory [28], and finite element model simulations [29]. Among them, equivalent circuits are one of the most frequently used calculation methods for PTs, which utilize mechanical properties to analogize circuit parameters. There are several types of circuit models, but the Mason circuit is one of the most popular. As a valuable tool for solving one-dimensional problems, network theory was utilized by Mason to create a more precise equivalent circuit based on this approach. This circuit divides the piezoelectric transducer into three ports, one electrical and two acoustic, using an ideal electromechanical transformer, as shown in Figure 2 [30,31]. The electrical port represented is made up of merely two capacitors, whose values are equal to C_0, but the electrodes are in opposite directions. N represents the electromechanical coupling, and the impedances (Z_T, Z_S) stand for the mechanical properties. The other parameters of a Mason circuit are shown in Table 2. Wu presented a novel equivalent circuit model for typical UWPT systems. They chose the T network as the electro–ultrasound–electric channel description, which significantly improved the calculation of circuit characteristics [32]. Mason equivalent circuits have also developed a slew of other formulations, including analogue networks [33], KLM circuits, and systems models [34]. However, there are only a few papers that investigate impedance mismatch using these equivalent models.

Table 2. The parameters of piezoelectric components.

Material Properties		Symbol	Meaning
$k_t^2 = \dfrac{e_{33}^2}{C_{33}^D \varepsilon_{33}^s} = \dfrac{h_{33}^2 \varepsilon_{33}^s}{C_{33}^D}$	$\Gamma = \dfrac{\omega}{v_D} = \omega\sqrt{\dfrac{\rho}{C_{33}^D}}$	ρ t	Density Thickness
$Z_0 = \rho A v_D = A\sqrt{\rho C_{33}^D}$	$N = C_0 h_{33}$	ε_{33}^s	Permittivity
$C_0 = \dfrac{\varepsilon_{33}^s A}{t}$	$h_{33} = k_t\sqrt{\dfrac{C_{33}^D}{\varepsilon_{33}^s}}$	k_t A	Electromechanical coupling Area
$Z_s = -jZ_0\csc(\Gamma t)$	$Z_r = jZ_0\tan\left(\dfrac{\Gamma t}{2}\right)$	C_{33}^D	Elastic stiffness

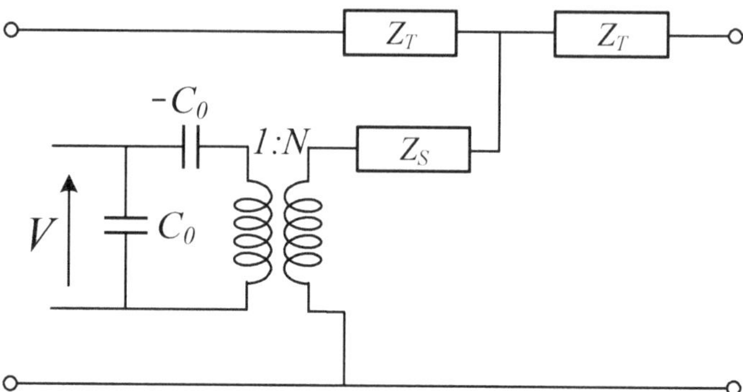

Figure 2. Mason's one-dimensional equivalent circuit.

2.1. Impedance Matching Equivalent Circuit

To verify that impedance matching can optimize system performance and that efficiency is an important expression of system performance, we chose efficiency to verify that impedance matching is effective for improving system performance. The method of this research is to calculate the partial derivative of the system efficiency and the impedance value of the matching layer in order to determine the value of impedance matching, which is used to improve the system performance. However, deriving formulas using the original Mason model would result in highly complex equations, impeding observation and practical applications. Therefore, the Mason one-dimensional equivalent circuit was first simplified, as shown in Figure 3. Figure 3a is a original Mason circuit, and Figure 3b is a simplified Mason circuit. Equation (1) shows the system efficiency equation derived from the simplified Mason circuit. However, the expansion of the efficiency formula is still very complicated; some parameters with small resistance in the circuit are ignored. The ultimate simplified efficiency formula is displayed in Equation (5), which can be derived by simplifying Equation (1). Equation (2) is the equivalent impedance parameter. Equations (3) and (4) are the currents I_1 and I_3 of the simplified Mason circuit in Figure 3b.

$$\eta_1 = \frac{Re\left[V_{out}\, I_3^*\right]}{Re\left[V\, I_1^*\right]} \tag{1}$$

$$
\begin{aligned}
Z_{11} &= Z_{s1} + Z_{t1} + Z_{tL} \\
Z_{12} &= Z_{t1} + Z_{tL} \\
Z_{22} &= Z_{t1} + Z_{tL} + Z_{t1} + Z_{tw} + Z_{sw} \\
Z_{21} &= Z_{t1} + Z_{tL} \\
Z_{23} &= Z_{sw} \\
Z_{33} &= Z_{tw} + Z_{sw} + Z_{leq} \\
Z_{32} &= Z_{sw}
\end{aligned}
\tag{2}
$$

$$I_1 = \frac{-(V \cdot (Z_{22} \cdot Z_{33} - Z_{23} \cdot Z_{32}))}{(Z_{11} \cdot Z_{23} \cdot Z_{32} - Z_{11} \cdot Z_{22} \cdot Z_{33} + Z_{12} \cdot Z_{21} \cdot Z_{33})} \tag{3}$$

$$I_3 = \frac{-(V \cdot (Z_{21} \cdot Z_{32}))}{(Z_{11} \cdot Z_{23} \cdot Z_{32} - Z_{11} \cdot Z_{22} \cdot Z_{33} + Z_{12} \cdot Z_{21} \cdot Z_{33})} \tag{4}$$

$$\eta_2 = \frac{R_{Leq}}{R_{Leq} + \left|\frac{Z_{33}}{Z_{32}}\right|^2 R_{t1} + \left|\frac{Z_{22}Z_{33} - Z_{23}Z_{32}}{Z_{21}Z_{32}}\right|^2 R_{s1}} \tag{5}$$

$$\eta_1 \approx \eta_2 \tag{6}$$

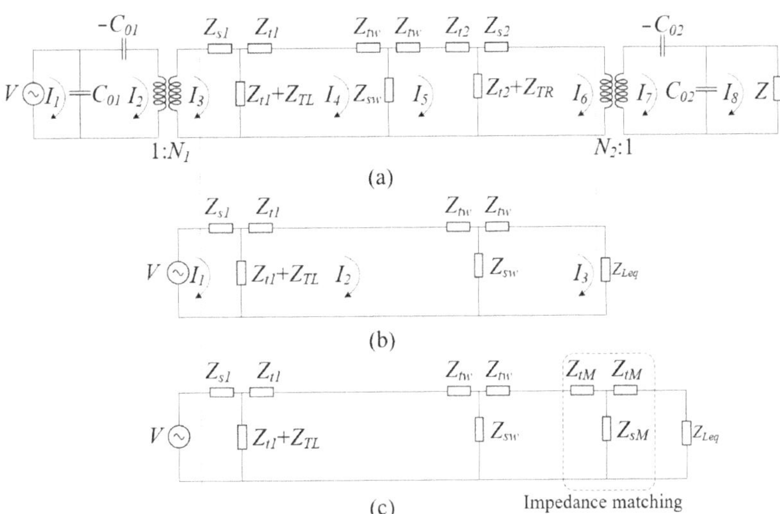

Figure 3. (a) Mason's one-dimensional equivalent circuit; (b) simplified diagram of Mason's one-dimensional equivalent circuit; (c) Mason's one-dimensional equivalent circuit with impedance matching.

From the derivation of Mason equation, it was discovered that there exists a minor difference in the system performance between Equations (1) and (5). Observing the simulation verification in Figure 4, Figure 4a is Equation (1) and Figure 4b is Equation (5). The imaginary part and real part of the matching impedance will appear at the peak point of the figure. The value of imaginary part and the real part of the impedance matching of Equation (1) and simplified Equation (5) are almost identical, as is shown in Equation (6). In addition, it shows that impedance matching is indeed effective in improving system performance and can be calculated via circuit simplification, thereby reducing the amount of calculation.

As shown in Figure 3c, the matched impedance circuit configuration is a T network. Where the imaginary and real parts of the impedance matching are equal to the real and imaginary parts of Z_{Leq}, resulting in the peak point of the figure. Therefore, the partial derivatives of the real part R_{Leq} and the imaginary part X_{Leq} of Z_{Leq} are calculated, respectively, based on Equation (5) to obtain the maximum efficiency value. Equation (7) is the partial derivatives of the real part R_{Leq} of Z_{Leq}. Equation (8) is the partial derivatives of the imaginary part X_{Leq} of Z_{Leq}. To simplify the lengthy formula, we replace the repeated equations within the procedure with variables A, B, C, D, E, F, and G, as indicated in Equation (9). Finally, the mathematical model is compared with the optimization module in MATLAB and found that the results are very close. The correctness of the mathematical model is proved. Additionally, it can be proved that the performance of the system can be improved through impedance matching.

$$
\begin{aligned}
\frac{\partial \eta}{\partial R_{leq}} &= \frac{1\sqrt{A}\sqrt{C}}{\sqrt{((-R_{21}^2 - X_{21}^2)(AR_{t1} + BR_{s1})C)}} \times (R_{21}^2 R_{sw}^2 R_{t1} + 2R_{21}^2 R_{sw} R_{t1} + R_{21}^2 R_{tw}^2 R_{t1} + X_{21}^2 R_{sw}^2 R_{t1} \\
&+ 2R_{sw} R_{t1} R_{tw} X_{21}^2 + R_{t1} R_{tw}^2 X_{21}^2 + R_{s1} R_{sw}^2 X_{22}^2 + 2R_{s1} R_{sw} R_{tw} X_{22}^2 + R_{s1} R_{tw}^2 X_{22}^2 \\
&- 2R_{32} R_{s1} R_{sw} X_{22} X_{23} - 2R_{32} R_{s1} R_{tw} X_{22} X_{23} + R_{32}^2 R_{s1} X_{23}^2 + X_{32}^2 R_{s1} X_{23}^2 + R_{23}^2 R_{s1} C \\
&- 2X_{22} R_{s1} X_{leq} X_{32} X_{23} + R_{21}^2 R_{t1} X_{leq}^2 + X_{22}^2 R_{s1} X_{leq}^2 - 2X_{22} R_{s1} X_{sw} X_{32} X_{23} + 2R_{21}^2 R_{t1} X_{leq} X_{sw} \\
&+ 2X_{21}^2 R_{t1} X_{leq} X_{sw} + 2R_{s1} X_{22}^2 X_{leq} X_{sw} + R_{21}^2 R_{t1} X_{sw}^2 + X_{21}^2 R_{t1} X_{sw}^2 + X_{22}^2 R_{s1} X_{sw}^2 \\
&+ 2(ADR_{t1} + R_{s1} X_{22}(-X_{32} X_{23} + DX_{22})) X_{tw} + (AR_{t1} + R_{s1} X_{22}^2) X_{tw}^2 - 2R_{22} R_{s1} X_{23}((-R_{sw} - R_{tw}) \\
X_{32} &+ R_{32} E) - 2R_{23} R_{s1}(R_{22} R_{32} F + F X_{22} X_{32} - X_{22} R_{32} E + R_{22} X_{32} E) + R_{22}^2 R_{s1}(F^2 + E^2))^{(\frac{1}{2})}
\end{aligned}
\tag{7}
$$

$$\frac{\partial \eta}{\partial X_{leq}} = \frac{1}{(R_{t1}A + R_{s1}B)} \times \left(R_{22}R_{32}R_{s1}X_{23} + X_{22}X_{32}R_{s1}X_{23} + R_{23}R_{s1}\left(R_{22}X_{32} - X_{22}R_{32} \right) - R_{s1}R_{22}^2 G \right.$$

$$\left. - R_{t1}AG - R_{s1}X_{22}^2 G \right) \tag{8}$$

$$\begin{aligned} A &= R_{21}^2 + X_{21}^2 \;\; B = R_{22}^2 + X_{22}^2 \;\; F = R_{sw} + R_{tw} \\ C &= R_{32}^2 + X_{32}^2 \;\; D = X_{leq} + X_{sw} \;\; G = X_{sw} + X_{tw} \\ E &= X_{leq} + X_{sw} + X_{tw} \end{aligned} \tag{9}$$

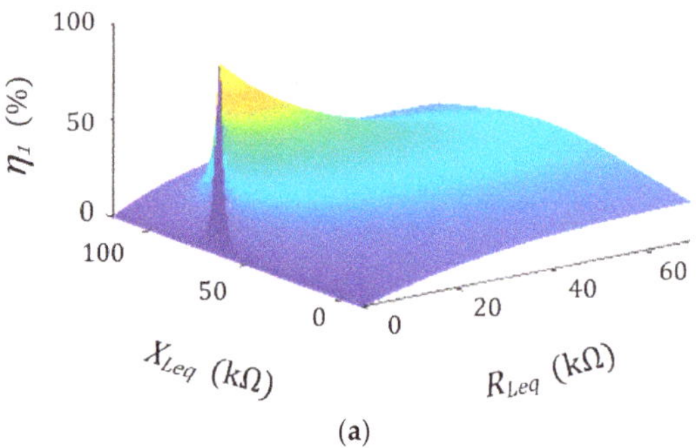

(a)

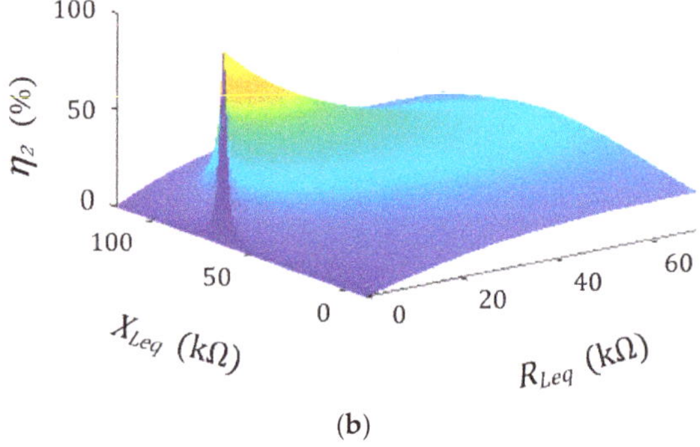

(b)

Figure 4. Efficiency comparison. (**a**) Original Mason circuit; (**b**) simplified circuit.

2.2. Proposed Approach for Impedance Matching—The Clamp of Piezoelectric Transducer

There are many methods of impedance matching [35–37], as shown in Figure 5. The method of traditional impedance matching uses special materials to coat the surface of the piezoelectric transducers, such as silver plating on the piezoelectric transducer, which has a low degree of freedom of operation, and it is difficult to control the thickness and uniformity of the application. A relatively new method of impedance matching is to use a meta-surface with a special structure to change the impedance [38]. However, the design and performance of an acoustic meta-surface are usually optimized for a specific application, so it is difficult to tune and change.

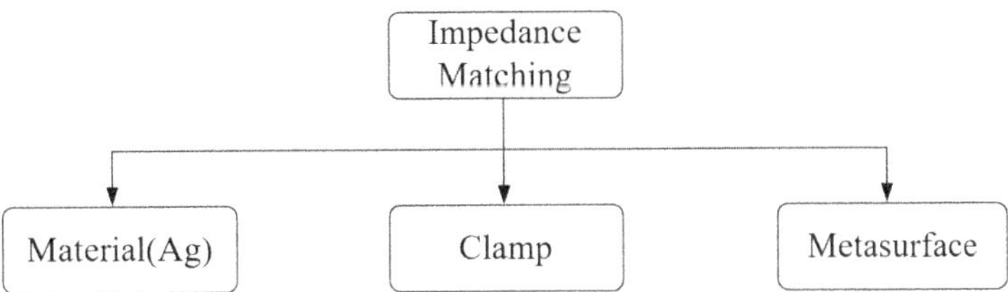

Figure 5. Method of impedance matching.

During this research, it was observed that the location of the fixed piezoelectric transducer influenced the vibration modes, which consequently impacted the impedance characteristics of the piezoelectric transducer. The traditional method is to fix the back of the piezoelectric transducer. Which is to glue it with the bracket to become a non-detachable, permanently bonded system. This method has many disadvantages, such as poor adjustability and stability. It was discovered that fixing the side of the piezoelectric transducer presents an effective means of adjusting impedance matching; however, this approach is scarcely mentioned in the existing literature. Even studies on detachable fixtures are focused on metals or solid media and mainly target the backside of the piezoelectric transducer, with no research on the side fixation, as is shown in Table 3. Therefore, this study demonstrates high innovation and practical applicability. A peripheral clamp consisting of three claws was proposed. The angle between each claw is 120 degrees, and the width of each claw corresponds to 30 degrees, as shown in Figure 6. This is a detachable clamp that can be adjusted for tightness, convenient, and flexible. A further advantage of the peripheral clamp is that they are all made using 3D printing, which reduces the price considerably.

Table 3. APT models based on separating media and excitation methods.

| Method | Ref | | Medium | Power/Sound | Detachable |
	Author	Year			
	[9] Freychet	2020	Wall	Power	Non-detachable
Special material	[10] Zhou	2022	Air	Power	Non-detachable
	[17] Rymantas	2017	Air	Sound	Detachable
Metasurface	[22] Eun	2018	Air/water	Sound	Detachable
Clamp	[25] Tseng	2022	Metal	Power	Detachable
This Research			Air	Power	Detachable

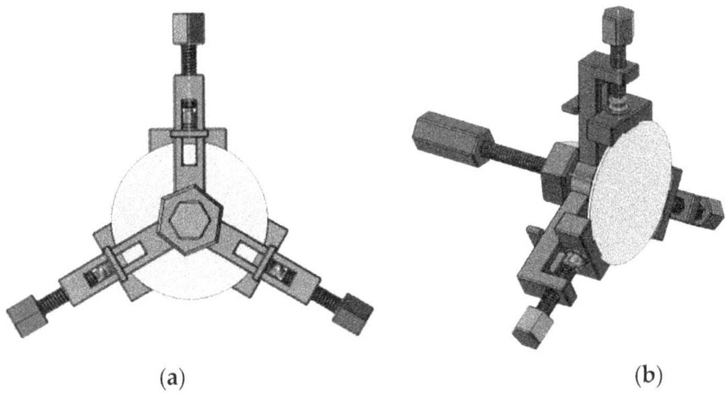

(a)   (b)

Figure 6. The model of the peripheral clamp. (**a**) Front view and (**b**) side view.

3. Simulation and Experimental Verification

The finite element simulation software COMSOL Multiphysics was used in this paper, which is for experimental preparation and verification of the effectiveness of the novel clamp. COMSOL Multiphysics is a finite element simulation software for multiphysics coupling. It is based on the finite element method and realizes the simulation of natural physical phenomena by solving partial differential equations. It uses mathematical techniques to solve physical phenomena in the real world.

In order to reduce the calculation, a 2D axisymmetric model was used for the simulation. A model of the mesh is shown in Figure 7. A represents the axis of symmetry, B is the input voltage side, C is the ground side, and D represents the fixed constraint. Ultrasonic waves are generated by applying a voltage to a piezoelectric transducer. Therefore, the boundary conditions must be considered simultaneously, as the mechanical and electrical parts are related to the vibration applied to the piezoelectric transducer in the simulation. Table 4 lists the settings of its boundary conditions. In addition, we have not set up any additional air damping and only accounted for the losses in the materials.

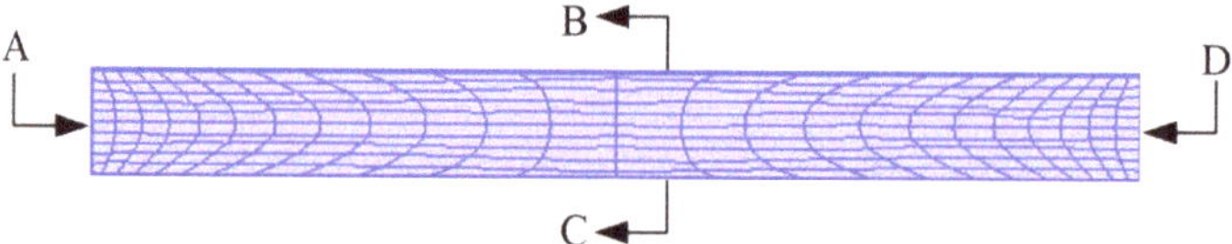

Figure 7. Mesh throwing of piezoelectric transducer in finite element simulation.

Table 4. The boundary state used for FEM model in simulation.

Boundary	Mechanical Status	Electrical Status
A	Axisymmetric	Axisymmetric
B	Free	Terminal Voltage
C	Free	Ground
D	Roller Support	Zero Charge

3.1. The Clamp of Piezoelectric Transducers

As shown in Table 5, the commonly used physical quantity for sound measurement is sound pressure, but usually, the sound pressure level describes the magnitude of sound pressure. The conversion equation between sound pressure and sound pressure level is shown in Equation (10). The variation in the vibration amplitude determines the sound intensity. Sound intensity is calculated using energy and sound pressure when expressed by pressure. Sound intensity is a vector; sound pressure is a scalar. The relationship between sound intensity (I) and sound pressure (p) is shown in Equation (11), where ρ is medium density and c is sound velocity. Sound power refers to the total energy the sound source radiates to space per unit of time; the unit is W. The relationship between sound power and sound intensity is shown in Equation (12), where A is the area through which the sound wave passes vertically. This implies that the acoustic pressure becomes stronger as the sound pressure level increases. A stronger acoustic pressure corresponds to a higher sound intensity, and the acoustic power within a given area increases with greater sound intensity. With the increase in acoustic power, the energy becomes more pronounced, indicating the effectiveness of the clamp.

$$L_p = 20 \times \log 10 \frac{p}{p_0} \tag{10}$$

$$I = \frac{p^2}{\rho c} \tag{11}$$

$$P = IA \tag{12}$$

Table 5. Common physical quantities and their units.

Physical Quantity	Unit
p (Sound pressure)	Pa
L_p (Sound pressure level)	dB
I (Sound intensity)	W/m^2
P (Sound power)	W
T (Stress)	Pa

The fundamental equations for piezoelectric transducers are shown in Equation (13), where T is the stress and E is the electric field, with both being independent variables, and where c^D, h, and β^S are the elastic stiffness coefficient, piezoelectric stiffness coefficient, and dielectric isolation rate, respectively. These equations state that the vibration strength and sound intensity increase with greater stress. Additionally, a higher sound intensity is indicative of a greater energy. Therefore, in COMSOL, we assess the effectiveness of different clamps by observing the sound pressure level and stress.

$$T = c^D S_3 - h D_3$$
$$E = -h S_3 + \beta^S D_3 \tag{13}$$

We used the acoustic–structure interaction module in the built-in acoustic module of COMSOL and added the Circuit module. The model of the APT system is established in COMSOL, the distance is about 2 mm, and the frequency is 51.5 kHz. As shown in Figure 8, the left is a schematic diagram, the dark colour is where the piezoelectric transducer is fixed, and the right is the sound pressure distribution obtained through COMSOL simulation. The dark part of Figure 8a is the back of the piezoelectric transducer. When the piezoelectric transducer is fixed on the back, the maximum sound pressure level of the total is 80 dB, which is considerably lower than that in Figure 8b. In the figure, the distribution in the middle is the sound pressure level on the piezoelectric transducers, and the surrounding distribution is the sound pressure level in air. The dark portion of Figure 8b represents the side of the piezoelectric transducer. When the piezoelectric transducer is fixed on the side, the maximum total sound pressure level is 100 dB, which is strong and uniformly distributed. This phenomenon shows that different fixed positions of piezoelectric transducers produce other effects. Setting the side of the piezoelectric transducer achieves a higher sound pressure than fixing the back.

3.1.1. No Clamp

The model of the piezoelectric transducer with a free state is built in COMSOL, as shown in Figure 9. It is the distribution of its total sound pressure level. In this context, we need to emphasize the "free state." Due to the inability of the piezoelectric transducer to float in the air without any support, we applied fixed constraints in the COMSOL model. However, compared with the constraint areas on the back and sides mentioned in the paper, the constraint area in the no-clamp model is negligible. It is found that the energy of the transmitting piezoelectric transducer is almost diffused to the surroundings when the piezoelectric transducer is in a free state, and there is not much sound pressure transmitted to the receiver. This situation poses significant challenges to efficient energy transmission within the system. Therefore, the design of the fixture is very important.

3.1.2. Back Clamp

The traditional method to fix a piezoelectric transducer is to glue it with the bracket to become a non-detachable, permanently bonded system. The holding clamp location is usually on the back of the piezoelectric transducer, as shown in Figure 10a. The model of the traditional clamp is established in COMSOL. The transmitter adds an input voltage of 8 V, and the output terminal is set to an open circuit state. The frequency is about 50 kHz. The distance between the transmitter and receiver is 2 mm. Figure 10b shows the stress transformation of the piezoelectric transducer, and its peak value is about 50 kPa. Figure 10c shows the change in its sound pressure level. The maximum value is about 80 dB. Compared with the free state, the stress has increased, but the peak value of the sound pressure level has not changed much. Nevertheless, this traditional clamp still has some influence on the distribution of sound pressure.

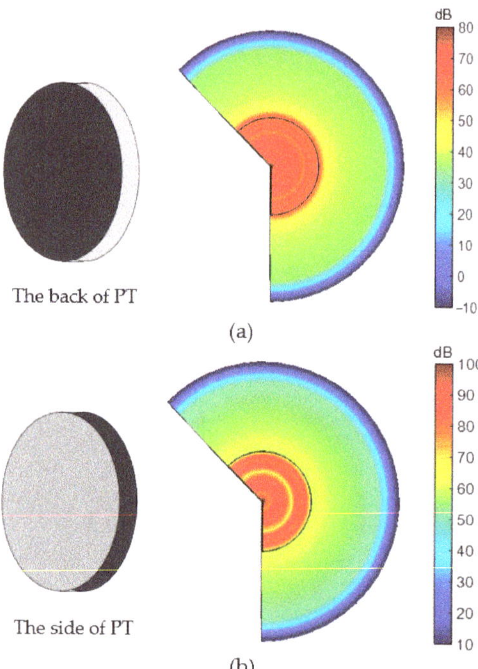

Figure 8. Total sound pressure level distribution of piezoelectric transducer. (**a**) Fixed piezo back; (**b**) fixed piezo side.

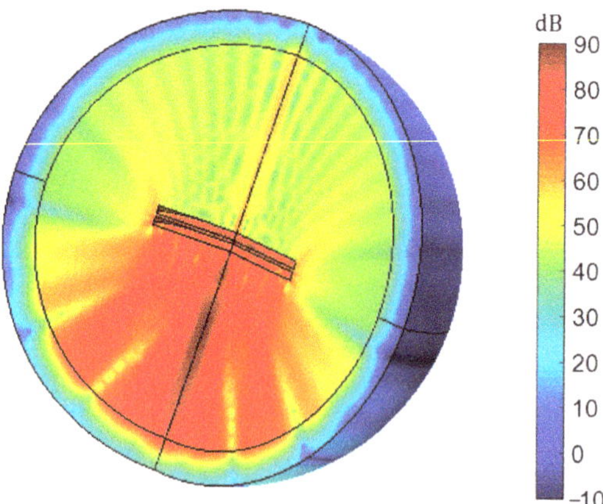

Figure 9. COMSOL model with no clamp and no fixed constraints.

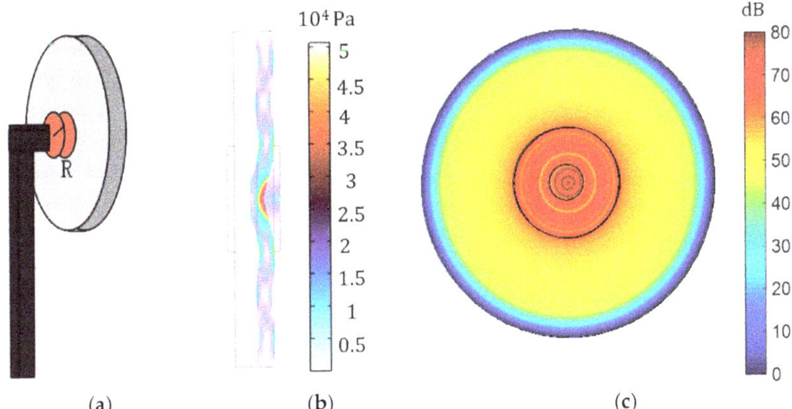

(a)   (b)   (c)

Figure 10. Fixed the back of piezo. (**a**) Schematic diagram of the back clamp; (**b**) the magnitude and deformation of the stress on the back clamp; (**c**) the distribution and size of the total sound pressure level on the back clamp.

3.1.3. Peripheral Clamp

A novel detachable clamp is designed in which the fixed position is on the side of the piezoelectric transducer, as shown in Figure 11a. The peripheral clamp has a total of three claws. The angle between each claw is 120 degrees, and the width of each claw corresponds to 30 degrees. The model of the peripheral clamp is established in COMSOL, and an input voltage of 8 V is added to the transmitter, and the output terminal is set to an open circuit state. The distance between the transmitter and receiver is 2 mm. The frequency is about 50 kHz. Figure 11b is the stress transformation of the piezoelectric transducer, and its peak value can reach 3 MPa. Figure 11c is the change in its sound pressure level; the maximum value can be up to 130 dB. The system performance is improved compared with the back clamp and no clamp. Noticeably, the sound pressure level has increased by nearly 50 dB.

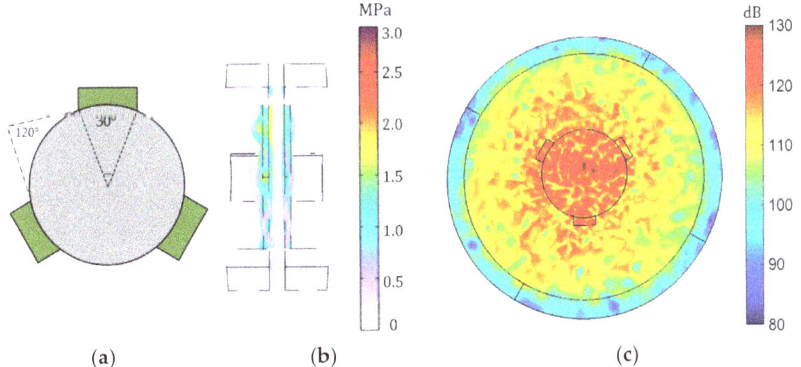

(a)   (b)   (c)

Figure 11. Fixed the side of piezo. (**a**) Schematic diagram of the peripheral clamp; (**b**) the magnitude and deformation of the side stress of the peripheral clamp; (**c**) the distribution and size of the total sound pressure level on the peripheral clamp.

3.1.4. Piezoelectric Transducer Array

Our current simulations and investigations are based on a single piezoelectric transducer, and the resulting values may be small compared with IPT. Additionally, because the acoustic impedance of the air does not match the acoustic impedance of the piezoelectric transducer seriously, its loss is very large. However, the research on APT systems in air is still very important, especially in biomedicine. For this reason, we made a simulation model of the PT array to prove that if the number of PTs increases, the result will obviously be improved. The model of the PT array is established in COMSOL, and the transmitter and receiver are composed of four piezoelectric transducers with a radius of 25 mm connected in parallel. The transmitter adds an input voltage of 8 V and sets the output to an open circuit state. The frequency is about 50 kHz. The distance between the transmitter and receiver is 2 mm. Figure 12a is the stress transformation of the piezoelectric array, and its peak value is up to 18 MPa. Figure 12b shows the change in its sound pressure level, with a maximum value up to 160 dB. The results can be nearly doubled compared with the results with no clamp. It proves that the number of PTs increases, and the experimental and simulation results will significantly improve.

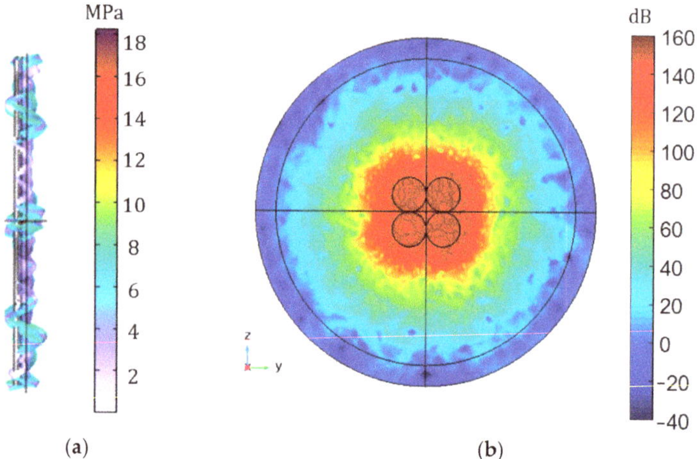

Figure 12. COMSOL model with PT array. (**a**) The magnitude and deformation of the stress of piezoelectric transducer array; (**b**) the distribution and size of the total sound pressure level of the piezoelectric transducer array.

Furthermore, the model in Figure 13 consists of piezoelectric transducer arrays with the same size and number of peripheral clamps. An input voltage of 8 V is applied to the transmitters, and the output is set to an open circuit state with a frequency of approximately 50 kHz. In Figure 13a, the stress transformation of the piezoelectric array with a peripheral clamp is presented with a peak value of up to 120 MPa. Figure 13b displays the change in sound pressure level, with a maximum value of 200 dB. These results are significantly higher than those of the piezoelectric array without a clamp, thus confirming the effectiveness of the clamp once again.

3.2. Experimental Verification

Under the condition that the resonant frequency of the system is 51.5 kHz, firstly, select the signal generator model Agilent 33250A and the ATA-122D Wide Band Amplifier as the system's power supply; the receiving end circuit is an open circuit. The transmitting end uses a differential probe Agilent N2772A probe to detect waveform. The receiver uses an oscilloscope probe to observe the waveform. The size of the piezoelectric transducer is a circular ceramic sheet with a radius of 25 mm, and the piezoelectric material is PZT-4. The components of the clamp are all manufactured by a FLASHFORGE 3D printer, and the 3D material is PLA. An experimental diagram of the system is shown in Figure 14.

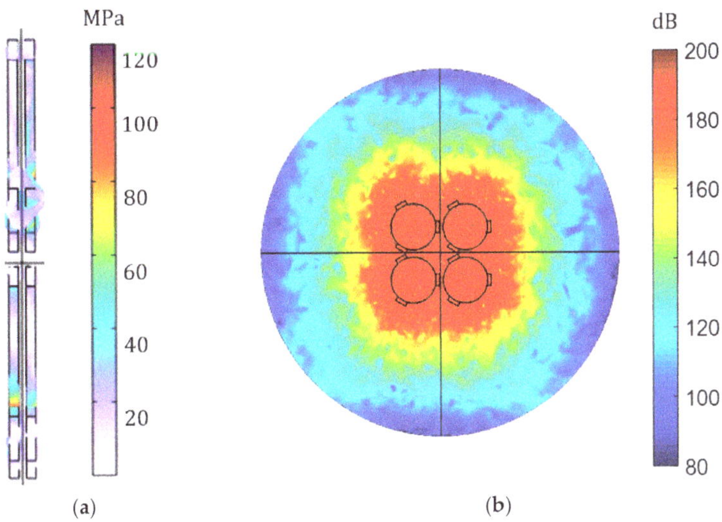

Figure 13. COMSOL model of a piezoelectric transducer array with peripheral clamp. (a) Magnitude and deformation of the piezoelectric transducer array with the peripheral clamp; (b) distribution and size of the total sound pressure level of the piezoelectric transducer array with the peripheral clamp.

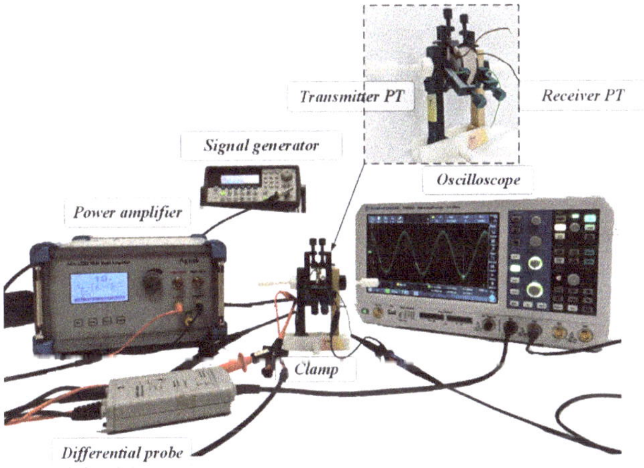

Figure 14. Experimental platform of APT system in air.

3.2.1. Impedance Characteristics

The impedance characteristics of a single piezoelectric transducer are measured in different cases, including no clamp, with a back clamp, and with a peripheral clamp. Figure 15a shows the impedance characteristic of a single piezoelectric transducer without any constraints and clamp. Figure 15b is the impedance characteristic of conditions on the back with the back clamp of a single piezoelectric transducer. Figure 15c is the impedance characteristic of a single piezoelectric transducer with constraints on its sides with the peripheral clamp. The Agilent E4980A Precision LCR Meter is used to measure the impedance of the piezoelectric transducer. The resonant frequency of a single piezoelectric transducer is 51.5 kHz, and its real resistance is 5.5 kΩ. The resonance frequency of the piezoelectric transducer fixed on the back is 51.5 kHz, and its real resistance is 16.3 kΩ; The resonance frequency of the piezoelectric transducer fixed on the side is 51.5 kHz, and its real resistance is 27.41 kΩ. According to the measurement results, the piezoelectric transducer's resistance is higher when using fixed constraints and a clamp, especially when the constraints with the peripheral clamp are added on the side. According to Equation (14), P is the sound pressure, Z is the acoustic impedance, and c is the volume velocity. When c is constant, the impedance is higher and the sound pressure is stronger.

$$P = c\,Z \tag{14}$$

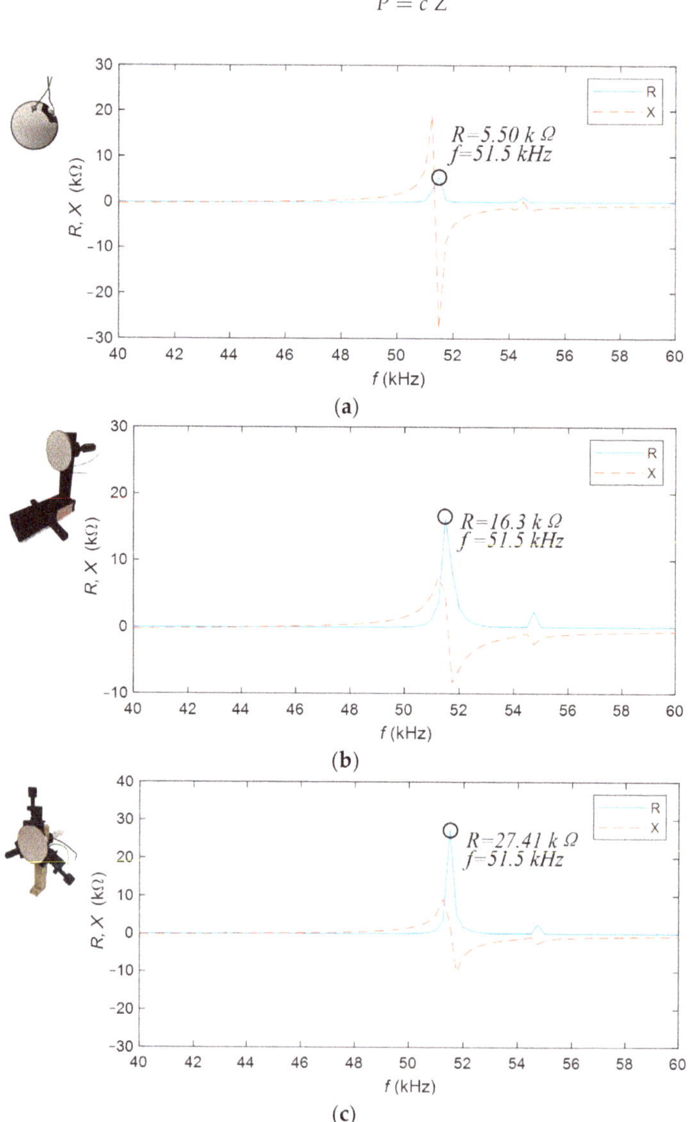

Figure 15. Impedance characteristics of a single piezoelectric transducer. (**a**) Without fixed constraints; (**b**) with fixed constraints on the back; (**c**) with fixed constraints on the sides.

3.2.2. Distance Characteristics

As shown in Figure 16, the red line is the experimental data about the relationship between the output voltage and the distance between the transmitter and receiver. The blue line is the simulation data about the relationship between the output voltage and the distance between the transmitter and receiver. These data were obtained under uniform conditions. The transmitter adds an input voltage of 8 V and sets the output to an open circuit state. The frequency is about 50 kHz. The trends of these two sets of values are very similar, with decreasing output voltage as distance increases. This result validates the simulation model.

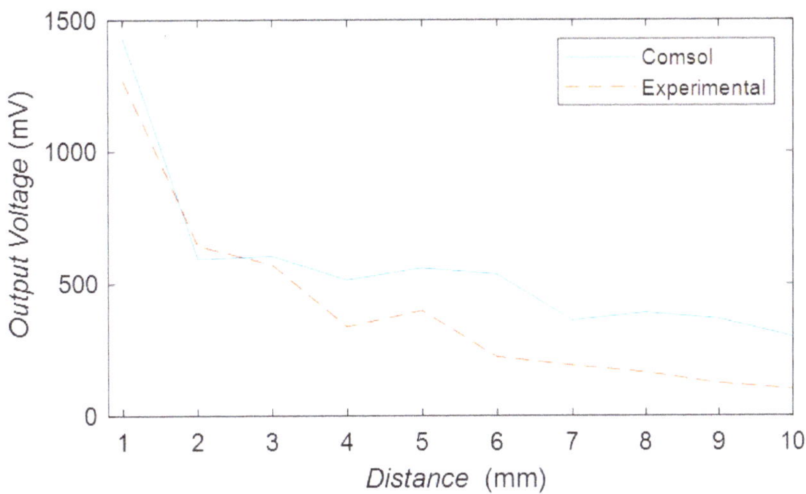

Figure 16. Simulation and experimental verification of the peripheral clamp.

The experiment compared the relationship between different distances and the output voltage with two additional clamps, as shown in Figure 17. The clamp that fixed the piezoelectric transducer on the back is a cylinder with a radius of 8 mm. The clamp that set the side of the piezoelectric transducer is distributed in an equilateral triangle with a fixed angle of 30 degrees. The distance between the transmitter and receiver is from 1 mm to 10 mm. An input voltage of 8 V is applied to the transmitters, and the output is set to an open circuit state with a frequency of approximately 50 kHz.

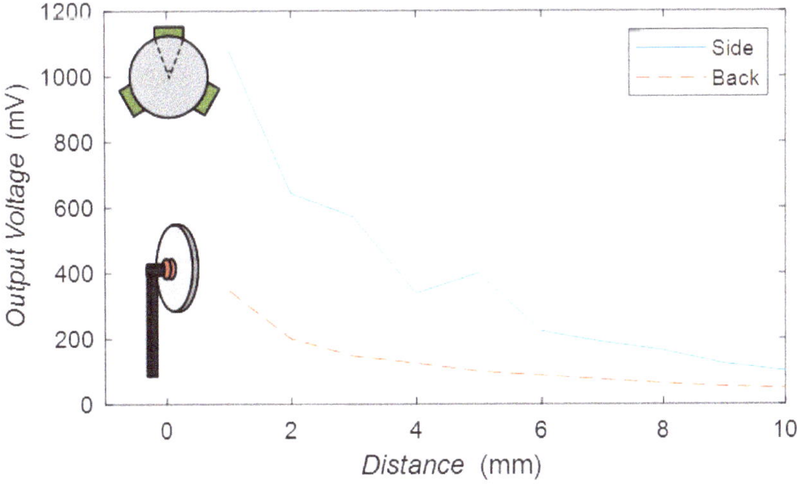

Figure 17. Distance characteristics of back clamp and peripheral clamp.

The blue curve in Figure 17 is the peripheral clamp, and the red curve is the back clamp. The trend of all curves decreases with increasing distance. It can be found that the output voltage of the novel clamp is always higher than that of the fixed back clamp at different distances. Therefore, the novel peripheral clamp proposed in this paper is effective.

4. Conclusions

In this paper, we introduced an impedance matching circuit to the Mason circuit and examined the correlation between input impedance and system efficiency. We identified a specific resistance value that maximizes efficiency, thus validating the efficacy of impedance matching. Furthermore, we compared different piezoelectric transducer configurations, including those with no fixed constraints, fixed constraints on the back, and fixed constraints on the side. Our findings revealed that the transducers with fixed constraints on the side exhibited higher sound pressure and output voltage compared with those without fixed constraints or with fixed constraints on the back. Based on these results, we proposed a novel peripheral clamp design, forming an equilateral triangle with a fixed width spanning 30 degrees at each point. This low-cost peripheral clamp can be entirely manufactured using 3D printers. We also investigated its impedance and distance characteristics, obtaining experimental results that aligned with our simulation outcomes, thus confirming the effectiveness of the novel peripheral clamp. While our studies primarily focused on a single piezoelectric transducer, we also conducted simulations involving piezoelectric arrays, demonstrating improved performance compared with performance with a single transducer. Therefore, if the measurement data from a single transducer are unsatisfactory, replacing it with a piezoelectric array can enhance the results.

Author Contributions: Conceptualisation, W.A.; Methodology, L.L.; Software, L.L.; Validation, L.L.; Writing—original draft preparation, L.L.; Writing—review and editing, W.A.; Supervision, W.A.; Research direction, W.A.; Resources, W.A. All authors have read and agreed to the published version of the manuscript.

Funding: This research received no external funding.

Institutional Review Board Statement: Not applicable.

Informed Consent Statement: Not applicable.

Data Availability Statement: Not applicable.

Conflicts of Interest: The authors declare no conflict of interest.

References

1. Rim, C.T.; Mi, C. *Wireless Power Transfer for Electric Vehicles and Mobile Devices*; John Wiley & Sons: New York, NY, USA, 2017.
2. Hui, S.Y.R.; Zhong, W.; Lee, C.K. A critical review of recent progress in mid-range wireless power transfer. *IEEE Trans. Power Electron.* **2013**, *29*, 4500–4511. [CrossRef]
3. Bhardwaj, A.; Allam, A.; Erturk, A.; Sabra, K.G. Enhancing underwater Acoustic Identification Tag design using concurrent wireless ultrasonic power transfer and backscatter communication. *J. Acoust. Soc. Am.* **2022**, *152*, A214. [CrossRef]
4. Roes, M.G.; Duarte, J.L.; Hendrix, M.A.; Lomonova, E.A. Acoustic energy transfer: A review. *IEEE Trans. Ind. Electron.* **2012**, *60*, 242–248. [CrossRef]
5. Basaeri, H.; Christensen, D.B.; Roundy, S. A review of acoustic power transfer for bio-medical implants. *Smart Mater. Struct.* **2016**, *25*, 123001. [CrossRef]
6. Kim, K.; Jang, S.G.; Lim, H.G.; Kim, H.H.; Park, S.M. Acoustic Power Transfer Using Self-Focused Transducers for Miniaturized Implantable Neurostimulators. *IEEE Access* **2021**, *9*, 153850–153862. [CrossRef]
7. Turner, B.L.; Senevirathne, S.; Kilgour, K.; McArt, D.; Biggs, M.; Menegatti, S.; Daniele, M.A. Ultrasound-Powered Implants: A Critical Review of Piezoelectric Material Selection and Applications. *Adv. Healthc. Mater.* **2021**, *10*, 2100986. [CrossRef]
8. Zaid, T.; Saat, S.; Jamal, N.; Husin, S.H.; Yusof, Y.; Nguang, S. A development of acoustic energy transfer system through air medium using push-pull power converter. *WSEAS Trans. Power Syst.* **2016**, *11*, 35–42.
9. Freychet, O.; Frassati, F.; Boisseau, S.; Garraud, N.; Gasnier, P.; Despesse, G. Analytical optimization of piezoelectric acoustic power transfer systems. *Eng. Res. Express* **2020**, *2*, 045022. [CrossRef]
10. Zhou, J.; Bai, J.; Liu, Y. Fabrication and modeling of matching system for air-coupled transducer. *Micromachines* **2022**, *13*, 781. [CrossRef]

11. Toda, M. New type of matching layer for air-coupled ultrasonic transducers. *IEEE Trans. Ultrason. Ferroelectr. Freq. Control.* **2002**, *49*, 972–979. [CrossRef]
12. Kelly, S.P.; Hayward, G.; Álvarez-Arenas, T.G. Characterization and assessment of an integrated matching layer for air-coupled ultrasonic applications. *IEEE Trans. Ultrason. Ferroelectr. Freq. Control.* **2004**, *51*, 1314–1323. [CrossRef]
13. Alvarez-Arenas, T.G. Acoustic impedance matching of piezoelectric transducers to the air. *IEEE Trans. Ultrason. Ferroelectr. Freq. Control.* **2004**, *51*, 624–633. [CrossRef]
14. Gomez Alvarez-Arenas, T.E. Air-coupled piezoelectric transducers with active polypropylene foam matching layers. *Sensors* **2013**, *13*, 5996–6013. [CrossRef] [PubMed]
15. Saito, K.; Nishihira, M.; Imano, K. Experimental Study on Intermediate Layer for Air-Coupled Ultrasonic Transducer with (0-3) Composite Materials. *Proc. Symp. Ultrason. Electron* **2006**, *27*, 213–214. [CrossRef]
16. Bovtun, V.; Döring, J.; Bartusch, J.; Beck, U.; Erhard, A.; Yakymenko, Y. Ferroelectret non-contact ultrasonic transducers. *Appl. Phys. A* **2007**, *88*, 737–743. [CrossRef]
17. Kazys, R.J.; Sliteris, R.; Sestoke, J. Air-coupled low frequency ultrasonic transducers and arrays with PMN-32% PT piezoelectric crystals. *Sensors* **2017**, *17*, 95. [CrossRef] [PubMed]
18. Kazys, R.; Sliteris, R.; Sestoke, J.; Vladisauskas, A. Air-coupled Ultrasonic Transducers based on an Application of the PMN-32% PT Single Crystals. *Ferroelectrics* **2015**, *480*, 85–91. [CrossRef]
19. Li, Y.; Yu, G.; Liang, B.; Zou, X.; Li, G.; Cheng, S.; Cheng, J. Three-dimensional ultrathin planar lenses by acoustic metamaterials. *Sci. Rep.* **2014**, *4*, 6830. [CrossRef]
20. Zhu, Y.; Fan, X.; Liang, B.; Cheng, J.; Jing, Y. Ultrathin acoustic metasurface-based Schroeder diffuser. *Phys. Rev. X* **2017**, *7*, 021034. [CrossRef]
21. Kim, J.; Park, S.; Anzan-Uz-Zaman, M.; Song, K. Holographic acoustic admittance surface for acoustic beam steering. *Appl. Phys. Lett.* **2019**, *115*, 193501. [CrossRef]
22. Bok, E.; Park, J.J.; Choi, H.; Han, C.K.; Wright, O.B.; Lee, S.H. Metasurface for water-to-air sound transmission. *Phys. Rev. Lett.* **2018**, *120*, 044302. [CrossRef] [PubMed]
23. Yang, T.; Lin, Z.; Yang, T. Experimental Evidence of High-Efficiency Nonlocal Waterborne Acoustic Metasurfaces. *Adv. Eng. Mater.* **2023**, *25*, 2200805. [CrossRef]
24. Leung, H.F.; Hu, A.P. Modeling the contact interface of ultrasonic power transfer system based on mechanical and electrical equivalence. *IEEE J. Emerg. Sel. Top. Power Electron.* **2017**, *6*, 800–811. [CrossRef]
25. Tseng, V.F.G.; Radice, J.J.; Drummond, T.E.; Goodrich, S.; Diamond, D.; Lazarus, N.; Bedair, S.S. Self-detachable through-metal acoustic wireless power transfer. *IEEE Trans. Ultrason. Ferroelectr. Freq. Control.* **2022**, *69*, 2381–2389. [CrossRef] [PubMed]
26. Shatalov, M.; Marais, J.; Fedotov, I.; Tenkam, M.D.; Schmidt, M. Longitudinal vibration of isotropic solid rods: From classical to modern theories. *Adv. Comput. Sci. Eng.* **2011**, *1877*, 187–216.
27. Fedotov, I.A.; Polyanin, A.D.; Shatalov, M.Y. Theory of Free and Forced Vibrations of a Rigid Rod Based on the Rayleigh Model. *Dokl. Phys.* **2007**, *52*, 607–612. [CrossRef]
28. Bishop, R.E.D. Longitudinal Waves in Beams. *Aeronaut. Q.* **1952**, *3*, 280–293. [CrossRef]
29. Allam, A.; Sabra, K.; Erturk, A. Comparison of various models for piezoelectric receivers in wireless acoustic power transfer. In Proceedings of the Active and Passive Smart Structures and Integrated Systems XIII, Denver, CO, USA, 27 March 2019; Volume 10967, pp. 198–205.
30. Mason, W.P. An electromechanical representation of a piezoelectric crystal used as a transducer. *Proc. Inst. Radio Eng.* **1935**, *23*, 1252–1263.
31. Mason, W.P. A dynamic measurement of the elastic, electric and piezoelectric constants of rochelle salt. *Phys. Rev.* **1939**, *55*, 775. [CrossRef]
32. Wu, M.; Chen, X.; Qi, C.; Mu, X. Considering losses to enhance circuit model accuracy of ultrasonic wireless power transfer system. *IEEE Trans. Ind. Electron.* **2019**, *67*, 8788–8798. [CrossRef]
33. Chubachi, N.; Kamata, H. Various equivalent circuits for thickness mode piezoelectric transducers. *Electron. Commun. Jpn. Part II Electron.* **1996**, *79*, 50–59. [CrossRef]
34. Thorvaldsson, T.; Nyffeler, F.M. Rigorous derivation of the Mason equivalent circuit parameters from coupled mode theory. In Proceedings of the IEEE 1986 Ultrasonics Symposium, Williamsburg, VA, USA, 17–19 November 1986; pp. 91–96.
35. Jin, P.; Fu, J.; Wang, F.; Zhang, Y.; Wang, P.; Liu, X.; Feng, X. A flexible, stretchable system for simultaneous acoustic energy transfer and communication. *Sci. Adv.* **2021**, *7*, eabg2507. [CrossRef] [PubMed]
36. Gokhale, V.J.; Downey, B.P.; Katzer, D.S.; Hardy, M.T.; Nepal, N.; Meyer, D.J. Engineering efficient acoustic power transfer in HBARs and other composite resonators. *J. Microelectromech. Syst.* **2020**, *29*, 1014–1019. [CrossRef]
37. Bakhtiari-Nejad, M.; Hajj, M.R.; Shahab, S. Dynamics of acoustic impedance matching layers in contactless ultrasonic power transfer systems. *Smart Mater. Struct.* **2020**, *29*, 035037. [CrossRef]
38. Song, K.; Kwak, J.H.; Park, J.J.; Hur, S. Acoustic metasurfaces for efficient matching of non-contact ultrasonic transducers. *Smart Mater. Struct.* **2021**, *30*, 085011. [CrossRef]

Article

All-Optical, Air-Coupled Ultrasonic Detection of Low-Pressure Gas Leaks and Observation of Jet Tones in the MHz Range

Kyle G. Scheuer [1],* and Ray G. DeCorby [2],*

[1] Ultracoustics Technologies Inc., Sherwood Park, AB T8A 3H5, Canada
[2] ECE Department, University of Alberta, 9211-116 St. NW, Edmonton, AB T6G 1H9, Canada
*   Correspondence: kyle@ultracoustics.ca (K.G.S.); rdecorby@ualberta.ca (R.G.D.)

Abstract: We used an ultrasensitive, broadband optomechanical ultrasound sensor to study the acoustic signals produced by pressurized nitrogen escaping from a variety of small syringes. Harmonically related jet tones extending into the MHz region were observed for a certain range of flow (i.e., Reynolds number), which is in qualitative agreement with historical studies on gas jets emitted from pipes and orifices of much larger dimensions. For higher turbulent flow rates, we observed broadband ultrasonic emission in the ~0–5 MHz range, which was likely limited on the upper end due to attenuation in air. These observations are made possible by the broadband, ultrasensitive response (for air-coupled ultrasound) of our optomechanical devices. Aside from being of theoretical interest, our results could have practical implications for the non-contact monitoring and detection of early-stage leaks in pressured fluid systems.

Keywords: jet tone; leak detection; air-coupled ultrasound; optical ultrasound detection; optomechanics; buckled dome microcavity

Citation: Scheuer, K.G.; DeCorby, R.G. All-Optical, Air-Coupled Ultrasonic Detection of Low-Pressure Gas Leaks and Observation of Jet Tones in the MHz Range. *Sensors* **2023**, *23*, 5665. https://doi.org/10.3390/s23125665

Academic Editors: Farook Sattar, Niladri Bihari Puhan and Reza Fazel-Rezai

Received: 26 May 2023
Revised: 15 June 2023
Accepted: 15 June 2023
Published: 17 June 2023

1. Introduction

Sounds produced by flowing liquids and gases [1] play a central role in a myriad of commonplace phenomena, including human speech [2], whistles produced by animals and engineered objects [3], and, of course, the rich sounds produced by many musical instruments (e.g., wind instruments and pipe organs [4,5]). In spite of their 'everyday' nature, the physics of flow-induced acoustics is quite complex [6] (and 'fundamentally non-linear' [1]), such that exact theoretical treatments (even for relatively simple geometries) are not routinely possible.

Nevertheless, the general features of flow-derived sound are well understood. Typically, acoustic signals arise due to turbulent conditions that correlate with vibrations (i.e., pressure waves) of the flow medium itself [3]. If appropriate feedback is present, periodically spaced vortexes can form in the turbulent flow and give rise to stable oscillations at resonant frequencies corresponding to the generation of 'flow tones', or, in more common terms, 'whistling'. For a given geometry, periodic vortex shedding and associated discrete tones typically arise for certain ranges of flow velocity [1]. A practical application of this phenomenon is the so-called vortex flow meter [7], in which the vortex shedding is induced via an engineered obstruction (i.e., a 'bluff body' [1]) placed in the flow path, and flow rates are extracted from measurement of the vortex-shedding frequencies.

In the present work, we describe a detailed experimental study of ultrasound produced by nitrogen gas jets emitted from a variety of syringes. Furthermore, we show that our observations are consistent with historical studies on gas jets emitted from much larger 'pipes' [8–15], albeit scaled to significantly higher frequencies in the present case. The results illustrate some unique capabilities of our recently reported [16] optomechanical ultrasound sensors, in particular their ultrasensitive and omnidirectional response to air-coupled ultrasound extending over a bandwidth of several MHz. Implications for practical applications, such as leak detection and metering of small-scale, high-pressure flows, are discussed.

The sensor we employ in this study is based on a buckled-microcavity Fabry–Perot resonator. We previously conducted several studies on the optical properties of these devices, including their suitability for quantum electrodynamics due to the high finesse (~10^3–10^4) and small mode volumes (as low as ~1.5 λ^3) routinely obtained [17–19], their applications in microfluidics as open-access cavities for liquids [20], and their ability to be fabricated with large birefringence and polarization mode non-degeneracy [21]. More recently, we demonstrated that our devices function as extremely sensitive optomechanical sensors for both static pressure differentials [22] and ultrasonic signals in air and water [16,23]. Notably, the ultrasonic force sensitivity is one or more orders of magnitude lower than other air-coupled ultrasound sensors, with a bandwidth spanning several MHz [24–26]. These properties prompted us to investigate the feasibility of using our devices for gas leak detection in industrial applications.

2. Materials and Methods

The sensor presented herein is based on a buckled-dome microcavity, where two Bragg mirrors, one planar and one concave, are separated by a partially evacuated and sealed cavity. Our previous work describes the buckled microcavity fabrication process in detail [16,23]. Nevertheless, we also provide an overview here. Briefly, a 3.5 period Si/SiO$_2$ Bragg stack centered at 1600 nm and terminated with Si was deposited on a quartz substrate via plasma-enhanced chemical vapor deposition (PECVD). Photolithography was performed with AZ1512 resist to pattern circular anti-features with a diameter of 100 μm. A ~15 nm fluorocarbon layer was then deposited, and lift-off was performed, leaving behind circular fluorocarbon pads that, ultimately, determined the dimensions of the devices. A second, identical Bragg mirror was deposited directly on top of the fluorocarbon layer and exposed bottom mirror. The substrate was then heated on a hotplate to induce bucking at the fluorocarbon sites, resulting in the formation of half-symmetric Fabry–Perot microcavities. Our previous work utilized a bulky optical table setup to conduct optical and ultrasonic measurements with our devices. This setup is not ideal for industrial applications, since rigid assembly is required to maintain optical alignment in extreme environmental conditions. To address these shortcomings, we designed and assembled a standalone probe unit using off-the-shelf optical components from Oz Optics and Thorlabs. A substrate containing our fabricated devices was affixed to threaded spacer (Thorlabs), which was then attached to a pigtail-style fiber focuser (Oz Optics). The fiber focuser was then connected to an optical circulator (Thorlabs) for interrogation and readout. A full description of the optical properties of our buckled dome microcavity devices is provided in a previous work [16], along with a detailed description of the optical interrogation and readout scheme. For the set of measurements presented herein, we used ~1 mW of optical power at ~1505 nm to couple to the dome mode and read out to the photodetector (Resolved Instruments). The photodetector was set to an 80 MHz sampling rate, and each measurement was averaged over 300 samples. No smoothing or additional post-processing was performed.

The gas system consisted of a nominally 2500 PSI N$_2$ tank (Linde) initially regulated to 40 PSI. The regulator was connected to a toggle valve that was normally open during experiments, and then to a needle valve (Swagelok) and pressure gauge (Baker Instruments), providing a pressure resolution of 0.1 PSI over the range of interest. The output of the pressure gauge was connected to the needle assembly under test. Needles of various gauges (Becton Dickenson PrecisionGlide, with dimensions shown in Table 1) were attached to syringes, which were then directly attached to a section of gas line tubing. A fitting was attached to the other end of the tubing, allowing each needle assembly to be easily attached or removed from the rest of the system. All measurements were performed in ambient laboratory conditions without any specialized acoustic treatment. Figure 1 shows a schematic representation of the gas handling system, along with a photograph of the probe/needle measurement configuration used throughout this manuscript. A photograph showing the needle assemblies is available in the supplementary information.

Table 1. Nominal dimensions of needles used in this work.

Gauge	Inner Diameter (mm)	Outer Diameter (mm)	Length (mm)
22	0.413	0.7176	38
26	0.260	0.4636	13
30	0.159	0.3112	25

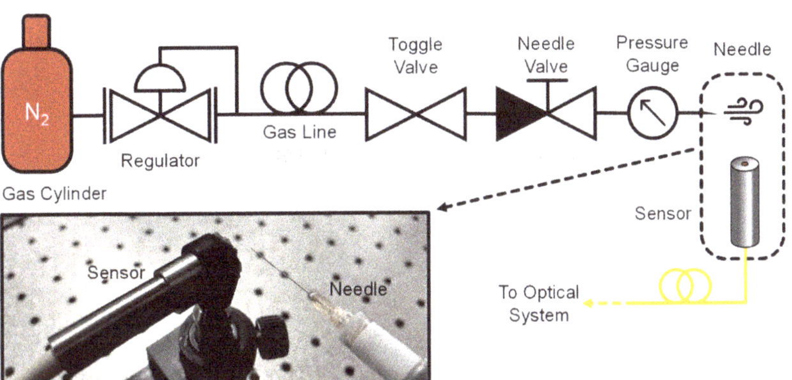

Figure 1. Experimental scheme showing gas handling setup. The setup served to both generate jet tones characteristic to the needle and simulate a controlled gas leak at an arbitrary pressure in range 0–40 PSI. The photograph shows the measurement configuration used throughout the manuscript, unless otherwise specified.

3. Results

3.1. Observation of Jet Tones in the MHz Frequency Range

Controllable pressures were achieved by first opening the toggle valve to allow the system to pressurize up to the supply side of the needle valve. The needle valve was subsequently adjusted while monitoring the pressure gauge, resulting in a stable leak through the needle orifice. Individual power spectral density (PSD) plots for the 30-gauge needle at different pressures are shown in Figure 2, along with a colormap that characterizes the PSD as a function of pressure. At low pressures, only low amplitude features that were invariant with pressure and characteristic to the gas handling system were observed. These features were also observed for the other needle gauges, as well as without a needle assembly present, as shown in the Supplementary Information. At ~8 PSI, jet tones that were evenly spaced in frequency began to emerge. Consistent with observations from Anderson [11], the spacing of these features is not initially representative of the fundamental tone, as the fundamental is often neither the highest amplitude nor the first-appearing resonance. The true fundamental frequency spacing begins to appear near 11 PSI, and is on the order of 60 KHz for the 30-gauge needle. These jet tones are positively correlated to pressure and continue to increase with pressure until eventually combining with the noise floor around 14 PSI, where the spectrum is dominated by white noise that extends into the MHz range. This rising noise floor can also be observed through comparing the individual PSD plots shown in Figure 2a,b.

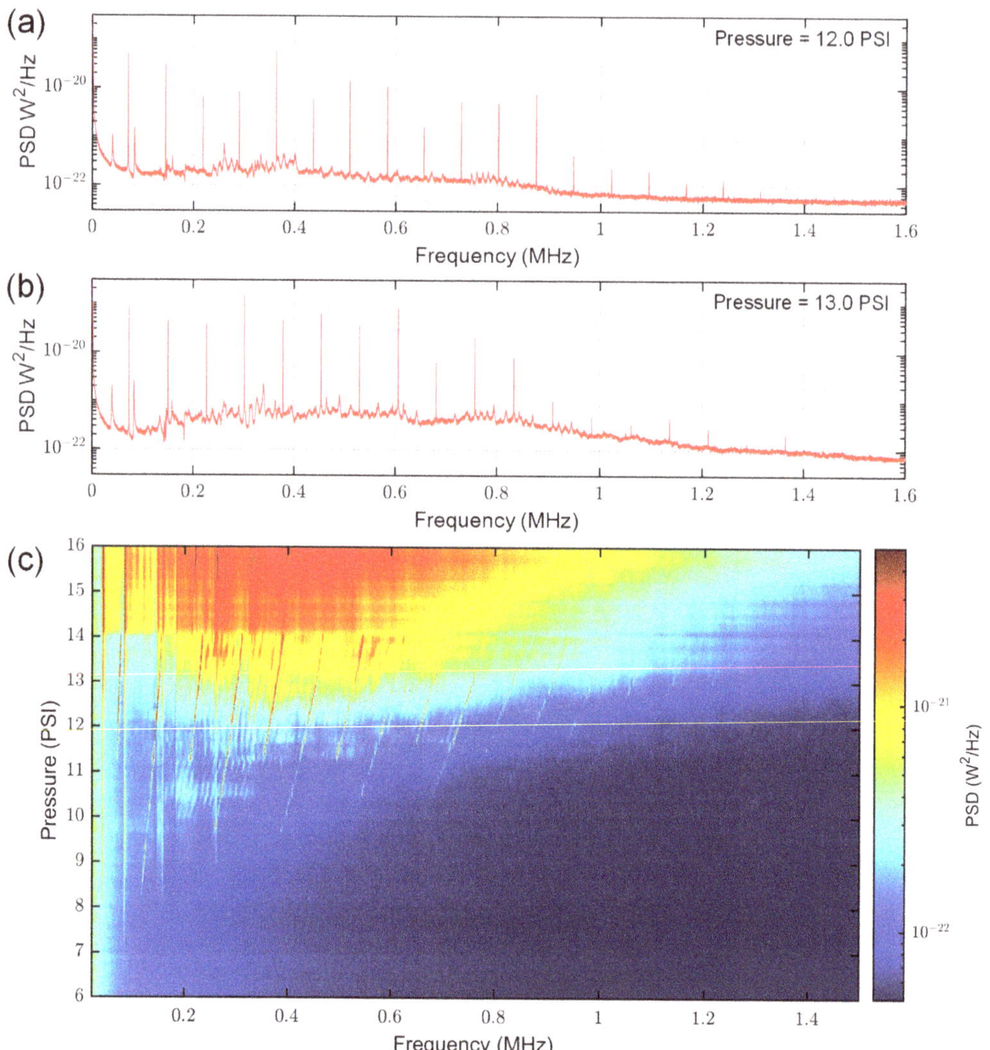

Figure 2. Generation of MHz frequency jet tones with a 30-gauge needle at various pressures. The sensor was placed 1 cm from needle at a 90° angle, as shown in Figure 1. (**a**) A power spectral density plot at 12.0 PSI. (**b**) A power spectral density plot at 13.0 PSI. (**c**) A colormap characterizing jet tones as a function of pressure. Jet tones are present in the region 8~14 PSI, while other static features inherent to gas handling system are present throughout a broader pressure range.

In addition to the 30-gauge needle, we performed similar characterization using both 22- and 26-gauge needles. Representative plots for each needle are shown in Figure 3, demonstrating frequency spacing that is a function of both the pressure and the dimensions of the needle. In general, the smaller the orifice diameter (i.e., higher needle gauge), the greater the spacing between jet tone harmonics and the higher they persist in frequency. As detailed in Section 4, our datasets are entirely consistent with historical observations [10,14] of acoustic signals emitted by gas jets emanating from pipe-like orifices.

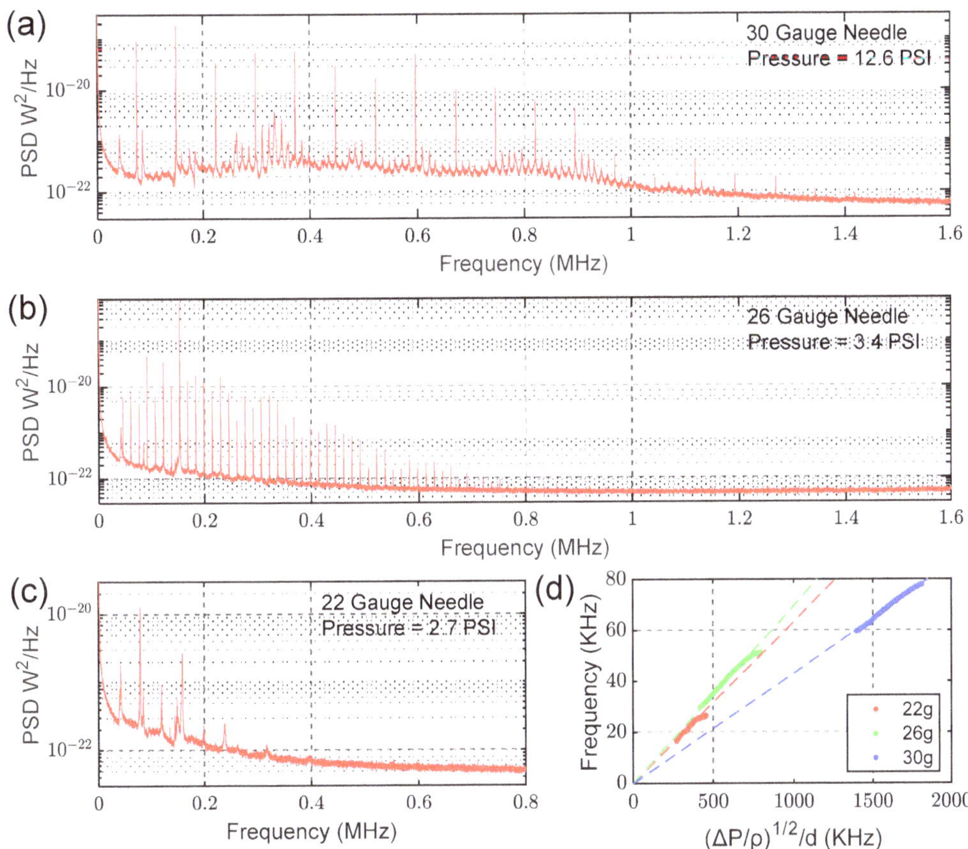

Figure 3. Generation of jet tones with a variety of needles, showing a clear dependence on dimensions of each needle. The sensor was placed 1 cm from needle at a 90° angle, as shown in Figure 1. (**a**) A power spectral density plot for the 30-gauge needle at 12.6 PSI. (**b**) A power spectral density plot for the 26 gauge needle at 3.4 PSI. (**c**) A power spectral density plot for the 22-gauge needle at 2.7 PSI. (**d**) A comparison of the fundamental jet tone frequency for all needle gauges as a function of pressure differential and orifice diameter.

3.2. Broadband Leak Detection

We now turn our attention to the characterization of high-pressure-differential signals where jet tones were not typically observed. Figure 4 shows the acoustic content of the 30-gauge needle in the pressure range 15.0–30.0 PSI. We observed a white noise contribution at high pressures, where the amplitude across the range 0~5 MHz was positively correlated with pressure. While the spectral content could be viewed as white noise in a flow rate sensing context, we note that the PSD is not entirely featureless, and could represent many densely spaced resonances. Regardless, the presence of spectral content in the MHz region, being orders of magnitude above the noise floor, clearly demonstrates the potential of our sensor for industrial applications, particularly in noisy settings where analysis in lower frequencies might not be possible.

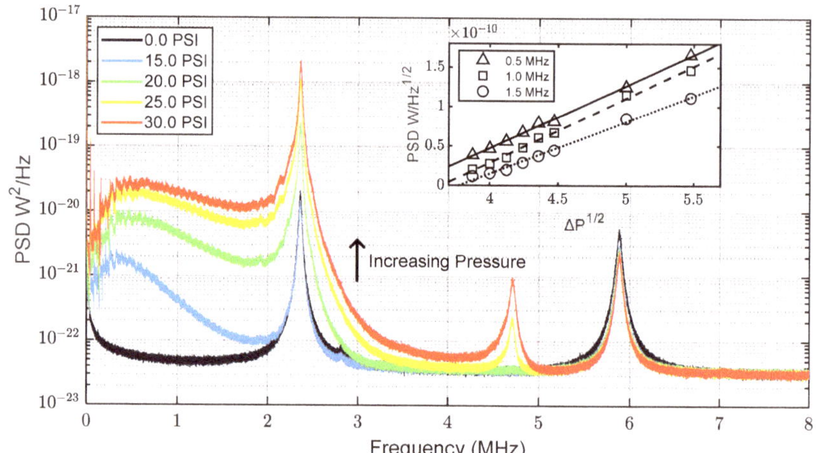

Figure 4. Sensing broadband frequency content of high-pressure nitrogen jets. Each trace shows measured PSD as the pressure of gas line was varied. The sensor was placed 1 cm from needle at a 90° angle, as shown in Figure 1. The plot illustrates that high pressure gas jets through small orifices possess spectral content well into the MHz frequency range, and that our sensor can detect such signals.

We found that while PSD generally increased across the frequency range 0~5 MHz, higher frequencies seemed to be particularly sensitive to pressure. The inset of Figure 4 shows how the PSD evolves with pressure at three discrete frequencies (0.5 MHz, 1.0 MHz, and 1.5 MHz). The power measured at a constant location is proportional to the acoustic power emitted by the source (i.e., the gas jet) [27]. It is also the case that the power emitted by the source is proportional to the mass flow rate of the gas, which scales with the square root of pressure. These quantities can be related to the sound pressure level (*SPL*) using the expression:

$$SPL \propto 10 \log(W/10^{-12}) \propto \log(\dot{m}RT/M). \tag{1}$$

Here, *W* is the sound power level at the source, $\dot{m}$ is the mass flow rate of the jet, *R* is the gas constant, *T* is the temperature, and *M* is the molecular weight [27,28]. We plotted the measured power at three distinct frequencies against the square root of the pressure differential applied to the needle. A linear fit was applied for each frequency, revealing excellent agreement ($R^2 > 0.99$), though there was increasing deviation at higher frequencies. We speculate that this result could be explained by a combination of attenuation in air and the possibility that higher pressures possess higher frequency content. We also recognize that understanding jet noise associated with highly turbulent flow is complicated in its own right, and that numerous theories were previously proposed [29–33].

We also observed additional non-linear contributions from our sensor at sufficiently high pressures in the form of a higher-order resonance feature near 4.8 MHz in some cases. This feature was present regardless of the needle used, as shown in Figure 5. We attribute this result to incoming pressure waves causing deflections that are a significant fraction of the linewidth of the optical resonance. In such cases, the relationship between pressure and optical power becomes non-linear, and harmonics of the natural vibrational modes of the dome appear in the noise spectrum (e.g., the feature at 4.8 MHz is a second-order harmonic of the dome fundamental vibrational resonance at 2.4 MHz). Figure 5 also illustrates that the broadband frequency content associated with higher pressures is not specific to the 30-gauge needle primarily studied in this work; rather, it is present regardless of the orifice size used. Additionally, the dynamic range between 0 and 20 PSI spans several orders of magnitude in all cases.

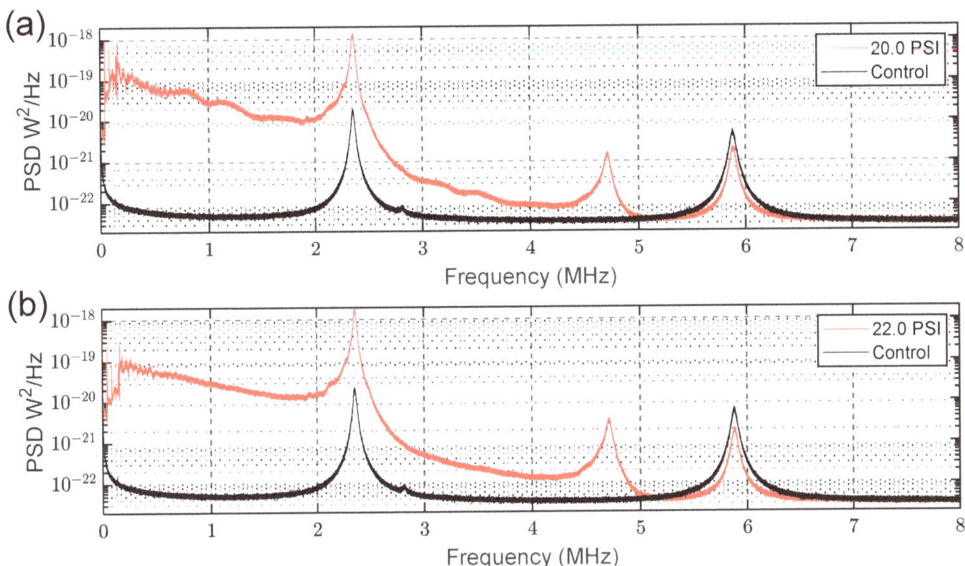

Figure 5. Broadband frequency content for additional needle gauges. Each needle was measured using the 90° needle–sensor configuration shown in Figure 1. A control measurement with the gas handling system depressurized is also presented for each case. (**a**) PSD plot for the 22-gauge needle at 20.0 PSI; (**b**) PSD plot for the 26-gauge needle at 22.0 PSI.

3.3. Omnidirectional Detection

The comparatively small active area of our sensors (~100 µm), combined with the nature of the buckled microcavity structure, provides inherent omnidirectionality [16]. To demonstrate this fact, we varied the lateral distance between the needle and sensor, while keeping the angle between them fixed at 0°, as shown in Figure 6c, as opposed to the 90° configuration used in previous measurements (Figure 1) The PSD plots in Figure 6a,b show results for two different lateral distances using the 30-gauge needle. We found that even in this extreme configuration, the spectral content was still visible in the <500 MHz region, above which air attenuation is suspected to be the limiting factor.

We also note that this distance is not representative of the ultimate device performance we project to be possible. The primary performance-limiting factor for our probe was the coupling between the interrogation laser and device, and better performance is anticipated with future iterations. This aspect could be addressed by designing a custom probe assembly with the ability to make small adjustments to correct for micron-scale misalignment. Aside from this issue, future work will involve further investigating the impact of the mirror layer structure and device size on the optomechanical performance.

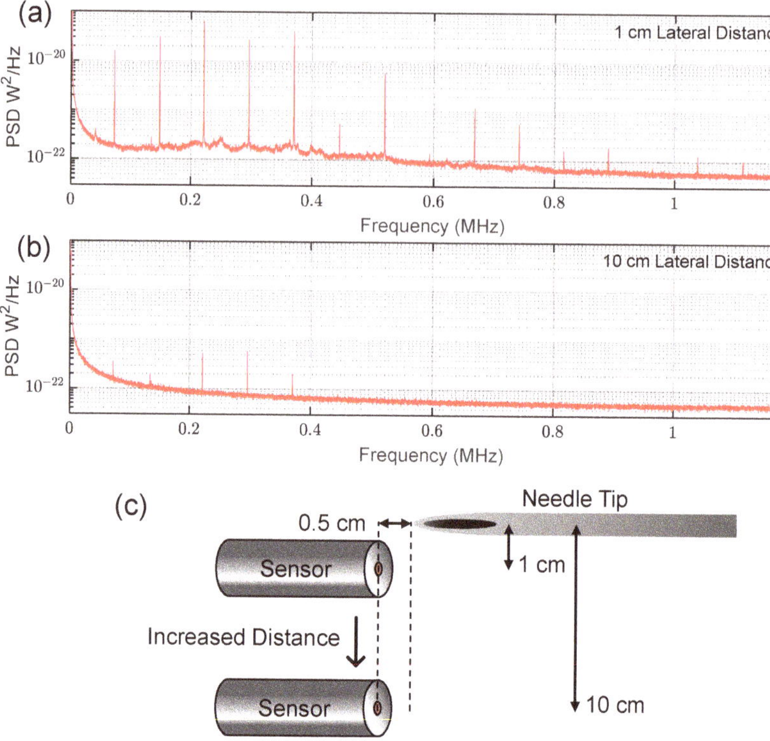

Figure 6. Omnidirectionality demonstration using the 30-gauge needle. PSD plots were recorded as sensor was moved laterally with respect to needle tip. The gas system was held at a constant pressure of 12.3 PSI. (**a**) PSD plot at a lateral distance of 1 cm; (**b**) PSD plot at a lateral distance of 10 cm; (**c**) Schematic showing the configuration for each measurement.

4. Discussion

As mentioned, our observations are consistent with the theoretical framework developed for gas jets emitted from pipe-like orifices. Here, the syringe needle itself plays the role of the pipe, and a gas flow through this needle is driven by a pressure differential between the internal body of the syringe and the external lab environment. Regimes of behavior can be understood using the well-known (and dimensionless) Reynolds (*Re*) and Strouhal (*St*) numbers. Here, $Re = \rho v d / \mu$ and $St = f d / v$, where ρ and μ are the density and dynamic viscosity of the gas, v is flow velocity, d is a characteristic dimension (approximated by the inner diameter of the needle here), and f is the 'vortex shedding' frequency. In many flow problems, *St* is approximately constant over a wide range of *Re*, implying that the observed vortex-shedding frequencies (and associated acoustic emissions) will scale as $f \sim v / d$. Thus, for the very small values of d studied here, we expect the acoustic noise and jet tones to extend to much higher frequencies than could be detected using conventional microphones in earlier studies [8–14], but which are well within the capabilities of our broadband ultrasound sensors.

For the pressurized syringe, high-level details of the flow properties and acoustic signals emitted can be understood as follows [9,10,12,15]:

i. Pressurized gas flows into the needle through an orifice termed the 'vena contracta' [12], which is an effective aperture of diameter δ, being slightly smaller than the inner diameter of the needle (e.g., $\delta \sim 0.63\, d$). This orifice represents the primary obstruction in the flow path and is, thus, the appropriate characteristic dimension to use in calculation of *Re* and *St*.

ii. For low *Re*, flow in the syringe needle is laminar, and negligible acoustic emissions are observed. Above some critical value of *Re*, however, eddies are expected to form near the needle entrance, resulting in a vortex trail forming along the inside walls of the 'pipe' and flowing in the downstream direction [10]. Within a certain range of *Re*, this vortex trail is approximately periodic, resulting in the emission of harmonically related jet tones at the exit aperture of the needle. Notably, both harmonics and sub-harmonics of the vortex-shedding frequency can be observed in the acoustic spectrum within this 'jet tone' regime, as observed and explained by Anderson [9].

iii. For sufficiently high Re, the vortex formation becomes increasingly chaotic, and the flow becomes increasingly turbulent. In this regime, jet tones are subsumed into a broad background of 'white' noise, and the power spectral density of this 'turbulent noise' continues to increase as the flow (i.e., *Re*) is increased.

The results reported in Section 3 (and the Supplementary Materials file) are in line with these expectations. Using the relationship $v = (2\Delta P/\rho)^{1/2}$ [5], where ΔP is the pressure differential across the syringe 'pipe', the curves plotted in Figure 3d verify the expected scaling $f \sim v/d$ discussed above. It should be noted that it is the fundamental jet tone frequency that is plotted. Since the tones observed at lower pressure spacings do not necessarily exhibit the full spectrum of harmonics [11], full colormaps for each needle were collected (and are available in the Supplementary Information), making it easier to identify the harmonic eigenfrequencies. However, it was somewhat difficult to extract the fundamental frequency for the 22-gauge needle due to the narrow pressure range where jet tones were observed as well as their comparatively low amplitude. We fit the data from each needle to a linear equation and found the slopes to be ~0.063 KHz/KHz, ~0.070 KHz/KHz, and ~0.043 KHz/KHz for the 22-gauge, 26-gauge, and 30-gauge needles, respectively ($R^2 > 0.99$ in all cases). It is interesting, though perhaps expected, that all three needles seem to exhibit similar slopes when the dependence on pressure and orifice diameter were taken into account. We suggest that the variation in slope observed might result from the fact that each gauge of needle has a different length, and that the characteristic dimension has some dependence on both d and L [9,11]. Moreover, the manufacturer does not specify a tolerance on any needle dimension, and we did not account for the needle bevel. Our data seem to suggest the absence of a jet tone in the case of a pressure differential below some threshold or for an infinitely large orifice diameter (in the form of an origin crossing), which is again consistent with Anderson's observations [9]. Anderson performed a similar experiment with orifice plates affixed to a tube of length L (where L was the characteristic dimension of the system) and noted nearly identical slopes. Another curiosity is that the slope for each needle deviated near the upper pressure range where jet tones were observed, at least in the case of the 22- and 26-gauge needles. We attribute this result to changes in the nature of vortex formation at high *Re* values [12].

Each needle was found to exhibit a distinct range of pressures where jet tones, and turbulent flow in general, were observed. This finding suggests the possibility of estimating the size of a pinhole leak in an industrial setting if the pressure of the system is known. The diameter of the orifice could be estimated using either the jet tone spacing or the amplitude of the white noise at a given distance. The ability to measure fluid leaks has wide appeal within the oil and gas industry, and significant effort was previously expended in developing early detection systems based on ultrasonic technology [34–38]. However, many of these systems are limited, at least in part, by their bandwidth and sensitivity. We believe that the ability to measure leaks from sub-millimeter holes at low pressures far into the MHz-frequency-range will be of particular interest for hazardous or explosive gases, where electronic components cannot be placed in close proximity. Additionally, many sources of ambient noise lie well below the MHz-range signals detected in the present work, which is a significant potential benefit of the broadband capabilities of our sensor. In future work, we hope to target the direct measurement of gas leaks in an industrial setting.

One additional intriguing possibility suggested by our results is that a small needle emitting a controlled gas jet could represent one way of generating a tunable acoustic frequency comb, especially since the pressures used here should be accessible to any space that already utilizes floating optical tables. This finding could represent a partial setup for photo-acoustic comb spectroscopy without requiring expensive acousto- or electro-optic modulators [39,40].

5. Conclusions

In summary, we have made three distinct contributions. First, as a compact probe for all-optical detection of MHz-frequency range, air-coupled ultrasound was constructed. Second, we generated MHz-range jet tones by passing pressurized nitrogen gas through a collection of small syringes and showed that previously established theory can be extended to this range. Finally, we showed that high-pressure gas leaks contain frequency content that extends far into the MHz range, lying orders of magnitude above the noise floor of our devices. Additionally, gas leaks can be sensed both off-axis and off-position. Our buckled microcavity-based devices function as uniquely enabling leak sensors due to their sensitivity, bandwidth, and omnidirectionality.

Supplementary Materials: The following supporting information can be downloaded at: https://www.mdpi.com/article/10.3390/s23125665/s1, Figure S1. A photograph of the needle assemblies from the main manuscript, Figure S2. Generation of MHz frequency jet tones with 22- and 26-gauge needles at various pressures, Figure S3. Extracted lineshape for the 30-gauge needle pressurized to 30.0 PSI, and Figure S4. Spectral content inherent to the gas system without a needle assembly in place. Reference [41] is cited in the Supplementary Materials.

Author Contributions: Conceptualization, K.G.S. and R.G.D.; methodology, R.G.D.; validation, K.G.S. and R.G.D.; formal analysis, K.G.S. and R.G.D.; investigation, K.G.S.; resources, R.G.D.; data curation, K.G.S.; writing—original draft preparation, K.G.S. and R.G.D.; writing—review and editing, K.G.S. and R.G.D.; visualization, K.G.S.; supervision, R.G.D.; project administration, R.G.D.; funding acquisition, R.G.D. and K.G.S. All authors have read and agreed to the published version of the manuscript.

Funding: This research was funded by the Government of Alberta (Innovation Catalyst Grant), Alberta Innovates, the Natural Sciences and Engineering Research Council of Canada (CREATE 495446-17), and the Alberta EDT Major Innovation Fund (Quantum Technologies).

Institutional Review Board Statement: Not applicable.

Informed Consent Statement: Not applicable.

Data Availability Statement: The data that support the findings of this study are available from the authors upon reasonable request.

Acknowledgments: We thank Graham Hornig for his assistance with the optical measurement system, and we thank Tim Harrison for his PECVD expertise.

Conflicts of Interest: Ultracoustics Technologies Ltd. (I,P) KGS, North Road Photonics Corp. (I,P) RGD.

References

1. Blake, W.K.; Powell, A. *Development of Contemporary Views of Flow-Tone Generation*; Springer: New York, NY, USA, 1986; pp. 247–325.
2. Shadle, C.H. Experiments on the Acoustics of Whistling. *Phys. Teach.* **1983**, *21*, 148–154. [CrossRef]
3. Chanaud, R.C. Aerodynamic Whistles. *Sci. Am.* **1970**, *222*, 40–47. [CrossRef]
4. Fletcher, N.H.; Thwaites, S. The Physics of Pipe Organs. *Sci. Am.* **1983**, *248*, 94–103. [CrossRef]
5. Fletcher, N.H.; Rossing, T.D. *The Physics of Musical Instruments*; Springer: New York, NY, USA, 1998.
6. Campos, L.M.B.C. On Waves in Gases. Part I: Acoustics of Jets, Turbulence, and Ducts. *Rev. Mod. Phys.* **1986**, *58*, 117. [CrossRef]
7. Venugopal, A.; Agrawal, A.; Prabhu, S.V. Review on Vortex Flowmeter—Designer Perspective. *Sens. Actuators A Phys.* **2011**, *170*, 8–23. [CrossRef]
8. Anderson, A.B.C. A Circular-Orifice Number Describing Dependency of Primary Pfeifenton Frequency on Differential Pressure, Gas Density, and Orifice Geometry. *J. Acoust. Soc. Am.* **1953**, *25*, 626–631. [CrossRef]

9. Anderson, A.B.C. A Jet-Tone Orifice Number for Orifices of Small Thickness-Diameter Ratio. *J. Acoust. Soc. Am.* **1954**, *26*, 21–25. [CrossRef]
10. Anderson, A.B.C. Metastable Jet-Tone States of Jets from Sharp-Edged, Circular, Pipe-Like Orifices. *J. Acoust. Soc. Am.* **1955**, *27*, 13–21. [CrossRef]
11. Anderson, A.B.C. Vortex-Ring Structure-Transition in a Jet Emitting Discrete Acoustic Frequencies. *J. Acoust. Soc. Am.* **1956**, *28*, 914–921. [CrossRef]
12. Thomas, N. On the Production of Sound by Jets. *J. Acoust. Soc. Am.* **1955**, *27*, 446–448. [CrossRef]
13. Jothi, T.J.S.; Srinivasan, K. Transonic Resonance Tones in Orifice and Pipe Jets. *Int. J. Aeroacoustics* **2013**, *12*, 103–121. [CrossRef]
14. Anderson, A.B.C. Dependence of *Pfeifenton* (Pipe Tone) Frequency on Pipe Length, Orifice Diameter, and Gas Discharge Pressure. *J. Acoust. Soc. Am.* **1952**, *24*, 675–681. [CrossRef]
15. Anderson, A.B.C. Structure and Velocity of the Periodic Vortex-Ring Flow Pattern of a *Primary Pfeifenton* (Pipe Tone) Jet. *J. Acoust. Soc. Am.* **1955**, *27*, 1048–1053. [CrossRef]
16. Hornig, G.J.; Scheuer, K.G.; Dew, E.B.; Zemp, R.; DeCorby, R.G. Ultrasound Sensing at Thermomechanical Limits with Optomechanical Buckled-Dome Microcavities. *Opt. Express* **2022**, *30*, 33083–33096. [CrossRef]
17. Potts, C.A.; Melnyk, A.; Ramp, H.; Bitarafan, M.H.; Vick, D.; LeBlanc, L.J.; Davis, J.P.; DeCorby, R.G. Tunable Open-Access Microcavities for on-Chip Cavity Quantum Electrodynamics. *Appl. Phys. Lett.* **2016**, *108*, 041103. [CrossRef]
18. Bitarafan, M.H.; DeCorby, R.G. Small-Mode-Volume, Channel-Connected Fabry–Perot Microcavities on a Chip. *Appl. Opt.* **2017**, *56*, 9992. [CrossRef]
19. Bitarafan, M.; DeCorby, R. On-Chip High-Finesse Fabry-Perot Microcavities for Optical Sensing and Quantum Information. *Sensors* **2017**, *17*, 1748. [CrossRef]
20. Maldaner, J.; Al-Sumaidae, S.; Hornig, G.J.; LeBlanc, L.J.; DeCorby, R.G. Liquid Infiltration of Monolithic Open-Access Fabry–Perot Microcavities. *Appl. Opt.* **2020**, *59*, 7125. [CrossRef]
21. Hornig, G.; DeCorby, R.; Bu, L.; Alsumaidae, S. Monolithic Elliptical Dome Fabry-Perot Microcavities Exhibiting Large Birefringence. *J. Opt. Soc. Am. B* **2022**, *39*, 884–890. [CrossRef]
22. Al-Sumaidae, S.; Bu, L.; Hornig, G.J.; Bitarafan, M.H.; DeCorby, R.G. Pressure Sensing with High-Finesse Monolithic Buckled-Dome Microcavities. *Appl. Opt.* **2021**, *60*, 9219. [CrossRef]
23. Hornig, G.J.; Scheuer, K.G.; DeCorby, R.G. Observation of Thermal Acoustic Modes of a Droplet Coupled to an Optomechanical Sensor. *arXiv* **2023**, arXiv:2305.13392.
24. Hansen, S.T.; Ergun, A.S.; Liou, W.; Auld, B.A.; Khuri-Yakub, B.T. Wideband Micromachined Capacitive Microphones with Radio Frequency Detection. *J. Acoust. Soc. Am.* **2004**, *116*, 828–842. [CrossRef]
25. Kuntzman, M.L.; Hall, N.A. A Broadband, Capacitive, Surface-Micromachined, Omnidirectional Microphone with More than 200 KHz Bandwidth. *J. Acoust. Soc. Am.* **2014**, *135*, 3416–3424. [CrossRef]
26. Basiri-Esfahani, S.; Armin, A.; Forstner, S.; Bowen, W.P. Precision Ultrasound Sensing on a Chip. *Nat. Commun.* **2019**, *10*, 132. [CrossRef]
27. Naranjo, E.; Baliga, S. Expanding the Use of Ultrasonic Gas Leak Detectors: A Review of Gas Release Characteristics for Adequate Detection. *Gases Instrum.* **2009**, *7*, 24–29.
28. Raichel, D.R. *The Science and Applications of Acoustics*, 2nd ed.; Springer: New York, NY, USA, 2006; ISBN 9780387260624.
29. Lighthill, M.J. Jet Noise. *AIAA J.* **1963**, *1*, 1507–1517. [CrossRef]
30. Karabasov, S.A. Understanding Jet Noise. *Philos. Trans. R. Soc. A Math. Phys. Eng. Sci.* **2010**, *368*, 3593–3608. [CrossRef]
31. Mu, G.; Lyu, Q.; Li, Y. An Analysis of Jet Noise Characteristics in the Compressible Turbulent Mixing Layer of a Standard Nozzle. *Machines* **2022**, *10*, 826. [CrossRef]
32. Masovic, D.; Sarradj, E. Derivation of Lighthill's Eighth Power Law of an Aeroacoustic Quadrupole in Acoustic Spacetime. *Acoustics* **2020**, *2*, 666–673. [CrossRef]
33. Mostafapour, A.; Davoudi, S. Analysis of Leakage in High Pressure Pipe Using Acoustic Emission Method. *Appl. Acoust.* **2013**, *74*, 335–342. [CrossRef]
34. Murvay, P.S.; Silea, I. A Survey on Gas Leak Detection and Localization Techniques. *J. Loss Prev. Process Ind.* **2012**, *25*, 966–973. [CrossRef]
35. Meribout, M.; Khezzar, L.; Azzi, A.; Ghendour, N. Leak Detection Systems in Oil and Gas Fields: Present Trends and Future Prospects. *Flow Meas. Instrum.* **2020**, *75*, 101772. [CrossRef]
36. Guenther, T.; Kroll, A. Automated Detection of Compressed Air Leaks Using a Scanning Ultrasonic Sensor System. In Proceedings of the 2016 IEEE Sensors Applications Symposium (SAS), Catania, Italy, 20–22 April 2016; pp. 116–121. [CrossRef]
37. Lee, J.C.; Choi, Y.R.; Cho, J.W. Pipe Leakage Detection Using Ultrasonic Acoustic Signals. *Sens. Actuators A Phys.* **2023**, *349*, 114061. [CrossRef]
38. Zuo, J.; Zhang, Y.; Xu, H.; Zhu, X.; Zhao, Z.; Wei, X.; Wang, X. Pipeline Leak Detection Technology Based on Distributed Optical Fiber Acoustic Sensing System. *IEEE Access* **2020**, *8*, 30789–30796. [CrossRef]
39. Wildi, T.; Voumard, T.; Brasch, V.; Yilmaz, G.; Herr, T. Photo-Acoustic Dual-Frequency Comb Spectroscopy. *Nat. Commun.* **2020**, *11*, 4164. [CrossRef]

40. Picqué, N.; Hänsch, T.W. Frequency Comb Spectroscopy. *Nat. Photonics* **2019**, *13*, 146–157. [CrossRef]
41. Bitarafan, M.H.; Ramp, H.; Allen, T.W.; PoBs, C.; Rojas, X.; MacDonald, A.J.R.; Davis, J.P.; DeCorby, R.G. Thermomechanical Characterization of On-Chip Buckled Dome Fabry–Perot Microcavities. *J. Opt. Soc. Am. B* **2015**, *32*, 1214. [CrossRef]

Article

Embedded Ultrasonic Inspection on the Mechanical Properties of Cold Region Ice under Varying Temperatures

Huimin Han [1], Li Wei [1], Nizar Faisal Alkayem [2,3] and Maosen Cao [1,*

[1] College of Mechanics and Materials, Hohai University, Nanjing 210024, China; hanhm9011@hhu.edu.cn (H.H.); weili0019@hhu.edu.cn (L.W.)
[2] College of Water Conservancy and Hydropower Engineering, Hohai University, Nanjing 210098, China; nizar-alkayem@hhu.edu.cn
[3] College of Civil and Transportation Engineering, Hohai University, Nanjing 210098, China
* Correspondence: cmszhy@hhu.edu.cn

Abstract: The mechanical properties of ice in cold regions are significantly affected by the variation in temperature. The existing methods to determine ice properties commonly rely on one-off and destructive compression and strength experiments, which are unable to acquire the varying properties of ice due to temperature variations. To this end, an embedded ultrasonic system is proposed to inspect the mechanical properties of ice in an online and real-time mode. With this system, ultrasonic experiments are conducted to testify to the validity of the system in continuously inspecting the mechanical properties of ice and, in particular, to verify its capabilities to obtain ice properties for various temperature conditions. As an extension of the experiment, an associated refined numerical model is elaborated by mimicking the number, size, and agglomeration of bubbles using a stochastic distribution. This system can continuously record the wave propagation velocity in the ice, giving rise to ice properties through the intrinsic mechanics relationship. In addition, this model facilitates having insights into the effect of properties, e.g., porosity, on ice properties. The proposed embedded ultrasonic system largely outperforms the existing methods to obtain ice properties, holding promise for developing online and real-time monitoring techniques to assess the ice condition closely related to structures in cold regions.

Keywords: ice; mechanical properties; embedded ultrasonic system; varying temperature; random pore model; porosity

Citation: Han, H.; Wei, L.; Alkayem, N.F.; Cao, M. Embedded Ultrasonic Inspection on the Mechanical Properties of Cold Region Ice under Varying Temperatures. *Sensors* **2023**, *23*, 6045. https://doi.org/10.3390/s23136045

Academic Editors: Farook Sattar, Niladri Bihari Puhan and Reza Fazel-Rezai

Received: 2 June 2023
Revised: 16 June 2023
Accepted: 19 June 2023
Published: 29 June 2023

1. Introduction

In recent decades, research on ice has grown remarkably. Intensive research is being performed on whether to examine global climate change [1–3], to break ice sheets in the arctic with icebreakers [4], to investigate ice-covered electric antennas [5], to design ships and offshore structures [6,7], to study the icing conditions in aviation [8], etc. In those investigations, it is essential to obtain the mechanical properties of ice, which will change with variations in ambient temperature.

Currently, diverse measurement methods of ice mechanical properties have been developed in an experimental way, including uniaxial compression [9–11], triaxial compression, and flexural strength tests [12]. Aksenov et al. [13] attained the preliminary temperature-related stress-strain properties of the freshwater ice by performing short-time uniaxial compression tests for cylindrical ice specimens at various specific temperatures. Moslet [14] conducted the uniaxial compression strength tests on columnar sea ice in the field on Svalbard, with the indication of a strong relationship between Young's modulus and ice porosity. Qiu et al. [15] obtained the compressive and tensile plastic properties of ice based on the triaxial compression test of columnar ice. However, due to the brittleness of the ice material itself, fracture is inevitable in the experiment. Moreover, the time between the testing starting and the fracture ending is very short. Therefore, in a single experiment, it is not possible to obtain changes in ice properties caused by continuous temperature changes. Additionally, those destructive tests are costly and inappropriate for structures in service. In order to obtain the mechanical properties of ice that continuously change with temperature without damaging the existing ice structure, a non-destructive testing method is needed that can achieve real-time monitoring and is suitable for ice materials. Accordingly, a non-destructive method for estimating the temperature-related mechanical properties of ice is demanded.

The ultrasonic technique is one of the most commonly used non-destructive methods in characterizations of material properties [16,17], derivations of dynamic responses [18], and condition monitoring [19–23]. Relevant sensing techniques for wave generation and measurement have been developed rapidly. Representative research works are as follows: Bayón et al. [24] employed the measured Rayleigh wave velocity and the aspect ratio of the elliptical trajectory amplitudes to obtain the Young's constants of isotropic linear materials. Further, Medina and Bayón [25] proposed a method for calculating the dynamic Young's constants of an anisotropic plate through the measured impact-echo resonance and Rayleigh wave velocity. Medina and Bayón [25] determined the mechanical properties and damage properties of a multilayer composite board by combining experimental and numerical data simultaneously. Ultrasonic technology has also been implemented in the estimation of ice properties. Bock and Polach [26] explored the applicability of the collected long-period surface wave dispersions in the inversion of ice shell thickness. Gagnon [27] presented an impulse-echo method to evaluate the longitudinal ultrasonic velocities of three different types of specimens to calculate Young's modulus. Noteworthy, a state-of-the-art literature review specified that only a limited number of studies dealt with the utilisation of ultrasonic techniques in the determination of ice properties, and there are still some crucial issues to be solved. In addition, after obtaining the monitoring data, this study used the methods mentioned in [28,29] to conduct the corresponding data modelling and analysis, including regression analysis and spectral analysis.

In addition to experimental testing methods, numerical simulation is a valuable alternative method for estimating the ice properties by taking advantage of wave propagation simulation techniques [22,30–33]. Various numerical models were established to represent the ice materials, including Young's non-linear behaviour model [34], the crushable foam model [35,36], and the user-defined elastic-plastic material model [37,38]. Liu et al. [6] proposed a quasi-static model on the basis of the strain rate-dependent plasticity theory to simulate ice structural behaviour by combining the Tsai Wu yield surface criteria and the associated flow rule. Bock and Polach [26] analysed the non-linear behaviour of the aqueous model ice, concluding that the non-linear behaviour of ice is independent of its crystal structure and chemical dopant. Gagnon et al. [27] adopted a "crushable foam" material type in LS-DYNA to model the ice behaviour, with the numerical results of the 'calibrated stress-volumetric strain relationship' showing a good agreement with the experimental ones. Noticeably, the existing numerical cases lacked studies on the influence of temperature variations and ice porosities.

From the surveyed literature, it is observed that the current prevalent methods to determine ice properties commonly rely on one-off and destructive compression and strength tests, which are incapable of acquiring the temperature variation-induced changes in ice properties. Moreover, ultrasonic technology, as one of the most commonly used non-destructive methods for characterising material properties, is rarely applied to ice mechanical properties. In addition, due to the unique nature of ice materials, it is necessary to study how ultrasonic sensors can be embedded in ice. Furthermore, there is a lack of numerical models for wave propagation in ice with different temperatures and porosities. To address these issues, an embedded ultrasound system is proposed to test the mechanical properties of ice, and its feasibility and effectiveness have been verified through ultrasound experiments and numerical models.

The rest of this paper is organised as follows: Section 2 formulates a novel ultrasonic method for determining the mechanical properties of ice. Section 3 builds the 2D and

3D ice models, with the numerical results at different temperatures and porosities being counted. Section 4 exhibits comparisons and discussions of experimental and numerical results. Section 5 presents the conclusions.

2. Test-Based Identification of the Mechanical Properties of Ice

The current prevalent methods to acquire the mechanical properties of ice are typically uniaxial compression, triaxial compression, and flexural strength experiments. However, due to the brittleness of the ice material itself, fracture is inevitable in the experiment. Moreover, the time between the testing starting and the fracture ending is very short. Therefore, in a single experiment, it is not possible to obtain changes in ice properties caused by continuous temperature changes. To address the deficiencies of existing methods, an embedded ultrasonic system is built as an alternative to compression tests to obtain the mechanical properties of ice at different temperatures.

2.1. Test Specimen

2.1.1. Actuators and Sensors

Piezoelectric material can be used as both actuators and transducers, which is beneficial for its direct and inverse piezoelectric effects [39]. Figure 1c shows a piezoelectric ceramic transducer (PZT), with its dimensions and material properties shown in Table 1. In the PZT, the electric wires are welded to the positive and negative poles on their different sides. Both sides are waterproof with the transparent insulating glues. The temperature sensor is made of a 4 mm-diameter, waterproof platinum probe (shown in Figure 1d). The measurement range is $-50-+200\ ^\circ$C.

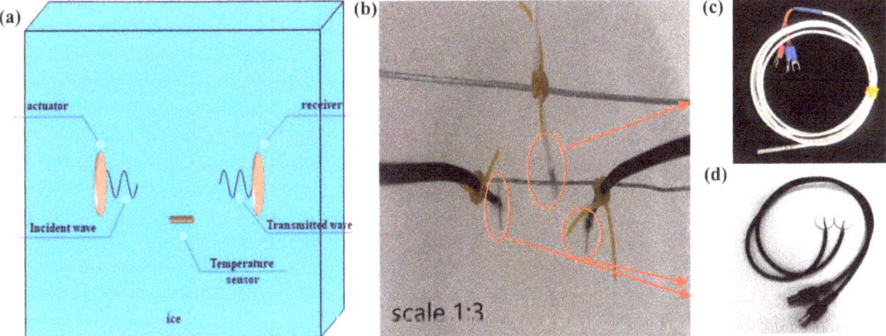

Figure 1. (**a**) Schematic diagram of ice sample structure; (**b**) ice specimen with sensors (scale 1:3); (**c**) the PZT; and (**d**) the temperature sensor.

Table 1. Dimensions and material properties of PZT.

Diameter	Thickness	d_{33}	K^{T3}
20 mm	1 mm	460×10^{-12} m/V	1700

2.1.2. Ice Specimen and Sensing Strategy

(1) Ice specimen

The ice specimen is fabricated using distilled water with the purpose of avoiding excessive impurities. The dimensions of the ice specimen are $150 \times 150 \times 200$ mm^3. Before freezing, all the required sensors during the ultrasonic tests are fixed in the middle of the mould with steel wires, as shown in Figure 1b.

The process for making ice samples is as follows: first, place the mould with distilled water and sensors into a refrigerator with an initial temperature of $-5\ ^\circ$C; second, when the surface water is frozen, progressively decrease the temperature of the refrigerator by 5 °C per two hours; next, when the refrigerator temperature reaches -35 °C, the ice specimen should be fully made; and finally, the newly-made ice specimen needs to be stored in a refrigerator at -35 °C for at least 48 h to ensure a uniform temperature distribution in the ice specimen. In this way, the ice cracks caused by big temperature variations could be avoided.

(2)   Sensing strategy

The PZTs and temperature sensor are set in the middle of the ice specimen. The distance between the two PZTs is considered to be 70 mm. Additionally, the temperature sensor is set at the uniform level of the PZTs due to the effect of temperature stratification in ice.

(3)   Considerations

Preliminary tests with a distance between the two PZTs are conducted to determine the appropriate properties. Comparing the results at 25 mm, 50 mm, and 70 mm, it is found that when the distance between the two PZTs is small, the wave packets of the longitudinal wave and the shear wave cannot be distinguished. The reason is that the propagation speed of the wave is very fast. When the shear wave propagates, the longitudinal wave has not been fully received, which causes the longitudinal wave and shear wave to be superimposed on each other, so it is difficult to distinguish a separate shear wave packet.

2.2. Velocity of the Transmitted Waves

Wave velocity, one of the key properties of wave propagation research, is defined as the velocity at which a disturbance propagates in specified materials. It mainly depends on the material properties, structural geometries, and external excitations. The longitudinal and shear wave velocities are the most widely used variables in ultrasonic-structural analysis. Their relationships with structural properties are formulated as follows:

$$V_l = \sqrt{\frac{E(1-\mu)}{\rho(1+\mu)(1-2\mu)}},\tag{1}$$

$$V_s = \sqrt{\frac{E}{2\rho(1+\mu)}}\tag{2}$$

According to Equations (1) and (2), the material properties (such as E, G, and μ) can be achieved with the measured longitudinal and shear wave velocities (V_l and V_s) (shown in Equations (3)–(6)). These relationships, which are the fundamentals of the characterizations of the ice properties, can be written as follows:

$$\mu = \frac{V_l^2 - 2V_s^2}{2\left(V_l^2 - V_s^2\right)},\tag{3}$$

$$E = \frac{V_s^2 \rho \left(3V_l^2 - 4V_s^2\right)}{V_l^2 - V_s^2},\tag{4}$$

$$G = \rho V_s^2,\tag{5}$$

$$K = \rho(V_l^2 - \frac{4}{3}V_s^2),\tag{6}$$

where ρ is the density, E is Young's modulus, μ is Poisson's ratio, and V_l and V_s represent the velocity of the longitudinal wave and the shear wave, respectively.

2.3. Experimental Setup and Methods

(1) Test process

The experimental apparatus for the ultrasonic test is presented in Figure 2.

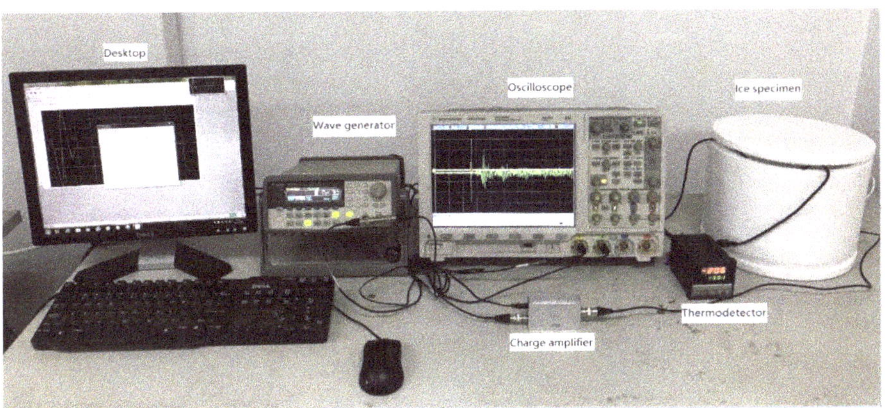

Figure 2. The experimental setup.

The excitation part includes the wave generator and desktop. Firstly, the waveform of excitation is generated on the desktop. Then, a tone burst signal, consisting of sinusoids modulated by the Hanning window [40], is utilised as the external excitation. This signal can be expressed as follows [41]:

$$x(t) = \frac{A}{2} \sin(2\pi f_c t) \left[1 - \cos\left(\frac{2\pi f_c t}{n} \right) \right],$$

(7)

where A is the amplitude, n denotes the number of signal cycles, and f_c represents the central frequency.

The designated excitation is then transferred into the Agilent 33250A arbitrary wave generator, which converts the excitation from a digital signal to an analogue one. The output terminal of the waveform generator is connected to the PZT through a Bayonet Nut Connector cable. Because of the positive piezoelectric effect, the PZT deforms in the longitudinal direction of the ice, leading to the generation of longitudinal waves. The output voltage of the waveform generator, which regulates the excitation amplitude, is set to be 10 V in order to ensure a fully deformed piezoelectric ceramic sheet.

In the receiving part, the PZT transducer generates current due to the inverse piezoelectric effect. Based on the characteristics of PZT, this generated current has a linear relationship with its deformation. Thus, the captured current from the receiver can be regarded as the propagated longitudinal wave in the ice. This received current is then amplified by a fixed-gain universal preamplifier, PXPA3, to increase its amplitude. The transfer gain of the charge amplifier is 10 mv/pc. Then the current is introduced into the Agilent DSO7034B oscilloscope to be converted into a digital signal. The sampling frequency of the oscilloscope is set to 50 MHz. In particular, the application of the waterproof glue on the surface of PZT not only makes the sensor insulated but also prevents the generated current from leaking into the ice. In addition, there is no extra circuit other than the experimental equipment. Therefore, the measured current from the receiver thoroughly originates from the inverse piezoelectric effect caused by the longitudinal wave.

During the ultrasonic test, the ice specimen is covered with a foam box, which is a common thermal insulation material in daily life and is used to slow down the impact of external temperature on water, thus slowing down the speed of water icing. If the freezing speed is too fast, it can cause the ice surface to expand and crack. Cracking ice will badly affect the experimental results. The internal temperature is monitored in real-time through the temperature sensor. By continuously recording these internal temperatures and the corresponding received signals, the non-destructive monitoring of wave propagation in ice under the condition of temperature variation is realised.

(2)   Determination of signal properties

Two properties (n and f_c) on the right-hand side of Equation (7) need to be determined in advance. Preliminary tests with different n and f_c are conducted to determine the appropriate properties. The measured ultrasonic signals with different n and f_c are shown in Figures 3 and 4, respectively.

In the preliminary test, the response signal results of 2 periods, 3 periods, 4 periods, and 5 periods are compared and analysed. The experimental results in Figure 3 indicate that the second peak of the measured signals can be more easily distinguished in cases of having a smaller number of cycles (n = 2, 3) than in cases of having a larger number of cycles (n = 4, 5).

On the other hand, the results in Figure 4 imply that the second peak of the measured signals is easier to identify in cases of having larger central frequencies (f_c = 250 kHz, 300 kHz, 350 kHz, 400 kHz, 450 kHz, and 500 kHz) than in cases of having smaller central frequencies (f_c = 150 kHz and 200 kHz).

Accordingly, a normalised 2-cycle 250 kHz tone burst signal (shown in Figure 5) is employed as the external excitation in the experimental tests below.

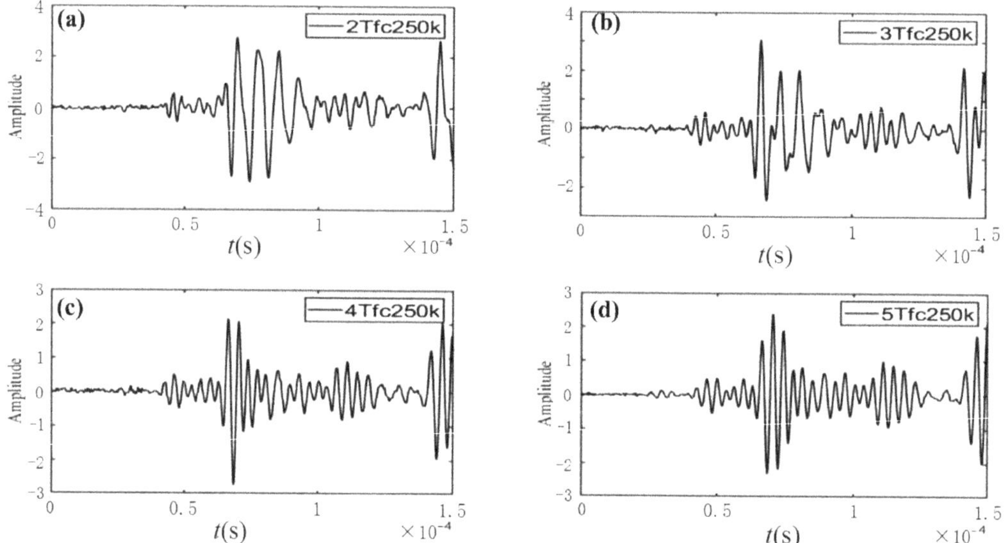

Figure 3. Measured ultrasonic signals with the same centre frequency (250 kHz) and various number of cycles: (**a**) n = 2, (**b**) n = 3, (**c**) n = 4, and (**d**) n = 5.

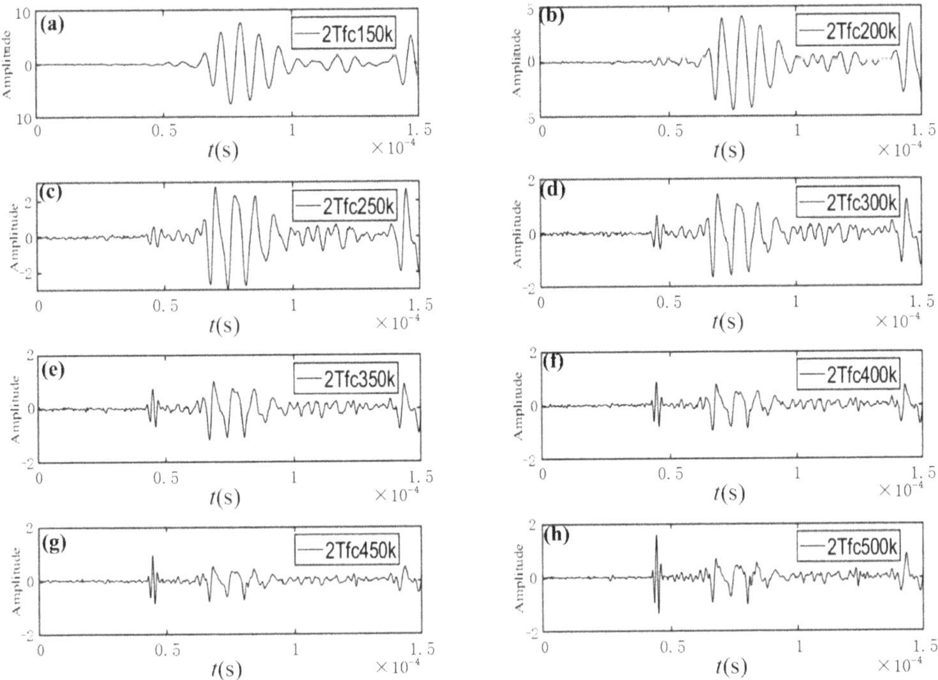

Figure 4. Measured ultrasonic signals with the same cycles (2T) and various central frequencies: (**a**) f_c = 150 kHz, (**b**) f_c = 200 kHz, (**c**) f_c = 250 kHz, (**d**) f_c = 300 kHz, (**e**) f_c = 350 kHz, (**f**) f_c = 400 kHz, (**g**) f_c = 450 kHz, and (**h**) f_c = 500 kHz.

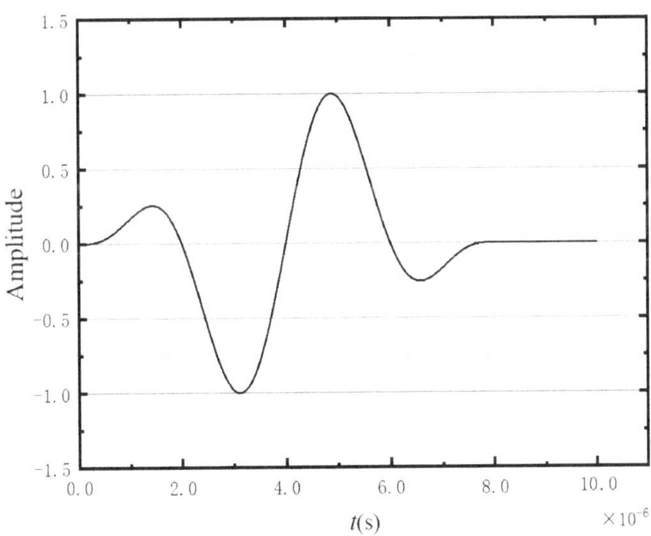

Figure 5. Waveform of the normalised 2-cycle 250 kHz tone burst signal.

2.4. Experimental Results

2.4.1. The Threshold Denoising Based on the Wavelet Transform

The mode reflection/conversion is inevitable at the external and internal boundaries of the ice specimen, contributing to a complex multi-mode wave signal. Moreover, the measured signals could be adversely affected by environmental noise. Therefore, it is necessary to rectify the measured signals to extract the inherent characteristics of the interested signal rather than those induced by mode reflection/conversion and noise. In this study, a wavelet transform-based threshold denoising process is adopted, with the results presented in Figure 6.

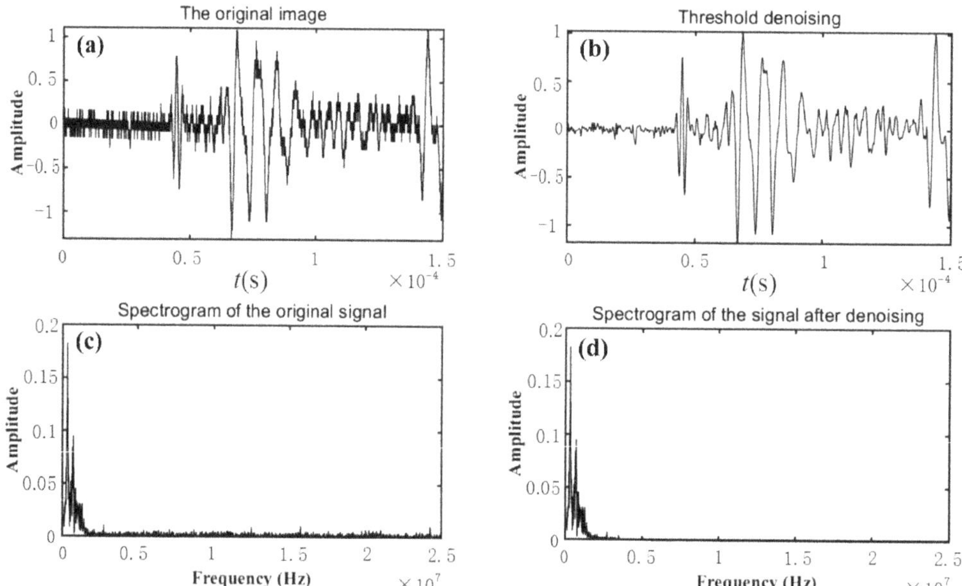

Figure 6. Time domain and spectrogram of the response signal: (**a**) original response signal, (**b**) signal after denoising, (**c**) spectrogram of the original signal, and (**d**) spectrogram of the signal after denoising.

2.4.2. Experimental Results

Considering the faster propagation of the longitudinal wave compared with the shear wave, the first and second peaks of the ultrasonic waveform can be regarded as the longitudinal wave and the shear wave, respectively. Figure 7 shows four denoised ultrasonic waveforms measured at different temperatures. According to Figure 7, the first peak of the presented waveforms cannot be clearly identified. Under these circumstances, the first wave trough of the excitation cycle is selected as a reference to calculate the propagation velocities of longitudinal and shear waves in the ice. Correspondingly, the selected troughs of the longitudinal and transverse waves are annotated by the green circleabaquss in Figure 7.

Figure 8 presents the calculated wave velocities with temperatures ranging from $-35\,^\circ\text{C}$ to $-0.5\,^\circ\text{C}$. Evidently, the velocities of longitudinal and shear waves are strongly related to temperature. To explicitly interpret the relationship between temperature and wave velocities, two quadratic functions are utilised to fit the velocity-temperature curves in Figure 8, respectively, which are:

$$V_l = -0.01139\,T^2 - 4.36647\,T + 3783.56,\ R = 0.99598, \tag{8}$$

$$V_s = -0.00401\ T^2 - 1.14069\ T + 1797.705,\ R = 0.99449, \tag{9}$$

where T represents the temperature and R is the correlation coefficient.

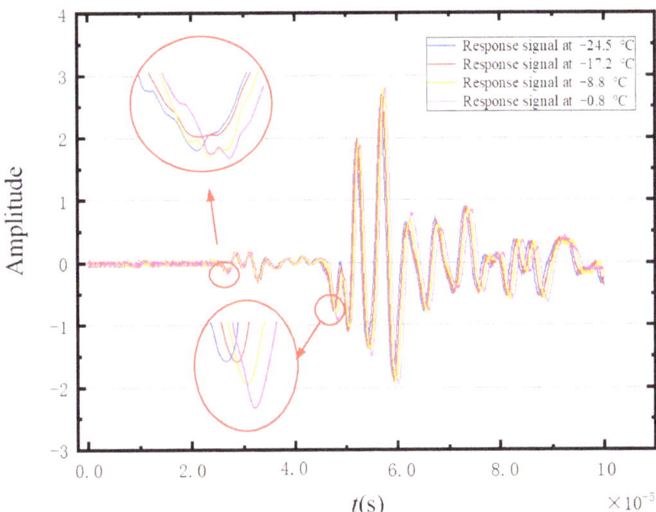

Figure 7. Four sets of measured waveforms.

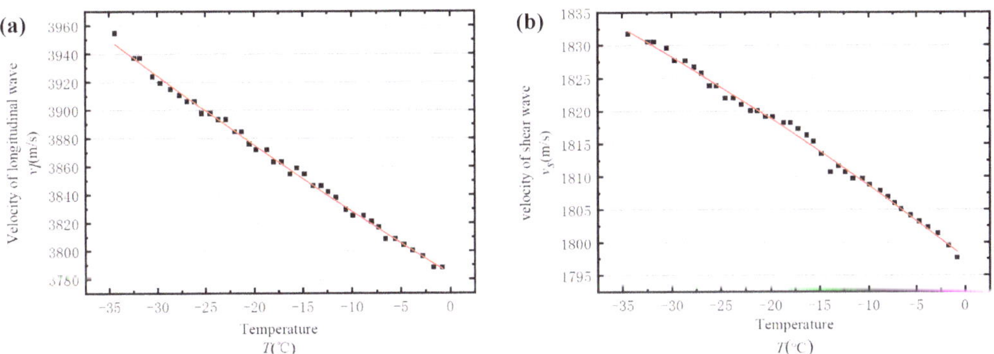

Figure 8. The relationship between wave velocity and temperature: (**a**) the relationship between longitudinal wave velocity and temperature and (**b**) the relationship between shear wave velocity and temperature.

Table 2 shows the longitudinal wave velocity in ice obtained from existing research. The velocity values are in good agreement, which validates the accuracy of experimental results in this study. The wave velocity and Young's modulus strongly depend on properties (such as salinity, impurity content, etc.) during the production of the ice, and the influences of temperature and time are not taken into account in these experiments. This may be the reason behind the deviation from the values, such as the distance of the transducers measured in the present experiment.

Table 2. Values of the longitudinal sound velocity in ice by different authors.

Studies	Kupperman et al. [42]	Gammon et al. [43]	Randhawa [44]	Present
Sound velocity (m/s)	3950	3892–4040	3980	3790–3960

In the experiments, the density of the utilised ice specimen was 890 kg/m^3, which is the average value determined by the drainage method [45–47]. Combined with Equations (3)–(6), the mechanical properties of ice at different temperatures can be achieved, including the Young's modulus, shear modulus, Poisson's ratio, and bulk modulus. The recognition results are exhibited in Figure 9, with the vertical axis representing the corresponding properties and the horizontal axis signifying the temperature variations. All the properties presented in Figure 9 decline continuously with increasing temperature.

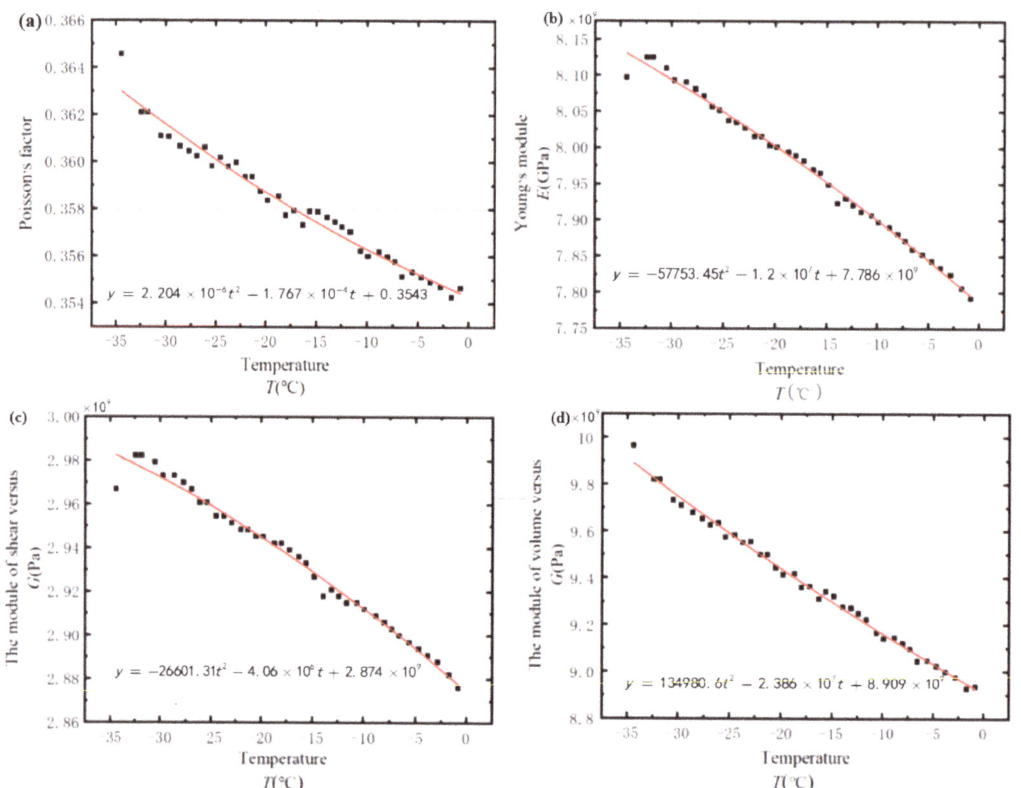

Figure 9. The changes of various properties with ice samples' temperatures: (**a**) the Poisson's factor of ice versus temperature of ice; (**b**) the Young's module versus temperature of ice; (**c**) the module of volume versus temperature of ice; and (**d**) the module of shear versus temperature of ice.

3. Physics-Directed Numerical Simulations

In nature, ice contains pores, which are related to the growth process of ice. The formation of ice from water is a process from the outside to the inside. After the surface freezes, the air in the water cannot pass through the ice, forming pores inside the ice. These pores vary in size and are randomly located, so randomly distributed circular pores are used in the manuscript to simulate the pores inside the ice. The dimension, distribution, and amount of the entrapped air bubbles highly affect the dynamic mechanical properties of ice. Hou et al. [48] indicted that those bubbles in ice could neither be manufactured artificially nor obtained in the desired amount and distribution. Therefore, the influence of the pores in the ice on the mechanical properties can only be studied by numerical methods. In this study, a stochastic algorithm is presented to produce the sparsely distributed air bubbles in the numerical model of ice.

3.1. Material Properties of the Ice Model

In order to better compare the numerical and experimental results, four sets of experimental results were selected as the model material's mechanical properties. The specific values are shown in Table 3.

Table 3. Material mechanical properties of ice.

Temperature T (°C)	Density ρ (kg·m^3)	Poisson's Factor μ	Young's Module E (GPa)	Shear Module G (GPa)	Volume Module K (GPa)
−24.5	890	0.360	8.0375	2.954	9.581
−17.2	890	0.358	7.982	2.939	9.363
−8.8	890	0.356	7.889	2.908	9.143
−0.8	890	0.355	7.792	2.876	8.935

3.2. Two-Dimensional Ice Random Pore Model

In the numerical model, ice is treated as a linear solid without considering its nonlinear behaviour or tensile or compressive damage. The dimensions of the tested specimens are 210×210 mm^2, and the pores (air bubbles) are concentrated within a central region of 180×180 mm^2. Additionally, the round pores in the 2D model are utilised. Before establishing the model, we observed and measured a large number of pores in natural and artificial ice. These pores, except for those larger than 5 mm on the surface of the ice, are very small in size inside the ice, some of which cannot be measured using conventional measurement methods. In order to restore the uniformly distributed pores accumulated in the ice, the porosity during simulation was fixed, the distribution was selected from the range of most pore sizes, and the appropriate pore distribution was calculated using MATLAB R2018a. They are classified into small pores having diameters of 0.5–1 mm and large pores having diameters of 1–1.5 mm, with the volume ratios being 6:4. This volume ratio was obtained through extensive testing, which not only meets the requirements of achieving porosity within the model but also achieves a uniform distribution of pores within the model. Moreover, the pores in the ice model are reciprocally independent, with distances between the pores greater than 1.5 mm. In the modelling process, random pores are first generated using MATLAB R2018a, and then the generated pore model is imported into the ice model built in ABAQUS 6.14. The final model is obtained by cutting out random pores from the original ice model. In the model, the excitation of the PZT actuator is simplified by applying a concentrated force at the same position as the PZT sensor [49]. The concentrated force is generated by the same 2-cycle 250 kHz tone burst signal as the experimental test (as shown in Figure 5). In addition, the distance between the actuator and receiver is 70 mm. In the model, the boundaries around the ice are the boundaries of wave propagation.

The porosity is obtained using the calculation method of Cox and Weeks [50]. The specific method can be expressed as:

$$v_a = 1 - \frac{\rho}{\rho_i(T_i)} + \rho S_i \frac{F_2(T_i)}{F_1(T_i)}, \tag{10}$$

The density of pure ice ρ_i (g/cm^3) is described as a function of temperature as [49]:

$$\rho_i(T_i) = 0.917 - 1.403 \times 10^{-4} T_i, \tag{11}$$

where v_a is the air volume ratio, ρ_i is the pure ice density, S_i is ice salinity, and T_i is the ice temperature. In this study, distilled water is used to make ice samples, so in the numerical model $S_i = 0$ is adopted. Therefore, Equation (10) becomes:

$$v_a = 1 - \frac{\rho}{\rho_i(T_i)}. \tag{12}$$

In the established numerical model, the density of the ice sample obtained from the experiment is employed to calculate the porosity at different temperatures following Equations (10)–(12).

Three 2D ice models (shown in Figures 10 and 11) with different porosities (0.5%, 1%, and 3%) were established to investigate the influence of porosities on the wave propagations in ice. The white dots in Figures 10 and 11 represent pores within the ice, while the green areas in Figure 10 represent the ice. The blue part in Figure 11 shows the grid inside ice. The mesh size is 1 mm, and the time step of numerical integration is 2×10^{-8} s. These small computational properties can ensure the accurate capture of the propagative behaviour of the ultrasonic wave.

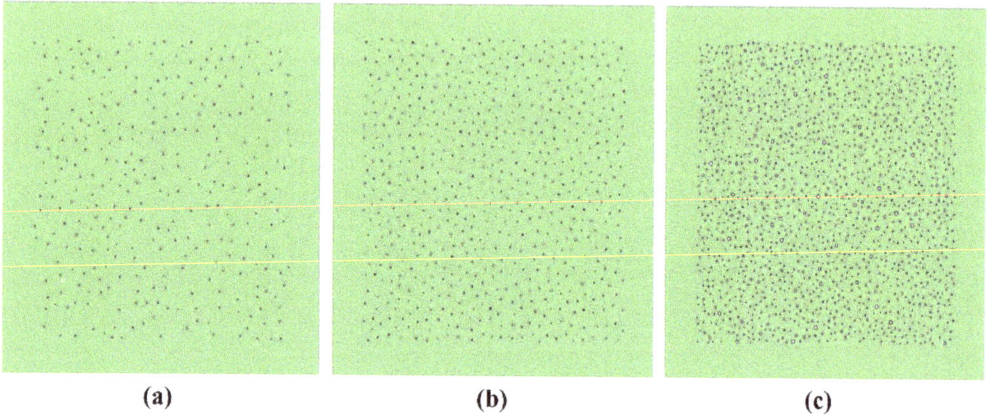

Figure 10. Numerical model of the ice specimen with different porosities: (**a**) 0.5%; (**b**) 1%; and (**c**) 3%.

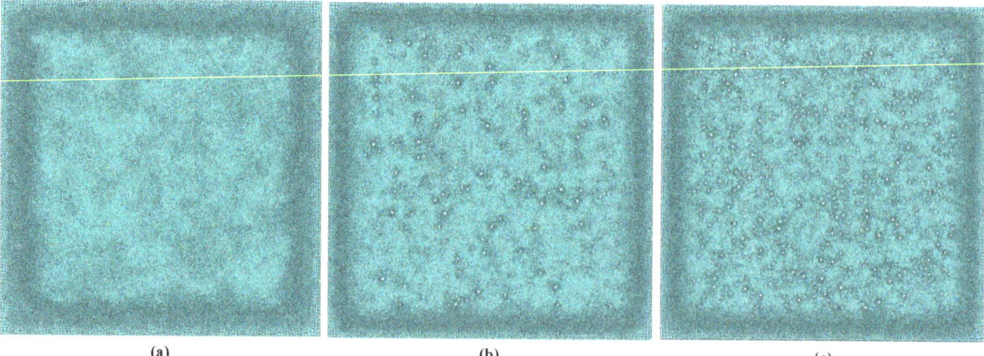

Figure 11. FEM meshes of the ice specimen with different porosities: (**a**) 0.5%; (**b**) 1%; and (**c**) 3%.

3.3. Numerical Simulations of 2D Ice Random Pore Model

3.3.1. Wave Propagation Analysis with 0.5% Porosity

Figure 12 shows the wave propagation process of the ice model with 0.5% porosity at different time values. Both the longitudinal and shear waves can be clearly distinguished within a short period of time after being excited. The red circle in Figure 12d shows that waves are superimposed on each other during propagation, and various waves cannot be separated. Figure 12e shows a horizontal rebound wave at the red circle. The whole specimen in Figure 12f is filled with mixed waveforms, and it is impossible to distinguish between shear waves and longitudinal waves.

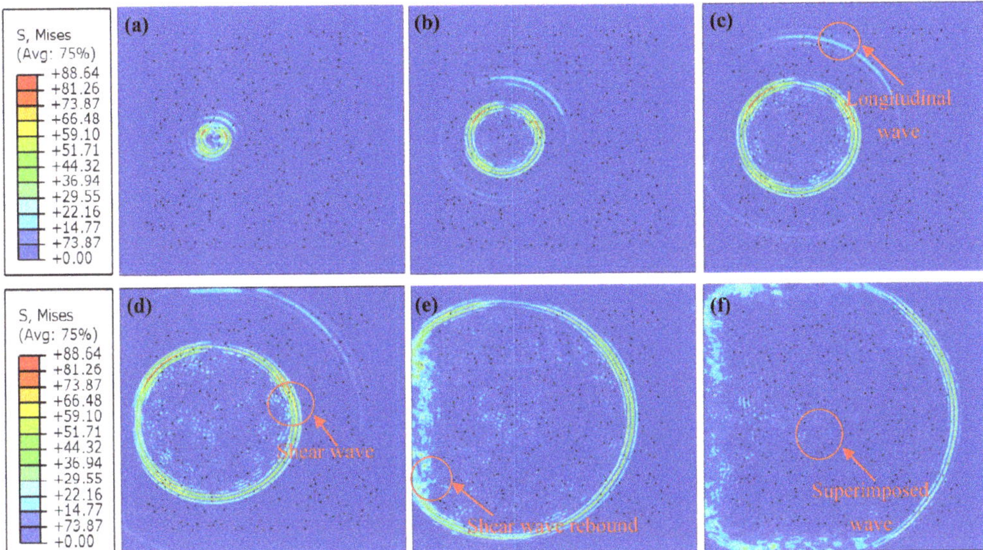

Figure 12. Wave propagation in a 2D ice model: (**a–f**) are the time slices of the wave propagation process: (**a**) time $= 1.0 \times 10^{-5}$ s, (**b**) time $= 2.0 \times 10^{-5}$ s, (**c**) time $= 3.0 \times 10^{-5}$ s, (**d**) time $= 4.0 \times 10^{-5}$ s, (**e**) time $= 6 \times 10^{-5}$ s, and (**f**) time $= 7 \times 10^{-5}$ s.

The waveforms received by the response point at four temperatures are extracted, and the drawn waveform is shown in Figure 13. As can be seen in Figure 13, there is no significant change in the morphology of the waves at the four temperatures. If the image is not enlarged, it is difficult to detect the difference between the four curves. Therefore, we believe that the change in temperature has little effect on the waveform. However, in the red dashed circle in Figure 13, the results of amplifying the peaks show that the time of the peaks appearing in the four curves is different, which means that the wave propagation speeds corresponding to the four temperatures are different (because the distance between the sensors remains constant).

With the application of ice properties from experiment tests, the numerical results of longitudinal and shear wave velocities at different temperatures can be obtained, as shown in Figure 14.

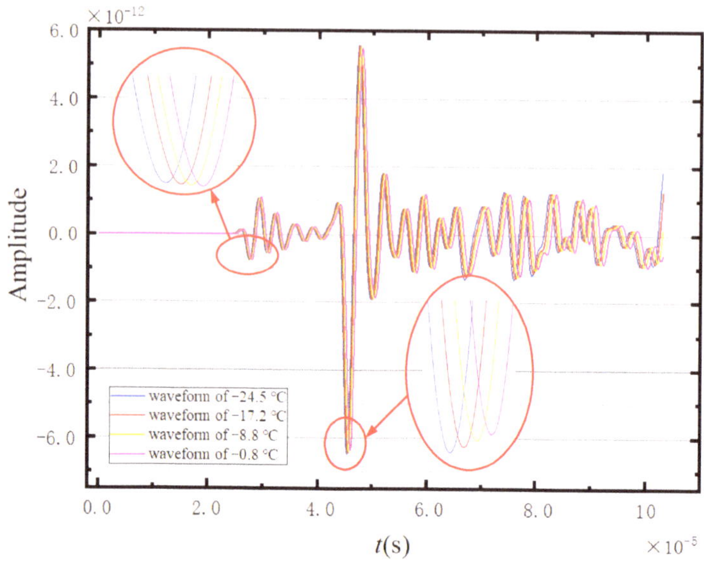

Figure 13. Four sets of numerical simulation waveforms.

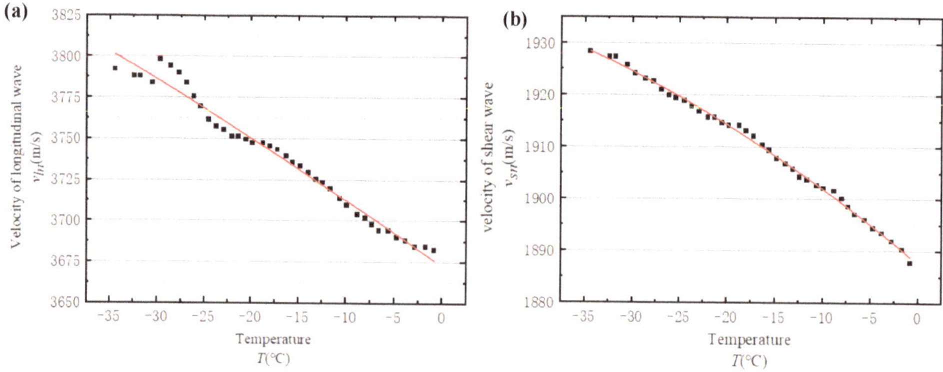

Figure 14. The relationship between wave velocities and temperatures: (**a**) the relationship between longitudinal wave velocity and temperature and (**b**) the relationship between shear wave velocity and temperature.

3.3.2. Non-Porous 2D Ice Model

To reveal the existence of pores on the wave propagation in ice, this section compares the numerical results attained by a non-porous ice model and a model with 3% porosity. The result of −17.2 °C is randomly selected to demonstrate the process of wave propagation in ice in numerical simulations. Of course, other temperature results can also be selected for display. The selection of different temperatures has no effect on the propagation process of waves in ice. The comparison results of wave propagations are depicted in Figures 15 and 16. In Figure 15b,c, the longitudinal and shear waves can be apparently inspected. According to the non-porous ice models in Figure 15b–f, no wave superpositions are observed near the excitation point, demonstrating the boundary of the critical rebound effect of the pore boundaries on the wave propagations in ice, even when the pore size is small.

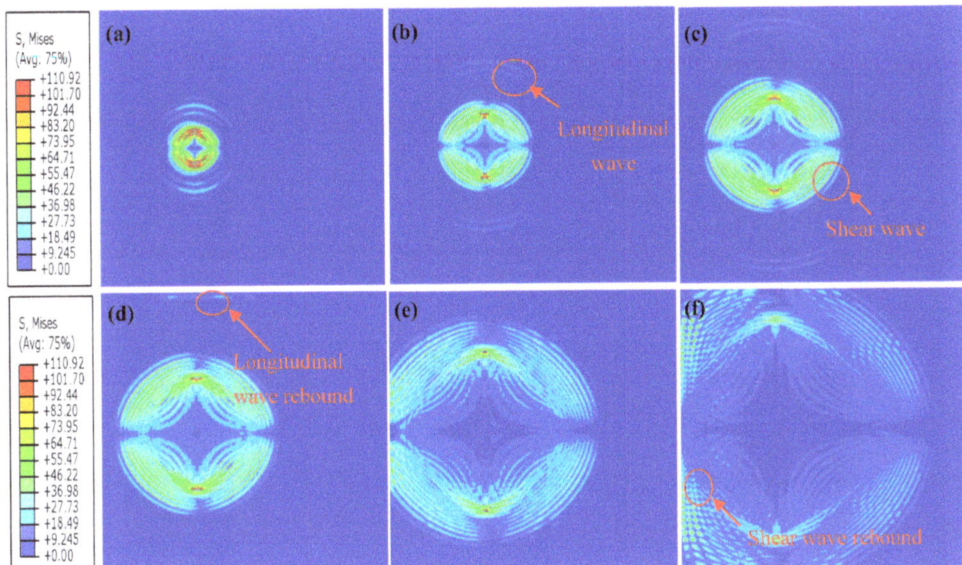

Figure 15. Wave propagation in a non-porous ice model and a 3% porosity ice model. (**a**–**f**) are the time slices of the wave propagation process: non-porous ice model (**a**) time = 2.0×10^{-5} s, (**b**) time = 3.0×10^{-5} s, (**c**) time = 4.0×10^{-5} s, (**d**) time = 4.5×10^{-5} s, (**e**) time = 6×10^{-5} s, and (**f**) time = 8×10^{-5} s.

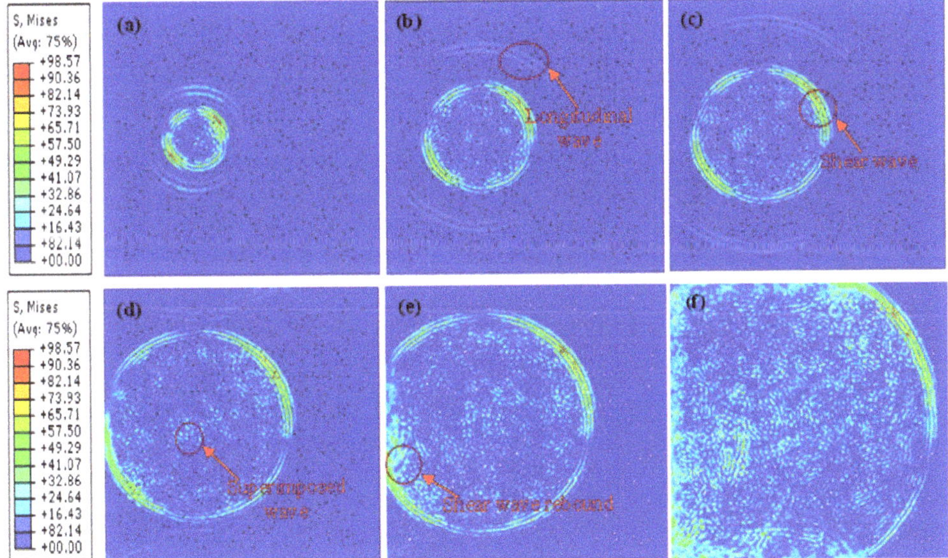

Figure 16. Wave propagation in a 3% porosity ice model, (**a**–**f**) are the time slices of the wave propagation process: (**a**) time = 2.2×10^{-5} s, (**b**) time = 3.2×10^{-5} s, (**c**) time = 4.0×10^{-5} s, (**d**) time = 5.0×10^{-5} s, (**e**) time = 5.8×10^{-5} s, and (**f**) time = 8×10^{-5} s.

Figure 17 presents the normalized waveforms of the response point. The circles represent the positions of the peaks used in velocity calculation, with the red circles indicating the peaks used for calculating the longitudinal wave velocity and the green circles used for calculating the shear wave velocity. According to Figure 17, the presence or absence of pores obviously affects the wave propagation speed. In Figure 17, the amplitude in the first wave packet of the non-porous ice model is greater than that of the porous ice model. However, in the range of 0.4–0.5 s, the amplitude of the porous ice model is greater than that of the non-porous ice model. This is because the pore size in the porous model is 0.5–1 mm, which is much smaller than the wavelength (about 1500 m). Therefore, waves will form multiple reflections, refractions, and diffraction superpositions under the rebound effect of the pore wall.

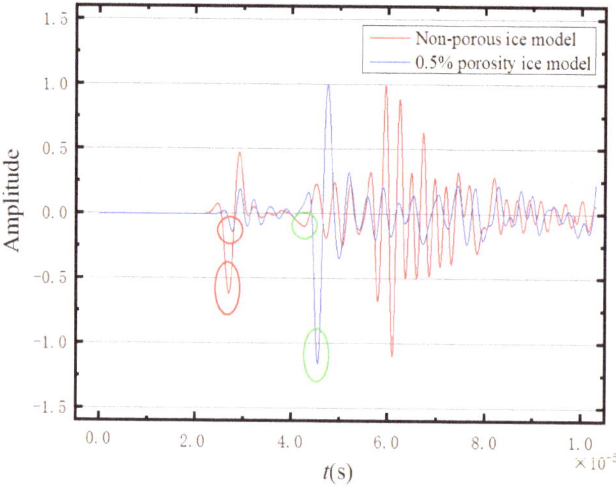

Figure 17. Comparison of the waveforms of the two models.

Figure 18 shows the normalised waveforms of the three porosities at −17.2 °C. The circles in Figure 18 indicate the values used for calculating the wave velocities, with the red circle is used for calculating the shear wave velocity and the green circles are used for calculating the longitudinal wave velocity. It can be seen from Figure 18 that the higher the porosity, the slower the wave propagation velocity.

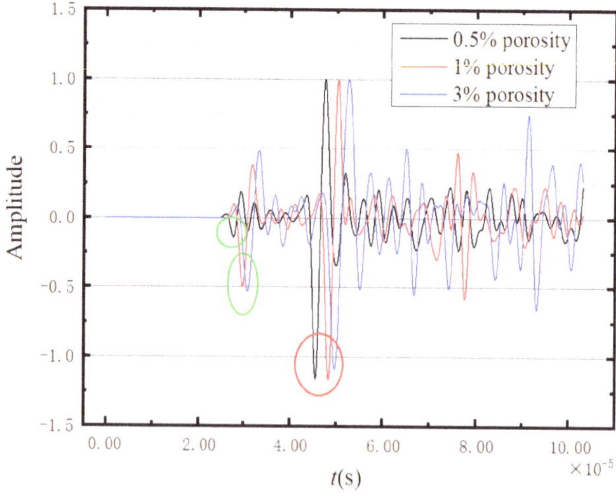

Figure 18. Comparison of three porosity waveforms.

3.4. Numerical Simulation Results of the 3D Ice Random Pore Model

In this section, the 3D random pore ice model is built by expanding the 2D model. The dimensions of the investigated 3D model are $150 \times 150 \times 50$ mm^3, and the pores are concentrated within a central cuboid range of $140 \times 140 \times 45$ mm^3. Similar to the 2D model, the random pores in the 3D model are spheres, which are classified into the small size with diameters of 0.5–1 mm and the large size with diameters of 1–1.5 mm. The ratio of small and large pores is 6:4. Each pair of pores has a distance larger than 3 mm. Additionally, the porosity of the 3D model can be designated in accordance with the porosity of the 2D model (shown in Equations (10)–(12)). Three 3D ice models with porosities of 0.5%, 1%, and 3% are counted below.

The coordinates of the centre and the diameter of the pores are randomly generated. When the diameter rotates 360 degrees around the centre, a spherical pore is developed. All pores of the entire model are shown in Figure 19a. By subtracting the pores from the whole ice, the final 3D ice model with random pores, as shown in Figure 19b, can be obtained. In addition, the three columns in Figure 19 represent three model diagrams with different porosities.

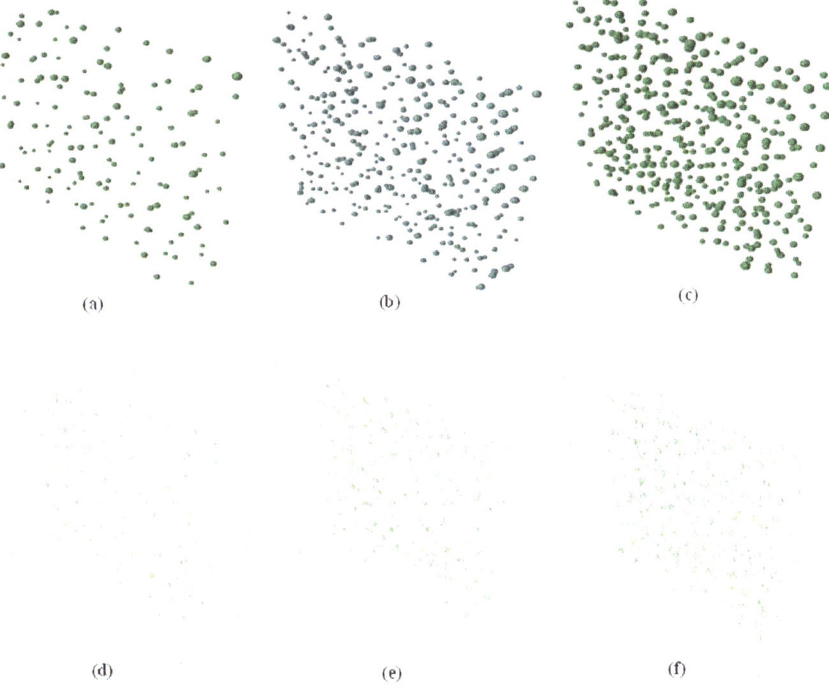

Figure 19. Three-dimensional numerical model: (**a–c**) the spherical pore of the three different porosities: (**a**) 0.5% porosity, (**b**) 1% porosity, and (**c**) 3% porosity; (**d–f**) three-dimensional ice random pore model of the three different porosities: (**d**) 0.5% porosity, (**e**) 1% porosity, and (**f**) 3% porosity.

The excitation signal in the 3D model adopts the waveform signal of Figure 5 and acts with concentrated force on the excitation point. The distance between the excitation and the response points is 70 mm. The centre frequency of the sine wave is $f_c = 250$ kHz. As the mesh size of the model is 1 mm, the time step is 2×10^{-8} s. The grid map is shown in Figure 20, where Figure 20b is a planned view of the model and Figure 20c is an enlarged picture of the red circle of Figure 20b, which is used to highlight the grid at the gap. The red circles in Figure 20c highlight the pores in the model.

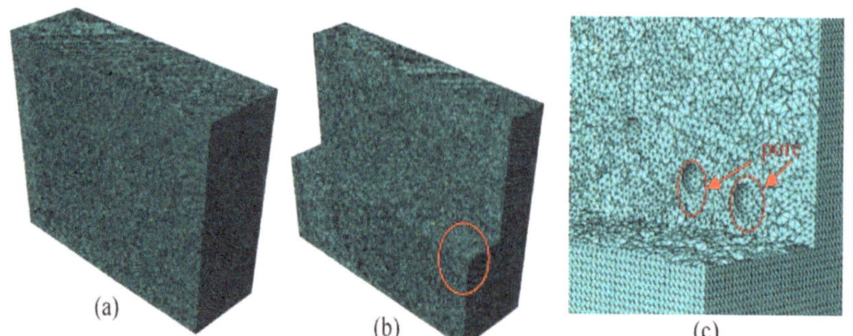

Figure 20. FEM meshes of the 3D ice model: (**a**) global view; (**b**) internal view; and (**c**) zoomed view of the red circle in (**b**).

3.5. Numerical Results from the 3D Ice Model

3.5.1. Wave Propagation in Ice with 1% Porosity

Figure 21 presents the propagation process of waves in the 3D model. According to Figure 21a–c, within a short period of time after excitation, the longitudinal and shear waves can be clearly distinguished. Nevertheless, as time goes on, the longitudinal and shear waves become bounced and superimposed on each other. Compared with the 2D model, these bounces and superpositions of ultrasonic waves in the 3D model are more serious and complex.

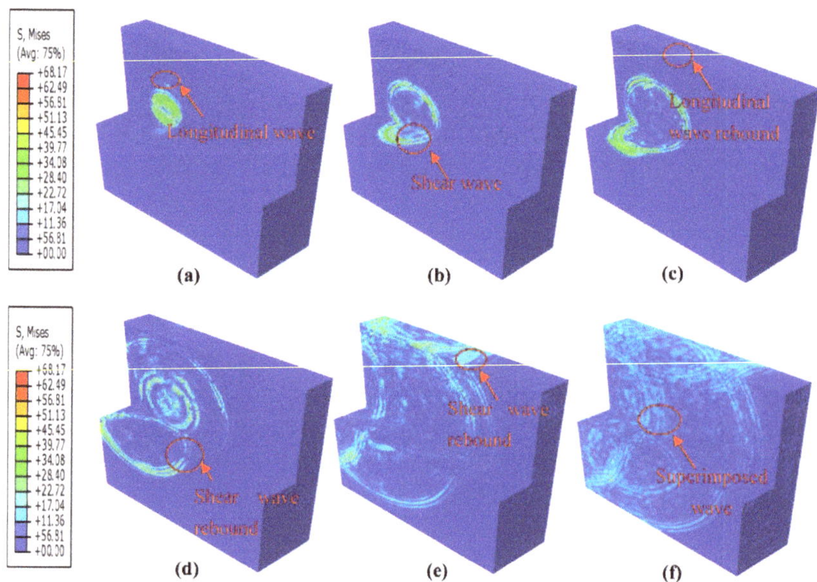

Figure 21. Wave propagation in the 3D ice model: (**a**–**f**) are the time slices of the wave propagation process: (**a**) time = 1.0×10^{-5} s, (**b**) time = 1.7×10^{-5} s, (**c**) time = 2.0×10^{-5} s, (**d**) time = 3.1×10^{-5} s, (**e**) time = 5.2×10^{-5} s, and (**f**) time = 6.3×10^{-5} s.

The mechanical properties of ice at various temperatures obtained from the experiment are input into the 3D numerical model as the material properties, and the shear wave together with the longitudinal wave velocities at different temperatures can be obtained. Finally, all wave velocities are plotted as a scatter diagram, as shown in Figure 22.

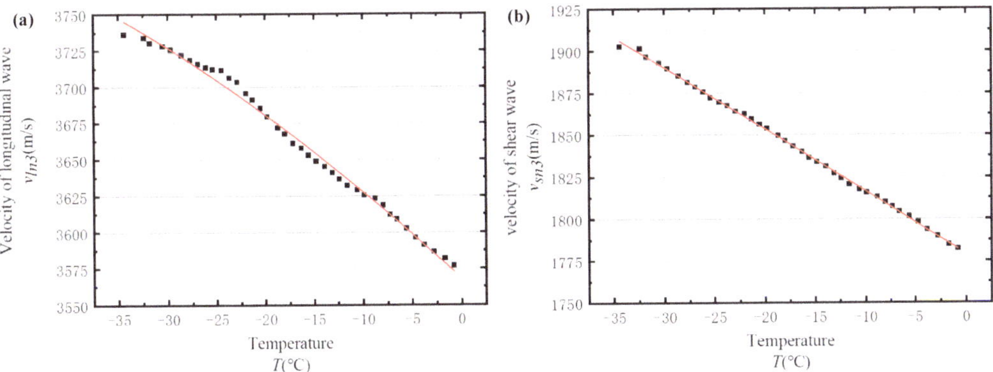

Figure 22. Three-dimensional numerical simulation of the relationship between wave velocities and temperatures: (**a**) the relationship between longitudinal wave velocity and temperature and (**b**) the relationship between shear wave velocity and temperature.

Four groups of data are selected from all these numerical outcomes. The waveforms of wave propagation in ice at these four temperatures are shown in Figure 23. It can be seen that the shape of the waveforms at each temperature is similar, and there are obvious differences in the speed of wave propagation at different temperatures. The positions of the red circles in Figure 23 correspond to the first troughs of the longitudinal wave and the shear wave, respectively.

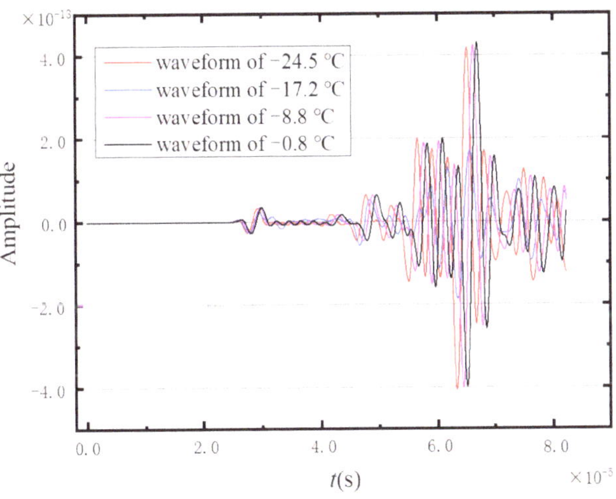

Figure 23. Combination chart of four sets of 3D numerical simulation.

3.5.2. Comparisons

This part investigates the influence of porosity on wave propagation in ice using a 3D model. The normalised waveforms at different porosities are shown in Figure 24. Similar to the results in 2D, the porosity only affects the speed of wave propagation, and the influence on the waveform can be ignored. The circles represent the peak positions used for velocity calculation. Among them, the red circle and the green circle represent the peak values used for calculating longitudinal wave velocity and shear wave velocity, respectively. The cluttered waveforms at the tail (the wave propagation time $t \geq 5.5 \times 10^{-5}$) in Figure 24 indicate the complex superposition and rebound of waves in the 3D model.

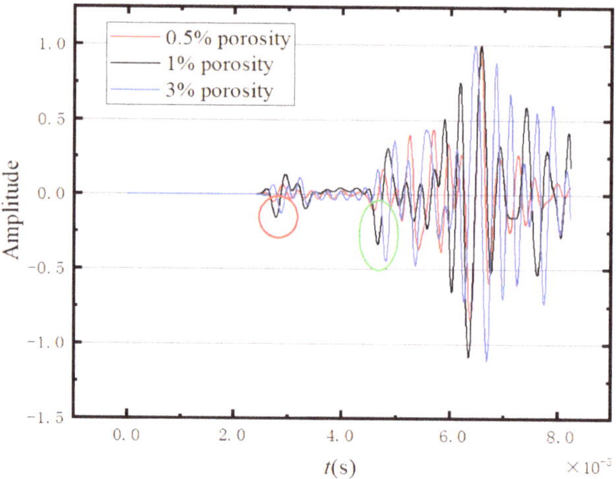

Figure 24. Comparison of three porosity waveforms of 3D models.

4. Discussion of Experimental and Numerical Results

This section compares the numerical and experimental results to explicitly verify the reliability of the established 2D and 3D numerical ice models. Taking the case of temperature $-17.2\ ^\circ\mathrm{C}$ as an example, the corresponding numerical and experimental results are presented in Figure 25. The figure shows that the waveform is normalised. The circles represent the peak positions used for velocity calculation. Among them, the red circle and the green circle represent the peak values used for calculating longitudinal wave velocity and shear wave velocity, respectively. According to Figure 25, both experimental and numerical results can distinguish the transverse wave from the longitudinal wave. Moreover, the waveforms are similar except for a certain difference in magnitude. Table 4 presents the simulation errors of the established 2D and 3D ice models at various temperatures. The relative errors are within the range of 0.15–5.7% according to the comparison of numerical and experimental results. The errors in the 3D model are less than those in the 2D model, which indicates that the 3D numerical ice model is more appropriate for wave propagation analysis of ice.

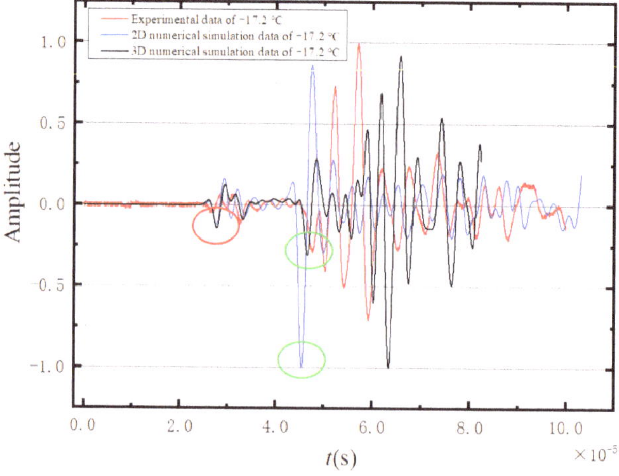

Figure 25. Comparison of experimental data and simulation.

Table 4. Wave velocity of tests and simulations and the errors between them.

Temperature T (°C)	Longitudinal Wave Velocity of the Test V_l (m/s)	Two-Dimensional Simulated Longitudinal Wave Velocity V_{ln2} (m/s)	Errors of 2D Simulation Results (%)	Three-Dimensional Simulated Longitudinal Wave Velocity V_{ln3} (m/s)	Deviation of 3D Simulation Results (%)	Two-Dimensional Shear Wave Velocity of the Test Vs (m/s)	Two-Dimensional Simulated Shear Wave Velocity V_{sn3} (m/s)	Deviation of 2D Simulation Results (%)	Three-Dimensional Simulated Shear Wave Velocity V_{sn3} (m/s)	Errors of 2D Simulation Results (%)
−34.4	3954.80	3791.98	4.12	3736.21	5.53	1825.77	1928.37	5.62	1902.83	4.22
−32.4	3937.01	3787.88	3.79	3733.69	5.16	1830.54	1927.31	5.29	1901.53	3.88
−31.8	3937.01	3787.88	3.79	3730.15	5.25	1830.54	1927.31	5.29	1896.71	3.61
−30.5	3923.77	3783.78	3.57	3728.05	4.98	1829.59	1925.72	5.25	1893.21	3.48
−29.7	3919.37	3798.16	3.09	3725.75	4.94	1827.68	1924.13	5.28	1889.56	3.39
−28.6	3914.99	3794.04	3.09	3721.98	4.93	1827.68	1923.08	5.22	1885.24	3.15
−27.7	3910.61	3789.93	3.09	3718.66	4.91	1826.72	1922.55	5.25	1881.36	2.99
−26.9	3906.25	3783.78	3.14	3715.92	4.87	1825.77	1920.97	5.21	1878.64	2.90
−26.1	3906.25	3775.62	3.34	3713.45	4.94	1823.87	1919.91	5.27	1875.64	2.84
−25.4	3897.55	3769.52	3.28	3712.08	4.76	1823.87	1919.39	5.24	1872.22	2.65
−24.5	3897.55	3761.40	3.50	3711.56	4.77	1821.97	1918.90	5.30	1869.66	2.62
−23.7	3893.21	3757.38	3.49	3706.31	4.80	1821.97	1917.81	5.26	1867.54	2.50
−22.9	3893.21	3755.36	3.54	3703.33	4.88	1821.02	1916.76	5.26	1864.26	2.37
−22	3884.57	3751.34	3.43	3695.85	4.86	1820.07	1915.71	5.25	1862.74	2.34
−21.3	3884.57	3751.34	3.43	3691.33	4.97	1820.07	1915.71	5.25	1859.69	2.18
−20.5	3875.97	3749.33	3.27	3685.46	4.92	1819.13	1914.66	5.25	1856.24	2.04
−19.8	3871.68	3747.32	3.21	3679.51	4.96	1819.13	1914.14	5.22	1853.96	1.91
−18.7	3871.68	3747.32	3.21	3671.69	5.17	1818.18	1914.14	5.28	1849.54	1.72
−18	3863.13	3745.32	3.05	3667.52	5.06	1818.18	1913.09	5.22	1846.33	1.55
−17.2	3863.13	3743.32	4.00	3661.09	5.23	1817.24	1912.05	4.94	1843.08	1.42
−16.3	3854.63	3739.32	2.99	3657.74	5.11	1816.29	1910.48	5.19	1839.85	1.30
−15.6	3858.88	3735.33	3.20	3652.86	5.34	1815.35	1909.44	5.18	1836.34	1.16
−14.8	3854.63	3733.33	3.15	3648.69	5.34	1813.47	1907.88	5.21	1833.85	1.12
−13.9	3846.15	3729.36	3.04	3645.21	5.22	1810.66	1906.84	5.31	1831.02	1.12
−13.1	3846.15	3725.39	3.14	3640.98	5.33	1811.59	1905.80	5.20	1826.98	0.85
−12.4	3841.93	3723.40	3.09	3636.54	5.35	1810.66	1904.24	5.17	1823.87	0.73
−11.6	3837.72	3719.45	3.08	3632.23	5.35	1809.72	1903.73	5.19	1820.33	0.59
−10.6	3829.32	3713.53	3.02	3628.96	5.23	1809.72	1902.69	5.14	1817.52	0.43
−9.9	3825.14	3709.59	3.02	3625.43	5.22	1808.79	1902.17	5.16	1815.65	0.38
−8.8	3825.14	3703.70	3.17	3623.19	5.28	1807.85	1901.66	5.10	1813	0.28
−8	3820.96	3701.75	3.12	3618.65	5.29	1806.92	1900.11	5.16	1809.89	0.16
−7.3	3816.79	3697.83	3.12	3612.12	5.36	1805.99	1898.56	5.13	1807.35	0.08
−6.6	3808.49	3693.93	3.01	3608.94	5.24	1805.05	1897.02	5.10	1804.22	0.05
−5.6	3808.49	3693.93	3.01	3602.54	5.41	1804.12	1895.99	5.09	1801.36	0.15
−4.7	3804.35	3690.04	3.00	3596.43	5.47	1803.19	1894.45	5.06	1797.95	0.29
−3.8	3800.22	3688.09	2.95	3591.23	5.50	1802.27	1893.43	5.06	1793.43	0.49
−2.8	3796.10	3684.21	2.95	3586.54	5.52	1801.34	1891.89	5.03	1789.53	0.66
−1.7	3787.88	3684.21	2.74	3581.93	5.44	1799.49	1890.36	5.05	1784.63	0.83
−0.8	3787.88	3682.27	2.70	3576.9	5.57	1797.64	1887.81	5.02	1782.08	0.87

Table 4 shows that the small deviation range demonstrates the reliability of both the experimental tests and the numerical simulations. Moreover, the mechanical properties of ice obtained from the experimental results are reasonable and applicable. In addition, the 2D and 3D numerical ice models can prove that this method of simulating ice with a random pore model is feasible. The ice model provides a reliable and feasible way for further investigations of ice-related engineering problems.

5. Conclusions

In this study, a sophisticated embedded ultrasonic system is proposed to inspect the mechanical properties of ice in real time and online. This embedded ultrasonic system provides a platform to continuously obtain the response of the ice. With this system, the ice properties under specific temperature conditions are identified based on the intrinsic relationships between the wave propagation velocities and the mechanical properties of ice. Furthermore, the feasibility and effectiveness of the proposed embedded ultrasonic system-based method are validated via ultrasonic experiments and numerical simulations. Both numerical and experimental results demonstrate the effectiveness of the proposed embedded ultrasonic system-based method to inspect the mechanical properties of ice. The conclusions can be summarised as follows:

(1) With the artificial ice specimen made of distilled water, the experiment results indicate the velocities of wave propagation in the ice decrease gradually with the elevated temperatures. Moreover, the velocity of the longitudinal wave is within the range of 3954.8–3787.88 m/s, while the velocity of the shear wave is within the range of 1831.54–1797.64 m/s;

(2) The experimentally obtained wave velocities have been brought into known formulas to calculate the material properties of ice at different temperatures, including Young's module, Poisson's ratio, shear modulus, and bulk modulus. The results have shown that the values of the four properties have a downward trend with the increase in temperature;

(3) With the help of the numerical models, the wave propagation in ice has been investigated, and the influence of varying temperatures on the wave propagation has been discussed in detail. The relative errors are within the range of 0.15–5.7% according to the comparison of numerical and experimental results, which proves the validity of the experimental results and reasonability of the calculated ice mechanics properties;

(4) By comparing the waveform diagrams of the non-porous model and different porosity models, it can be found that the porosity affects the wave propagation velocity. By which, the greater the porosity, the slower the wave propagation velocity;

(5) This research proves that the random pore model can simulate ice well. Additionally, by modifying the properties of the input model, more different ice models at different temperatures can be obtained, which provides a reliable modelling method for later research on ice.

Author Contributions: Methodology, software, formal analysis, writing—original draft, H.H.; data curation, validation, writing—original draft, L.W.; data curation, validation, writing—original draft, N.F.A.; conceptualisation, supervision, writing—review and editing, funding acquisition, M.C. All authors have read and agreed to the published version of the manuscript.

Funding: The authors are grateful for the Fundamental Research Funds for the Central Universities: No.B220204002; the 2022 National Young Foreign Talents Program of China: No. QN2022143002L; Jiangsu International Joint Research and Development Program: No. BZ2022010; and the Nanjing International Joint Research and Development Program: No. 202112003.

Institutional Review Board Statement: Not applicable.

Informed Consent Statement: Not applicable.

Data Availability Statement: The datasets used and/or analysed during the current study are available from the corresponding author on reasonable request.

Conflicts of Interest: All authors certify that they have no affiliations with or involvement in any organisation or entity with any financial or non-financial interest in the subject matter or materials discussed in this manuscript.

References

1. Screen, J.A.; Simmonds, I. The central role of diminishing sea ice in recent Arctic temperature amplification. *Nature* **2010**, *464*, 1334–1337. [CrossRef]
2. Kumar, A.; Perlwitz, J.; Eischeid, J.; Quan, X.; Xu, T.; Zhang, T.; Hoerling, M.; Jha, B.; Wang, W. Contribution of sea ice loss to Arctic amplification. *Geophys. Res. Lett.* **2010**, *37*, 1–17. [CrossRef]
3. Gebre, S.; Boissy, T.; Alfredsen, K. Sensitivity to climate change of the thermal structure and ice cover regime of three hydropower reservoirs. *J. Hydrol.* **2014**, *510*, 208–227. [CrossRef]
4. Zhou, L.; Riska, K.; Ji, C. Simulating transverse icebreaking process considering both crushing and bending failures. *Mar. Struct.* **2017**, *54*, 167–187. [CrossRef]
5. Balkhanov, V.K.; Bashkuev, Y.B.; Khaptanov, V.B. Deformation of freshwater ice cover measured by a horizontal electric antenna. *Tech. Phys.* **2007**, *52*, 120–122. [CrossRef]
6. Liu, Z.; Amdahl, J.; Løset, S. Plasticity based material modelling of ice and its application to ship–iceberg impacts. *Cold Reg. Sci. Technol.* **2011**, *65*, 326–334. [CrossRef]
7. Iliescu, D.; Baker, I. The structure and mechanical properties of river and lake ice. *Cold Reg. Sci. Technol.* **2007**, *48*, 202–217. [CrossRef]
8. Wang, Y.; Xu, Y.; Huang, Q. Progress on ultrasonic guided waves de-icing techniques in improving aviation energy efficiency. *Renew. Sustain. Energy Rev.* **2017**, *79*, 638–645. [CrossRef]
9. Sun, X.; Li, S.; Dun, X.; Li, D.; Li, T.; Guo, R.; Yang, M. A novel characterization method of piezoelectric composite material based on particle swarm optimization algorithm. *Appl. Math. Model.* **2019**, *66*, 322–331. [CrossRef]
10. Wang, Q.; Li, Z.; Lei, R.; Lu, P.; Han, H. Estimation of the uniaxial compressive strength of Arctic sea ice during melt season. *Cold Reg. Sci. Technol.* **2018**, *151*, 9–18. [CrossRef]
11. Han, H.; Li, Z.; Huang, W.; Lu, P.; Lei, R. The uniaxial compressive strength of the Arctic summer sea ice. *Acta Oceanol. Sin.* **2015**, *34*, 129–136. [CrossRef]
12. Karulina, M.; Marchenko, A.; Karulin, E.; Sodhi, D.; Sakharov, A.; Chistyakov, P. Full-scale flexural strength of sea ice and freshwater ice in Spitsbergen Fjords and North-West Barents Sea. *Appl. Ocean Res.* **2019**, *90*, 101853. [CrossRef]
13. Aksenov, V.I.; Gevorkyan, S.G.; Iospa, A.V. Temperature Dependence of Stress-Strain Properties of Freshwater Ice. *Soil Mech. Found. Eng.* **2019**, *56*, 366–370. [CrossRef]
14. Moslet, P. Field testing of uniaxial compression strength of columnar sea ice. *Cold Reg. Sci. Technol.* **2007**, *48*, 1–14. [CrossRef]
15. Qiu, W.-L.; Peng, R.-X. Research on the numerical simulation for plastic model of ice as building materials under triaxial compression. *Constr. Build. Mater.* **2020**, *268*, 121183. [CrossRef]
16. Tinard, V.; François, P.; Fond, C. The Potential Scope of the Ultrasonic Surface Reflection Method Towards Mechanical Characterisation of Isotropic Materials. Part 2. Experimental Results. *Exp. Mech.* **2021**, *61*, 1161–1170. [CrossRef]
17. Tinard, V.; François, P.; Fond, C. The Potential Scope of the Ultrasonic Surface Reflection Method Towards Mechanical Characterisation of Isotropic Materials. Part 1. A Theoretical Analysis. *Exp. Mech.* **2021**, *61*, 1153–1160. [CrossRef]
18. Zhu, Z.H.; Post, M.A.; Meguid, S.A. The Potential of Ultrasonic Non-Destructive Measurement of Residual Stresses by Modal Frequency Spacing using Leaky Lamb Waves. *Exp. Mech.* **2012**, *52*, 1329–1339. [CrossRef]
19. Mustansar, Z.; Shaukat, A.; Margetts, L. Effect of using different types of methods for the derivation of elastic modulus of bone—A critical survey. *MATEC Web Conf.* **2017**, *108*, 13002. [CrossRef]
20. Lin, S.; Shams, S.; Choi, H.; Azari, H. Ultrasonic imaging of multi-layer concrete structures. *NDT E Int.* **2018**, *98*, 101–109. [CrossRef]
21. Yang, X.; Verboven, E.; Ju, B.-F.; Kersemans, M. Comparative study of ultrasonic techniques for reconstructing the multilayer structure of composites. *NDT E Int.* **2021**, *121*, 102460. [CrossRef]
22. Cao-Rial, M.; Moreno, C.; Quintela, P. Determination of Young modulus by using Rayleigh waves. *Appl. Math. Model.* **2019**, *77*, 439–455. [CrossRef]
23. Maguire, R.R.; Schmerr, N.C.; Lekić, V.; Hurford, T.A.; Dai, L.; Rhoden, A.R. Constraining Europa's ice shell thickness with fundamental mode surface wave dispersion. *Icarus* **2021**, *369*, 114617. [CrossRef]
24. Bayón, A.; Gascón, F.; Nieves, F.J. Estimation of dynamic elastic constants from the amplitude and velocity of Rayleigh waves. *J. Acoust. Soc. Am.* **2005**, *117*, 3469–3477. [CrossRef]
25. Medina, R.; Bayón, A. Elastic constants of a plate from impact-echo resonance and Rayleigh wave velocity. *J. Sound Vib.* **2010**, *329*, 2114–2126. [CrossRef]
26. Polach, R.U.F.V.B.U.; Ettema, R.; Gralher, S.; Kellner, L.; Stender, M. The non-linear behavior of aqueous model ice in downward flexure. *Cold Reg. Sci. Technol.* **2019**, *165*, 102775. [CrossRef]
27. Gagnon, R. Results of numerical simulations of growler impact tests. *Cold Reg. Sci. Technol.* **2007**, *49*, 206–214. [CrossRef]
28. Wang, Y.; Ni, Y. Bayesian dynamic forecasting of structural strain response using structural health monitoring data. *Struct. Control Health Monit.* **2020**, *27*, e2575. [CrossRef]

29. Wang, Q.-A.; Zhang, C.; Ma, Z.-G.; Ni, Y.-Q. Modelling and forecasting of SHM strain measurement for a large-scale suspension bridge during typhoon events using variational heteroscedastic Gaussian process. *Eng. Struct.* **2022**, *251*, 113554. [CrossRef]
30. Invernizzi, S.; Lacidogna, G.; Carpinteri, A. Particle-based numerical modeling of AE statistics in disordered materials. *Meccanica* **2012**, *48*, 211–220. [CrossRef]
31. Markovic, N.; Nestorovic, T.; Stojic, D. Numerical modeling of damage detection in concrete beams using piezoelectric patches. *Mech. Res. Commun.* **2015**, *64*, 15–22. [CrossRef]
32. Cao, M.; Ren, Q.; Wang, H. A method of detecting seismic singularities using combined wavelet with fractal. *Chin. J. Geophys.* **2005**, *48*, 672–679. [CrossRef]
33. Alkayem, N.F.; Cao, M. Damage identification in three-dimensional structures using single-objective evolutionary algorithms and finite element model updating: Evaluation and comparison. *Eng. Optim.* **2018**, *50*, 1695–1714. [CrossRef]
34. Bonath, V.; Edeskär, T.; Lintzén, N.; Fransson, L.; Cwirzen, A. Properties of ice from first-year ridges in the Barents Sea and Fram Strait. *Cold Reg. Sci. Technol.* **2019**, *168*, 102890. [CrossRef]
35. Gharamti, I.; Dempsey, J.; Polojärvi, A.; Tuhkuri, J. Fracture of warm S2 columnar freshwater ice: Size and rate effects. *Acta Mater.* **2020**, *202*, 22–34. [CrossRef]
36. Ji, S.; Yue, Q.; Bi, X. Probability distribution of sea ice fatigue parameters in JZ20-2 sea area of the Liaodong Bay. *Ocean Eng.* **2002**, *20*, 6.
37. Cai, W.; Zhu, L.; Yu, T.; Li, Y. Numerical simulations for plates under ice impact based on a concrete constitutive ice model. *Int. J. Impact Eng.* **2020**, *143*, 103594. [CrossRef]
38. Zhang, J.-P.; Zhou, D. Numerical modeling for strain rate effect and size effect of ice under uniaxial tension and compression. *Commun. Nonlinear Sci. Numer. Simul.* **2021**, *96*, 105614. [CrossRef]
39. Wang, Z.; Wei, L.; Cao, M. Damage Quantification with Embedded Piezoelectric Aggregates Based on Wavelet Packet Energy Analysis. *Sensors* **2019**, *19*, 425. [CrossRef]
40. Li, W.; Wang, Z.; Cao, M.; Fu, R. Concrete strength monitoring based on the variation of ultrasonic waveform acquired by piezoelectric aggregates. *Struct. Eng. Mech.* **2020**, *76*, 591–598.
41. Zhang, W.; Hao, H.; Wu, J.; Li, J.; Ma, H.; Li, C. Detection of minor damage in structures with guided wave signals and nonlinear oscillator. *Measurement* **2018**, *122*, 532–544. [CrossRef]
42. Kupperman, D.S.; Reimann, K.J.; Abrego-Lopez, J. Ultrasonic NDE of cast stainless steel. *NDT Int.* **1989**, *22*, 309. [CrossRef]
43. Gammon, P.H.; Klefte, H.; Clouter, M.J. Elastic Constants of Ice Samples by Brillouin Elastic Constants of Ice Samples by Brillouin Spectroscopy. *J. Phys. Chem.* **1983**, *87*, 4025–4029. [CrossRef]
44. Randhawa, K.S. The Measurement of the Young's Modulus of Ice with Ultrasonic Waves. Ph.D. Thesis, Technische Universität Hamburg, Hamburg, Germany, 2018.
45. Pustogvar, A.; Kulyakhtin, A. Sea ice density measurements. Methods and uncertainties. *Cold Reg. Sci. Technol.* **2016**, *131*, 46–52. [CrossRef]
46. Timco, R.F.G.W. A review of sea ice density. *Cold Reg. Sci. Technol.* **1996**, *24*, 1–6. [CrossRef]
47. Timco, G.W.; Frederking, R.M.W. Compressive strength of sea ice sheets. *Cold Reg. Sci. Technol.* **1990**, *17*, 227–240. [CrossRef]
48. Song, Z.; Hou, L.; Whisler, D.; Gao, G. Mesoscopic numerical investigation of dynamic mechanical properties of ice with entrapped air bubbles based on a stochastic sparse distribution mechanism. *Compos. Struct.* **2020**, *236*, 111834. [CrossRef]
49. Chen, H.; Xu, B.; Zhou, T.; Mo, Y.-L. Debonding detection for rectangular CFST using surface wave measurement: Test and multi-physical fields numerical simulation. *Mech. Syst. Signal Process.* **2018**, *117*, 238–254. [CrossRef]
50. Cox, G.F.N.; Weeks, W.F. Equations for Determining the Gas and Brine Volumes in Sea-Ice Samples. *J. Glaciol.* **1983**, *29*, 306–316. [CrossRef]

Article

Estimation Method of an Electrical Equivalent Circuit for Sonar Transducer Impedance Characteristic of Multiple Resonance

Jejin Jang [1], Jaehyuk Choi [1], Donghun Lee [2] and Hyungsoo Mok [3,*]

1   Research and Development Division, Tin Technology Co., Ltd., Seongnam 13212, Republic of Korea; jjjang@tin-technology.com (J.J.); jhchoi@tin-technology.com (J.C.)
2   Agency for Defense Development, Changwon 51678, Republic of Korea; leedhun@add.re.kr
3   Department of Electrical Engineering, Konkuk University, Seoul 05029, Republic of Korea
*   Correspondence: hsmok@konkuk.ac.kr; Tel.: +82-2-450-3479

Abstract: Improving the operational efficiency and optimizing the design of sound navigation and ranging (sonar) systems require accurate electrical equivalent models within the operating frequency range. The power conversion system within the sonar system increases power efficiency through impedance-matching circuits. Impedance matching is used to enhance the power transmission efficiency of the sonar system. Therefore, to increase the efficiency of the sonar system, an electrical-matching circuit is employed, and this necessitates an accurate equivalent circuit for the sonar transducer within the operating frequency range. In conventional equivalent circuit derivation methods, errors occur because they utilize the same number of RLC branches as the resonant frequency of the sonar transducer, based on its physical properties. Hence, this paper proposes an algorithm for deriving an equivalent circuit independent of resonance by employing multiple electrical components and particle swarm optimization (PSO). A comparative verification was also performed between the proposed and existing approaches using the Butterworth–van Dyke (BVD) model, which is a method for deriving electrical equivalent circuits.

Keywords: sound navigation and ranging; particle swarm optimization; electrical equivalent circuit; multiple resonant characteristics

Citation: Jang, J.; Choi, J.; Lee, D.; Mok, H. Estimation Method of an Electrical Equivalent Circuit for Sonar Transducer Impedance Characteristic of Multiple Resonance. *Sensors* **2023**, *23*, 6636. https://doi.org/10.3390/s23146636

Academic Editors: Farook Sattar, Niladri Bihari Puhan and Reza Fazel-Rezai

Received: 4 July 2023
Revised: 21 July 2023
Accepted: 23 July 2023
Published: 24 July 2023

1. Introduction

Sound navigation and ranging (sonar) systems detect underwater objects by utilizing electrical energy to sound energy conversion. Sonar power systems consist of the following components, as shown in Figure 1 [1–3]: (1) a direct current (DC) power supply, which serves as the electrical energy source required for generating acoustic energy; (2) a power converter, which converts the DC voltage into alternating current (AC) to provide the desired electrical energy supply for the required sound signal intensity; (3) an LC filter (or low-pass filter), which removes unnecessary frequencies for acoustic detection; (4) an impedance-matching transformer, which eliminates the reactive power generated by the material characteristics of the sonar sensor; and (5) a sonar sensor, which converts the input electrical signal into an acoustic signal.

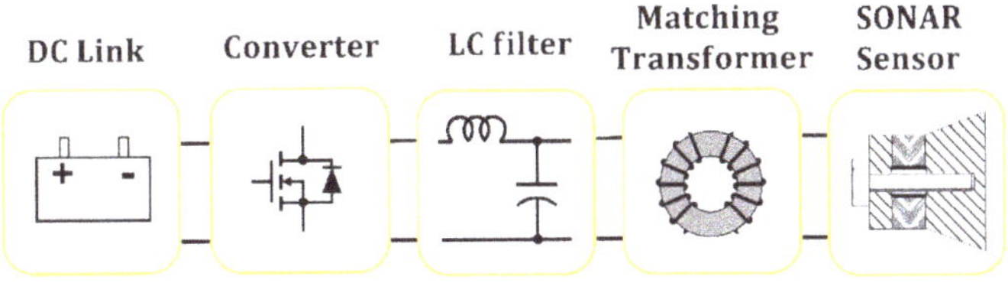

Figure 1. Sonar transducer power system.

Sonar systems have been developed for precise and wide-ranging exploration aimed at broadband operation [4,5], with high capacity, and high efficiency [6]. When designing high-power and high-efficiency power converters for sonar systems, reactive power and active power must be minimized and converted into acoustic energy [6]. After designing a sonar power system that meets these requirements, it is essential to verify the sonar sensor's performance. However, the electrical characteristics of sonar sensors can vary depending on the installation environment, such as terrestrial or submarine, the sensor type, and the sensor array structure. Therefore, while designing and validating high-power and high-efficiency power converters, accurate electrical equivalent models of the sonar sensor are required. Equivalent circuit models can be of various types and reflect the physical characteristics or operational features of the sonar sensors.

The equivalent circuit model originated from Mason's one-dimensional (1D) model of a piezoelectric transducer proposed in 1942 [7]. In this model, the physical movement of the piezo-ceramic was compared with that of a spring, and an ideal transformer was used to represent the electrical circuit. In 1961, Redwood proposed a method that used electrical transmission lines to model the transient behavior of sonar sensors [7]. In 1970, the Krimtholz, Leedom, and Mattthaei (KLM) model was proposed to simulate sonar sensors operating in the high-frequency range [7–12]. In 1994, Leach proposed the Butterworth–van Dyke (BVD) model as a replacement for the conventional Mason model and provided the simplest approach for simulating multilayered sensor structures. Although this methodology is not well suited for high-frequency characteristics and multiple resonances [7], its application as a modeling circuit remains predominant owing to its remarkable accuracy at the resonance points.

An analytical method using approximation techniques in a certain range of sonar sensor impedances was employed to estimate the parameters of equivalent circuits [7,13]. However, this approach is ineffective when the impedance variation at the resonant frequencies is small. To overcome this limitation, an alternative approach that calculates the equivalent circuit parameters based on the resonance frequencies within the actual impedance of the sensor was proposed. However, this method yielded inaccurate results when deriving the parameters of an equivalent circuit [14]. Building on this calculation method, Ramesh and Ebenezer [15] proposed a parameter estimation technique using the least squares method. They assumed that each resonance point is independent and treated the equivalent circuit for each resonance as a parallel RLC branch. They essentially proposed a method for deriving an equivalent circuit and its parameters for a sensor with two resonance frequencies [15]. A drawback of this method is that the estimation errors increase when the resonance frequencies are close to each other. Consequently, to accurately model sensors with abrupt impedance variations is challenging. To address these limitations, recent studies have proposed parameter estimation approaches using particle swarm optimization (PSO) [16]. However, owing to the limitations of the frequency degrees of freedom of the equivalent circuit, even using such methods, the modeling of rapidly changing impedance characteristics remains difficult. Therefore, in this study, we devised an electrical equivalent circuit derivation approach by constructing an equivalent circuit composed of multiple RLC components configured in series and parallel. The PSO algorithm was employed to determine the physically realizable values of the equivalent circuit component parameters. To validate the proposed approach, electrical equivalent circuits of sonar sensors with single, dual, and multiple resonances were derived using both conventional and proposed methods. The accuracy of the impedance estimation was evaluated with respect to the change in the number of resonances.

2. Electrical Equivalent Circuit for Sonar Sensors

2.1. Conventional Electrical Equivalent Circuits

The BVD equivalent circuit, shown in Figure 2, is composed of an RLC branch consisting of resistance (R_1), inductance (L_2), and capacitance (C_3) to mimic the electrical resonance frequency characteristics, along with a parallel capacitor (C_0) representing the electrical capacitance properties of the piezoelectric component. Therefore, in this study, to reduce sensor errors caused by physical and material factors, an enhanced equivalent circuit with increased electrical resonance modes and RLC branches was employed within the BVD model.

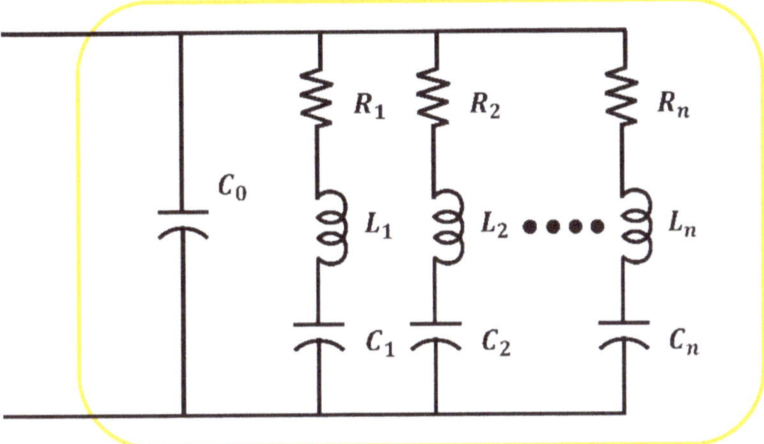

Figure 2. BVD equivalent circuit.

The electrical admittance of Figure 2 is determined as follows:

$$Y = \frac{I}{V} = j\omega C_0 + \sum_{i=1}^{n} \frac{1}{R_i + j\omega L_i + \frac{1}{j\omega C_i}} \tag{1}$$

2.2. Proposed Electrical Equivalent Circuits

The equivalent electrical circuit in Figure 2 represents the resonances within the operating frequency range by using an RLC branch for the nth resonance. However, such equivalent models face challenges with regard to selecting different models based on the observed resonances in the measured sonar impedance data and estimating the parameters within the selected model.

Furthermore, based on the measured characteristics of the sonar sensor, an equivalent circuit was determined. When using a large number of components, parameter estimation required a relatively long time, resulting in more accurate results. Conversely, for a small number of components, the estimation time was reduced, although the accuracy decreased, thus exhibiting a trade-off relationship. Additionally, after deriving the corresponding parameters, a trial-and-error process was performed under human intervention for additional calibration. Therefore, in this paper, we propose a high-degree model composed of 54 RLC components, as depicted in Figure 3, which allows for the selection of different equivalent circuits for each resonance frequency. Thus, the proposed model seeks to overcome the conventional limitations of the sensor characteristic simulation due to the constraints on the number of RLC components.

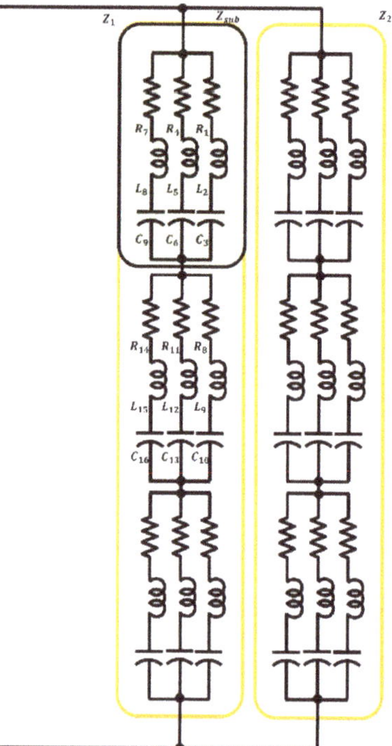

Figure 3. Proposed equivalent circuits, high-degree model.

3. Estimation of Electrical Equivalent Circuit Using Particle Swarm Optimization

To derive the electrical equivalent model of the sonar sensor, the electrical characteristics of the sensor must be analyzed and an appropriate type of equivalent model must be selected. Once the equivalent model is determined, optimization algorithms are used with the measured impedance magnitude and phase characteristics of the actual sonar sensor in its operating frequency range as reference values. This allows for the extraction of parameter values at the point where the combination of parameter values minimizes errors in the impedance magnitude and phase.

As shown in Table 1, the least squares method, genetic algorithm (GA) and PSO have their own characteristics. The least squares method faces the difficulty of achieving optimal results in the presence of large errors. Meanwhile, the GA suffers from the lack of diversity among individuals, leading to convergence to nonoptimal solutions.

Table 1. Characteristic of each type of parameter estimation algorithm.

Algorithm Type	Characteristic
Least square	The optimal solution is derived in the direction that minimizes the total sum of errors between the reference and estimated values.
Genetic algorithm	The optimal solution is found within a group, and the search is repeated based on the selected solution to derive the optimal solution.
PSO algorithm	Individual particles and clusters exchange error information based on one reference value to derive the optimal solution.

In this study, the PSO algorithm was employed to derive the equivalent circuit parameters of a sonar sensor [17]. The PSO algorithm is based on swarm intelligence, which is inspired by the collective behavior of flocks of birds and schools of fish. It has the advantage of obtaining results quickly because it only transmits the best global optimum information instead of requiring overlap or mutation operations [18]. However, the PSO algorithm may suffer from convergence to the local optima rather than the global optimum if the speed and direction of the particles are inaccurate. To address this issue, the inertia weight (w) in the PSO algorithm was examined so as to improve its performance [19].

3.1. Conventional Method

The electrical equivalent model of the sonar sensor was derived using the PSO algorithm as shown in Figure 4, and related variables are shown in Table 2.

(1)　Initialization phase: The necessary variables for the PSO algorithm are initialized, and the initial values are set.

(2)　Fitness calculation of the particles: The impedance characteristics of the sensor are calculated based on the estimated parameter values of the equivalent circuit and the variables obtained from the previous step. The calculated and measured impedance characteristics are compared to evaluate the error level. The evaluation is performed using Equation (2), where e represents the error value, k_{mag} and k_{phase} are the weighting factors for the magnitude and phase errors, respectively, and Z_{mag} and Z_{phase} represent the errors in the estimated impedance magnitude and phase, respectively.

$$e = k_{mag}Z_{mag} + k_{phase}Z_{phase} \tag{2}$$

(3)　$pbest_{id}^{k}$ and $gbest_{id}^{k}$ update phase: In this phase, the evaluation values of each particle, determined by the fitness function are compared to the previously recorded best evaluation value of $pbest_{id}^{k}$. If the current evaluation value is lower than the previous best evaluation value, the parameter values are updated and assigned as the new $pbest_{id}^{k}$. Subsequently, among all the updated $pbest_{id}^{k}$ values, the value with the best evaluation result of $gbest_{id}^{k}$ is compared with the previous best evaluation result within the swarm. If the new evaluation result is superior, the derived parameters are updated. Once all the updates are completed, the algorithm checks whether the maximum number of iterations has been reached or if the evaluation result satisfies the termination criterion. If either condition is satisfied, the PSO algorithm terminates. This process ensures that the particles continuously update their personal best positions and velocities, while also considering the global best position within the swarm.

(4)　x_{id}^{k+1} calculation phase: In this phase, when the maximum number of iterations is not exceeded and the evaluation result does not satisfy the termination criterion, new parameter values (x_{id}) are assigned for the next optimization operation. The new parameter values are determined according to the PSO algorithm, as shown in Equation (3). The velocity change (v_{id}), representing the direction and magnitude of the particle movement, is added to the current x_{id} value, resulting in the derivation of the new x_{id} value, as shown in Equation (4). This calculation phase ensures that each particle updates its position based on its previous position and the velocity change determined by the PSO algorithm, allowing for continuous exploration and refinement of the parameter space in the search for an optimal solution.

$$V_{id}^{k+1} = wV_{id}^{k} + c_1 r_i^{k}\left(pbest_{id}^{k} - x_{id}^{k}\right) + c_2 r_2^{k}\left(gbest_{d}^{k} - x_{id}^{k}\right) \tag{3}$$

$$x_{id}^{k+1} = x_{id}^{k} + V_{id}^{k+1} \tag{4}$$

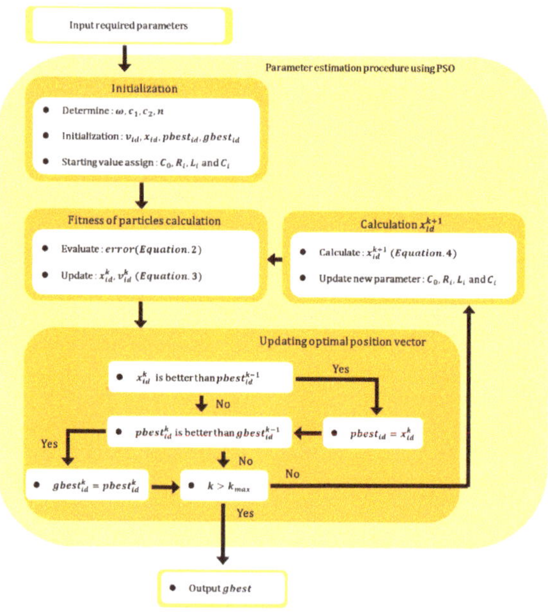

Figure 4. Conventional procedure for parameter estimation using PSO.

Table 2. Variable definitions for Equations (2)–(4).

Variables Name	Definition
c_1, c_2	Acceleration constants
d	Number of swarm
n	Total number of parameters to derive PSO results
r_1, r_2	Uniformly distributed random numbers
ω	Inertia weight factor
i	Number of particles
p_{num}	Total number of particles for the swarm
V_{id}^{t}	Present velocity vector of the swarm
V_{id}^{t+1}	Next velocity vector of the particle
x_{id}^{k}	Present position vector of the swarm
x_{id}^{k+1}	Next position vector of the particle
$gbest_d$	Optimal position vector of the swarm
$pbest_d$	Optimal position vector of the particle

3.2. Proposed Method

As shown in Figure 5, The process of deriving the parameters to simulate multiple resonance characteristics and rapidly changing electrical impedance characteristics using multiple electrical RLC components is as follows:

(1) Initialization phase: the necessary variables are initialized, and initial values are set for the PSO algorithm.

(2) Multiple-component parameter derivation is through the PSO process, as shown in Figure 4, and the electrical parameters of the entire set of RLC components are derived, as shown in Figure 3.

(3) Selection of restricted component range: component parameters that are physically difficult to implement are selected as the basis for defining the restricted range.

(4) Selection of feasible components considering manufacturability: based on the upper and lower limits of the impedance determined in step 3, the RLC component parameters that exceed the impedance thresholds are selected.

(5) PSO reiteration using the modified equivalent circuit: using the modified equivalent circuit from step 4, the components that were not selected are excluded, and the PSO derivation process is performed, as shown in Equations (3) and (4), to derive the electrical equivalent circuit parameters.

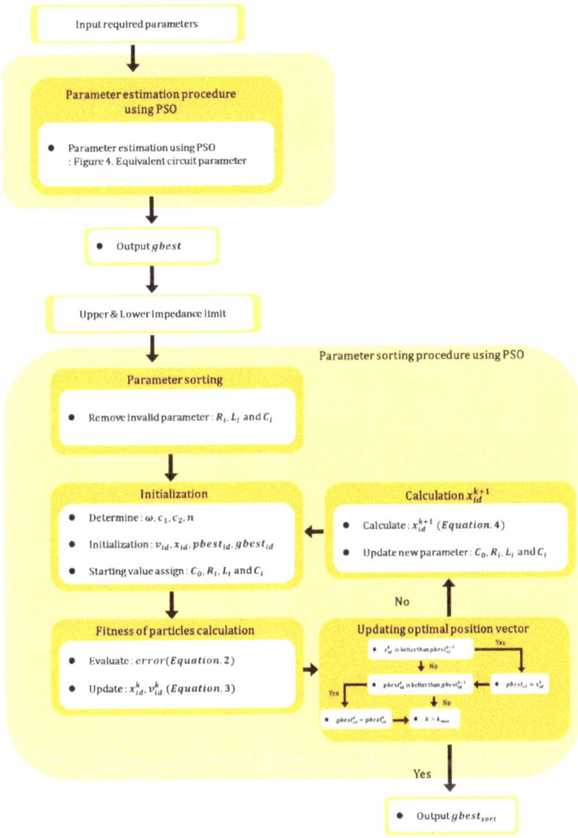

Figure 5. Proposed procedure of PSO algorithm for parameter estimation of high-degree electrical equivalent circuit.

4. Results

The results of the equivalent circuit parameter estimation for single, double, and multiple resonant sonar sensors were compared to validate the accuracy of the proposed equivalent circuit and parameter estimation algorithm. The average error rates between the measured impedance magnitude and phase values (using Equations (5) and (6)) and the estimated impedance characteristics were utilized to compare the accuracy of the impedance characteristics. Here, $Z_{mag\_error\_avg}$ and $Z_{ph\_error\_avg}$, respectively, represent the average error rates of the impedance magnitude and phase. Further, Z_{mag_real}, Z_{mag_est}, Z_{ph_real}, and Z_{ph_est} denote the measured and estimated values of the impedance magnitude and phase, respectively.

In this paper, the results of parameter derivation for circuit modeling are presented by comparing the conventional equivalent circuit and PSO algorithm-based parameter derivation with the proposed method using an equivalent circuit and the PSO algorithm. The results of the proposed algorithm for the parameters obtained in procedure (3) of Section 4.2 are denoted by "·" for electrical shorts and "X" for electrical opens in the table.

$$Z_{mag\_error\_avg}[\%] = \frac{1}{n}\sum_0^n \sqrt{\left(Z_{mag\_real} - Z_{mag\_est}\right)^2} \Big/ \left(\frac{1}{n}\sum_0^n Z_{mag\_real}\right) \times 100 \qquad (5)$$

$$Z_{ph\_error\_avg}[\%] = \frac{1}{n}\sum_0^n \sqrt{\left(Z_{ph_{real}} - Z_{ph_{est}}\right)^2} \Big/ \left(\frac{1}{n}\sum_0^n Z_{ph_{real}}\right) \times 100 \qquad (6)$$

4.1. Single Resonance Characteristic Results

As shown in Figure 6a, for a sonar sensor with a single resonant frequency within the operating frequency range, a conventional equivalent circuit with an RLC branch was constructed to simulate the single resonant mode in the BVD model. The equivalent circuit parameters were derived by utilizing the PSO algorithm, illustrated in Figure 4, in which the measured impedance data are represented by the red line, and the characteristics of the equivalent circuit are denoted by the line of crosses in Figure 6b. The values obtained for the four parameters of the equivalent circuit are presented in Table 3.

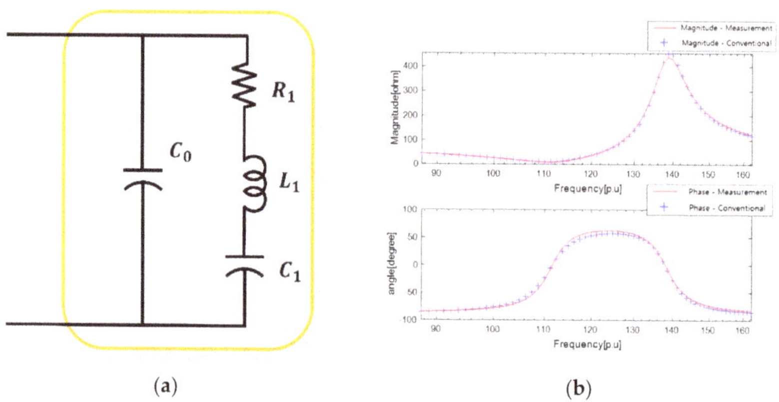

Figure 6. Results of deriving the equivalent circuit of a single resonant sensor using the conventional method: (**a**) conventional equivalent circuit; and (**b**) electrical impedance characteristics.

Table 3. Results of deriving equivalent circuit parameters of a single resonant sonar sensor using the conventional method.

Parameter Name	Value	Parameter Name	Value
C_0	19.91 nF	L_1	287.58 μH
R_1	11.57 Ω	C_1	11.15 nF

To compare the proposed equivalent circuit and derivation approach for a single-resonant sonar sensor, the results of deriving the equivalent circuit using the PSO algorithm shown in Figure 5 are summarized in Figure 7. Specifically, Figure 7a,b show the configurations of the equivalent circuits with 54 RLC components [EA] each. After excluding components with excessively large or small parameter values, which may appear electrically open or short, the equivalent circuit was sorted, as shown in Figure 7b. The parameter values for each component constituting the equivalent circuit are listed in Tables 4 and 5. The component arrangement is organized in the order of subscripts from the top-left RLC component of Z_1 in the table, which corresponds to the 27 top-left RLC components, as shown in Figure 7a. The electrical characteristics of these components are shown in Figure 7c, wherein the blue and black lines represent the characteristic results.

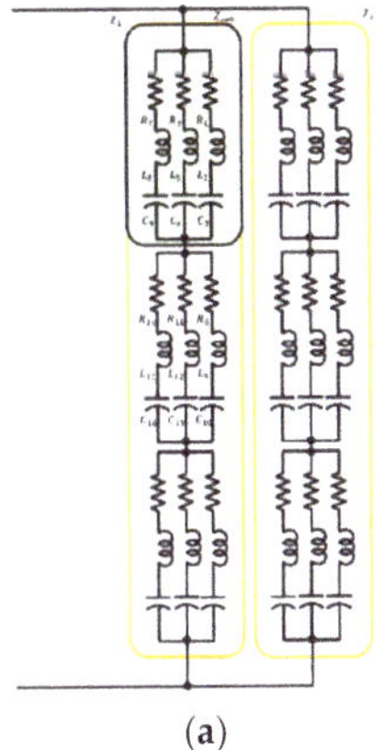

(a)

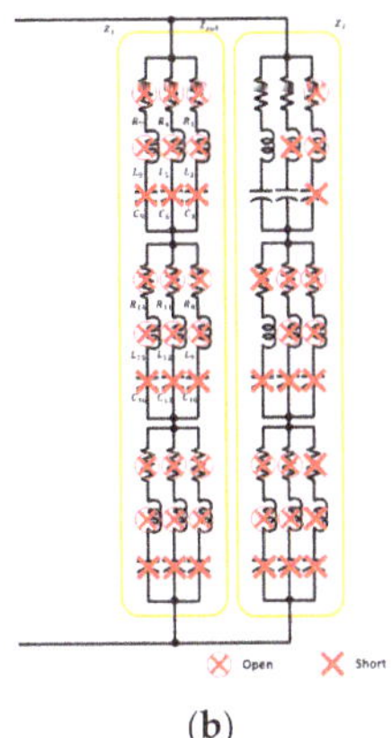

(b)

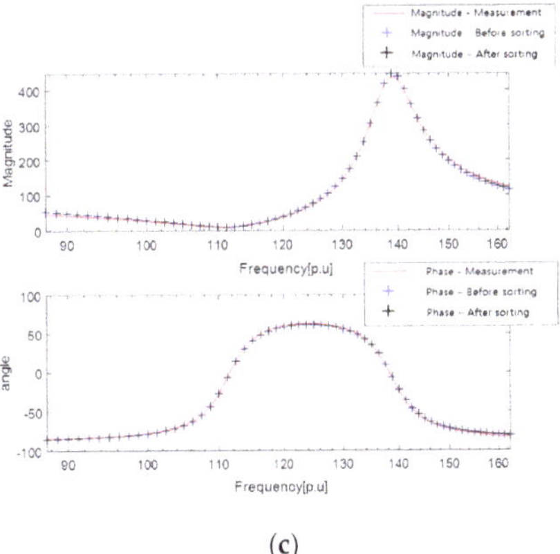

(c)

Figure 7. Results of deriving the equivalent circuit of a single resonant sensor using the proposed method: (**a**) high-degree equivalent circuit; (**b**) equivalent circuit after sorting unnecessary elements; (**c**) electrical impedance characteristics.

Table 4. Results of deriving equivalent circuit parameters of a single resonant sonar sensor using the proposed method before sorting.

	R [Ω]	L [mH]	C [nF]	R [Ω]	L [mH]	C [nF]	R [Ω]	L [mH]	C [nF]
Z_1	1.30×10^{31}	2.49×10^{38}	4.13×10^{35}	1.34×10^{21}	6.27×10^{32}	7.34×10^{32}	2.51×10^{30}	9.87×10^{36}	1.37×10^{35}
	3.09×10^{34}	1.35×10^{28}	6.33×10^{33}	7.05×10^{40}	2.08×10^{34}	4.55×10^{41}	3.00×10^{23}	1.45×10^{32}	1.31×10^{33}
	2.43×10^{33}	8.11×10^{39}	3.75×10^{38}	8.72×10^{27}	2.37×10^{38}	3.74×10^{34}	2.30×10^{28}	2.59×10^{38}	9.62×10^{32}
Z_2	3.22×10^{31}	2.86×10^{37}	7.54×10^{38}	6.76×10^{26}	1.97×10^{34}	1.68×10^{29}	0.0013	0.00013	7.14×10^{31}
	2.99	7.25×10^{-6}	17.1	1.81×10^{31}	6.38×10^{36}	1.45×10^{34}	4.14×10^{29}	2.29×10^{40}	2.04×10^{33}
	12.93	0.37	8.10	0.0025	0.018	8.30×10^{27}	4.76×10^{32}	1.16×10^{31}	3.12×10^{37}

Table 5. Results of deriving equivalent circuit parameters of a single resonant sonar sensor using the proposed method after sorting.

	R [Ω]	L [mH]	C [nF]	R [Ω]	L [mH]	C [nF]	R [Ω]	L [mH]	C [nF]
Z_1	X	X	·	X	X	·	X	X	·
	X	X	·	X	X	·	X	X	·
	X	X	·	X	X	·	X	X	·
Z_2	X	X	·	X	X	·	·	·	·
	2.99	·	17.1	X	X	·	X	X	·
	12.93	0.37	8.10	·	0.02	·	X	X	·

4.2. Dual Resonance Characteristic Results

Following the same procedure as that used for the comparison and validation of the equivalent circuit for single-resonant sonar sensors, we also compared the precision of the equivalent circuit results for sonar sensors with double resonant frequencies. The results of PSO obtained in Figure 4 were utilized to derive the equivalent circuit of a dual-resonant SONAR sensor, as presented in Figure 8 and Table 6 in the paper.

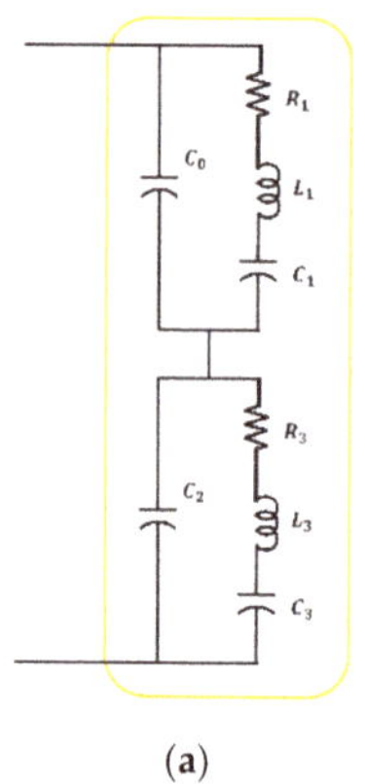

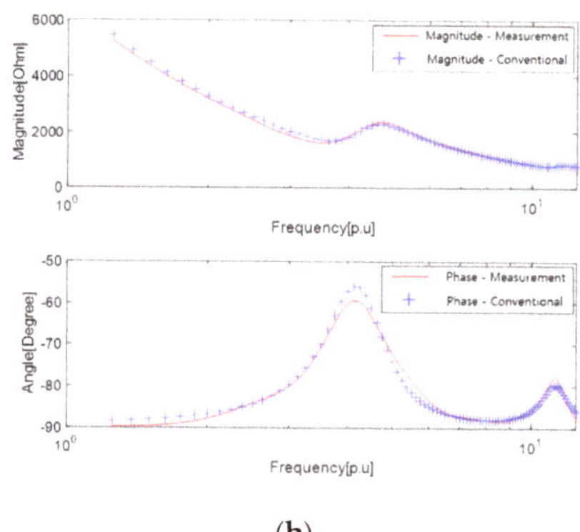

(**a**) (**b**)

Figure 8. Results of deriving the equivalent circuit of the dual resonance sensor using the conventional method: (**a**) equivalent circuit; and (**b**) electrical equivalent characteristics.

Table 6. Results of deriving equivalent circuit parameters of the dual resonance sonar sensor using the conventional method.

Parameter Name	Value	Parameter Name	Value
C_0	22 nF	C_2	390 nF
R_1	3.3 kΩ	R_3	12 Ω
L_1	425 mH	L_3	1.22 mH
C_1	5.6 nF	C_3	680 nF

The results of the parameter estimation using PSO are presented in Tables 7 and 8. The characteristics of the equivalent circuit based on the parameters listed in Tables 7 and 8 are shown in Figure 9, wherein the reference sensor characteristics are in red and the equivalent circuit characteristics are in blue.

Table 7. Results of deriving equivalent circuit parameters of the dual resonant sonar sensor using the proposed method before sorting.

	R [Ω]	L [mH]	C [nF]	R [Ω]	L [mH]	C [nF]	R [Ω]	L [mH]	C [nF]
	6.78×10^{34}	8.39×10^{28}	7.80×10^{30}	0.00012	0.019	1813.33	0.71	1.19×10^{-6}	3637.09
Z_1	1.37	8.47×10^{-7}	15,210.46	8.18×10^{-5}	9.64	2.59×10^{34}	8.49×10^{38}	1.60×10^{35}	6.86×10^{32}
	4.38×10^{36}	7.73×10^{34}	1.39×10^{30}	64.89	9.88	595.98	0.00023	7.59	1.16×10^{32}
	6.23	0.69	1910.65	0.00016	5.56	46,620.45	2.39×10^{37}	3.95×10^{35}	1.23×10^{41}
Z_2	5.62×10^{29}	2.31×10^{36}	7.21×10^{40}	64.78	21.10	525.72	9.72×10^{37}	1.16×10^{46}	1.42×10^{33}
	0.81	1.34	4.75×10^{8}	4.64×10^{45}	3.51×10^{34}	3.49×10^{29}	2.12×10^{-5}	0.35	1.03×10^{44}

Table 8. Results of deriving equivalent circuit parameters of the dual resonant sonar sensor using the proposed method after sorting.

	R [Ω]	L [mH]	C [nF]	R [Ω]	L [mH]	C [nF]	R [Ω]	L [mH]	C [nF]
	X	X	·	X	X	·	X	X	·
Z_1	·	·	·	·	9.64	·	X	X	·
	X	X	·	·	9.88	595.98	·	7.59	·
	·	·	X	X	·	46,620.45	X	X	·
Z_2	X	X	·	21.1	526	525.72	X	X	·
	·	·	·	X	X	·	·	·	·

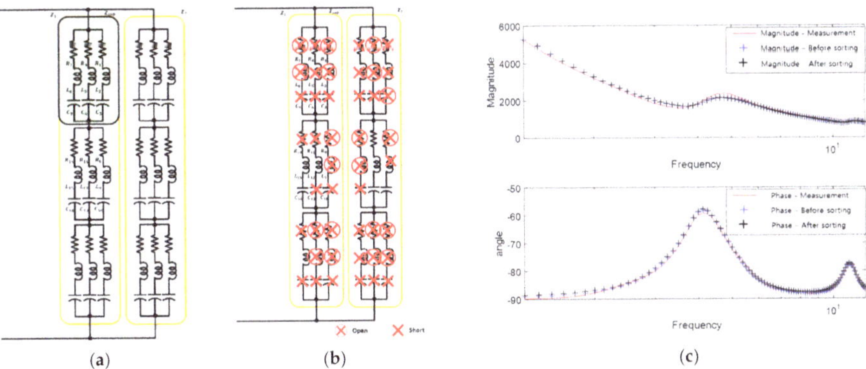

Figure 9. Results of deriving the equivalent circuit of a single resonant sensor using the proposed method: (**a**) high-degree equivalent circuit; (**b**) equivalent circuit after sorting unnecessary elements; and (**c**) electrical impedance characteristics.

4.3. Multiple Resonance Characteristic Results

To verify the feasibility of simulating multiple resonant characteristics and rapid impedance changes with varying frequencies, we compared the existing and proposed approaches based on the measured data for sensors with multiple resonant frequencies and characteristics exhibiting rapid variations.

Owing to the phase characteristics of the multi-resonant sonar sensor, an additional parallel L_1 was included in the existing BVD model to construct an equivalent circuit. The results of the parameter estimation using PSO are presented in Table 9. The characteristics of the equivalent circuit based on the parameters listed in Table 9 are shown in Figure 10, wherein the reference sensor characteristics are in red and the equivalent circuit characteristics are in blue.

Table 9. Results of deriving equivalent circuit parameters of the multiple resonance sonar sensor using the conventional method.

Parameter Name	Value	Parameter Name	Value
L_1	5.356 mH	L_7	11.844 mH
C_2	806.5 nF	C_8	318.989 nF
R_3	10 Ω	R_9	129.631 Ω
L_4	6.344 mH	L_{10}	29.999 mH
C_5	149.46 nF	C_{11}	380.17 nF
R_6	90.877 Ω	–	–

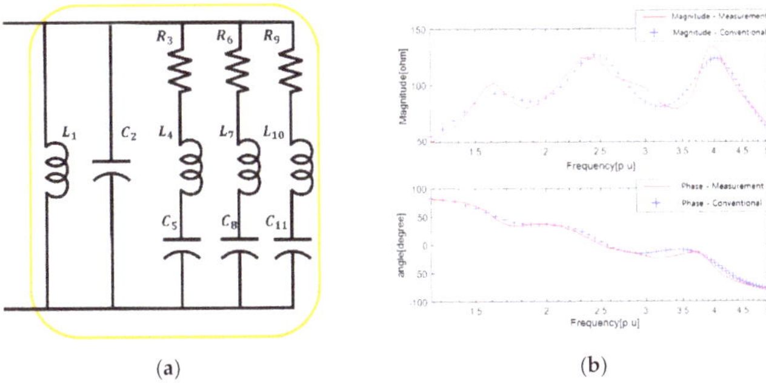

Figure 10. Results of deriving the equivalent circuit of the multiple resonance sensor using the conventional method: (**a**) equivalent circuit; and (**b**) electrical equivalent characteristics.

To further compare and validate the proposed approach, we employed the PSO algorithm to derive the equivalent circuit and parameters, as shown in Figure 11. The results are presented in Tables 10 and 11. The electrical impedance characteristics of each equivalent circuit in Figure 11c are represented by blue and black lines, as aforementioned.

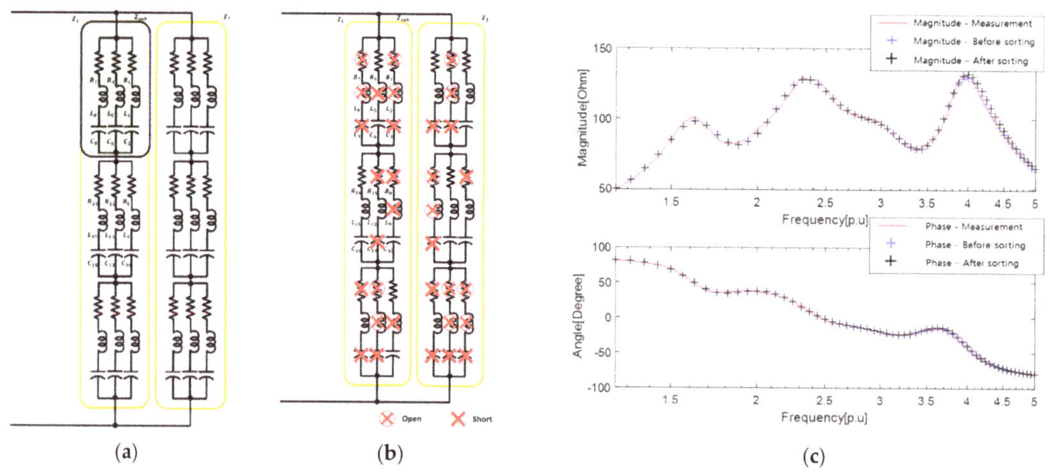

Figure 11. Results of deriving the equivalent circuit of the multiple resonant sensor using the proposed method: (**a**) high-degree equivalent circuit; (**b**) equivalent circuit after sorting unnecessary elements; and (**c**) electrical impedance characteristics.

Table 10. Results of deriving equivalent circuit parameters of the multiple resonant sonar sensor using the proposed method before sorting.

	R [Ω]	L [mH]	C [nF]	R [Ω]	L [mH]	C [nF]	R [Ω]	L [mH]	C [nF]
Z_1	6.78×10^{34}	8.39×10^{28}	7.80×10^{30}	0.00012	0.019	1813.33	0.709	1.19×10^{-6}	3637.09
	1.37	8.47×10^{-7}	1.52×10^{4}	8.18×10^{-5}	9.64	2.59×10^{34}	8.49×10^{38}	1.60×10^{35}	6.86×10^{32}
	4.38×10^{36}	7.73×10^{34}	1.39×10^{30}	64.89	9.88	595.98	0.00023	7.59	1.16×10^{32}
Z_2	6.23	0.69	1.91×10^{3}	0.00016	5.56	46,620.45	2.39×10^{37}	3.95×10^{35}	1.23×10^{41}
	5.62×10^{29}	2.31×10^{36}	7.21×10^{40}	64.78	21.1	525.72	9.72×10^{37}	1.16×10^{46}	1.42×10^{33}
	0.81	1.34	4.75×10^{8}	4.64×10^{45}	3.51×10^{34}	3.49×10^{29}	2.12×10^{-5}	0.35	1.03×10^{44}

Table 11. Results of deriving equivalent circuit parameters of the multiple resonant sonar sensor using the proposed method after sorting.

	R [Ω]	L [mH]	C [nF]	R [Ω]	L [mH]	C [nF]	R [Ω]	L [mH]	C [nF]
Z_1	X	X	·	·	·	1813.33	0.71	·	3637.09
	1.37	Short	15,210.46	·	9.64	·	X	X	·
	X	X	·	64.89	9.88	595.98	·	7.59	·
Z_2	6.23	0.69	1910.65	·	5.56	46,620.45	X	X	·
	X	X	·	64.78	21.1	525.72	X	X	·
	0.81	1.34	·	X	X	·	·	0.35	·

Table 12 lists the average error rates relative to the reference impedance characteristics when different equivalent circuits were employed for each type of sensor. "Conventional" refers to the approach of deriving an equivalent circuit for each sonar sensor by modifying it based on the number of resonant frequencies. The terms "Before sorting" and "After sorting" define the stages in the process of utilizing the proposed 54 [EA] components-based equivalent circuit, where "Before" represents the initial result derived through the PSO algorithm, and "After" denotes the exclusion of certain components that are physically unsuitable.

Table 12. Average error rate according to the equivalent circuit by sonar sensor type.

Equivalent Type / Sensor Characteristic	Impedance Magnitude Error Ratio [%]			Impedance Phase Error Ratio [%]		
	Conventional	Before Sorting	After Sorting	Conventional	Before Sorting	After Sorting
Single resonant sensor	0.063	0.055	0.055	0.11	0.031	0.031
Dual resonant sensor	1.31	1.43	1.51	3.93	0.061	0.078
Multi-resonant sensor	5.64	0.018	0.82	12.34	0.62	2.58

Based on the average error rates for each equivalent circuit, the existing approaches were shown to have limitations in accurately estimating the impedance characteristics when the impedance exhibited rapid variations at the resonant points and frequencies, with maximum error rates of 5.64 and 12.34%, respectively. However, the proposed approach utilizing multiple components for the equivalent circuit and optimization algorithm demonstrated a relatively accurate simulation of the impedance characteristics, with maximum error rates of 1.43 and 2.52%, respectively, achieving average error rates within 3%. This validates the capability of accurately simulating the impedance characteristics and confirms that the proposed equivalent circuit and algorithm enable the derivation of the equivalent circuit of the sensor, regardless of the number of resonant frequencies.

5. Conclusions

In existing sonar sensor equivalent circuits, the circuits are modified based on the resonant characteristics of the sensor, and the limitations of the number of passive components within the circuit make it difficult to accurately reflect the impedance characteristics and rapidly changing impedance characteristics at adjacent resonant frequencies. To overcome these challenges, we developed an electrical equivalent circuit for sonar sensors based on their impedance characteristics and devised an estimation approach to derive the equivalent circuit.

To validate the proposed approach, we compared the average errors between the measured reference values and the estimated results of the equivalent circuit parameters for sensors with single, dual, and multiple resonances. The comparison results showed that, when using the conventional approach, the maximum average errors were 5.64 and 12.34% for the impedance magnitude and phase, respectively. However, when utilizing the proposed approach, the maximum errors were reduced to 0.8 and 2.58% for the impedance magnitude and phase, respectively, demonstrating higher precision compared to the conventional approach. Moreover, the proposed approach maintained high precision with average error rates below 1%, even with variations in the number of resonances and in the impedance characteristics of the sensor.

Furthermore, recent sonar sensors are not only operated at fixed frequencies but over a wide frequency range to enhance detection performance. Impedance matching is essential for overall system efficiency when operating a sensor across a wide frequency range. It is crucial to reflect the frequency characteristics of the load accurately when designing an impedance-matching circuit. The precision of the derived equivalent circuit, as previously described, is expected to enable the accurate reflection of load characteristics when designing amplifiers for sonar operation.

Author Contributions: Conceptualization, J.C.; methodology, J.C.; software, J.C. and J.J.; validation, J.C. and J.J.; formal analysis, J.J.; investigation, J.J.; resources, J.J.; data curation, J.J.; writing—original draft preparation, J.J.; writing—review and editing, H.M.; visualization, J.J. and J.C.; supervision, H.M.; project administration, D.L.; funding acquisition, D.L. and H.M. All authors have read and agreed to the published version of the manuscript.

Funding: This research was supported by Agency for Defense Development (The Study of high power amplification techniques for FFR suitable for CAS transmission).

Conflicts of Interest: The authors declare no conflict of interest.

References

1. Choi, J.H.; Mok, H.S. Simultaneous Design of Low-Pass Filter with Impedance Matching Transformer for SONAR Transducer Using Particle Swarm Optimization. *Energies* **2019**, *12*, 4646. [CrossRef]
2. Song, S.M.; Kim, I.D.; Lee, B.H.; Lee, J.M. Design of Matching Circuit Transformer for High-Power Transmitter of Active Sonar. *J. Electr. Eng. Technol.* **2020**, *15*, 2145–2155. [CrossRef]
3. Choi, J.; Lee, D.H.; Mok, H. Discontinuous PWM Techniques of Three-Leg Two-Phase Voltage Source Inverter for Sonar System. *IEEE Access* **2020**, *8*, 199864–199881. [CrossRef]
4. Kye, J.E.; Cho, J.I.; Yoo, W.P.; Choi, S.L.; Park, J.H. Trends and Applications on Multi-beam Side Scan Sonar Sensor Technology. *Electron. Telecommun. Trends* **2013**, *28*, 167–179.
5. Oh, Y.-S. Development of High Resolution Side Scan Sonar System Using Multi Beam Array Porcessing. Ph.D. Thesis, Department of Ocean Engineering, Graduate School of Korea Maritime University, Busan, Republic of Korea, 2012.
6. Song, S.M.; Kim, I.D.; Lee, B.H.; Lee, J.M. A Study on Feedback and Feedforward Control of High Power Transmitters for Active Sonar. In Proceedings of the 2019 IEEE 4th International Future Energy Electronics Conference (IFEEC), Singapore, 25–28 November 2019.
7. Shen, Z.; Xu, J.; Li, Z.; Chen, Y.; Cui, Y.; Jian, X. An Improved Equivalent Circuit Simulation of High Frequency Ultrasound Transducer. *Front. Mater.* **2021**, *8*, 109. [CrossRef]
8. Oakley, C.G. Calculation of ultrasonic transducer signal-to-noise ratios using the KLM model. *IEEE Trans. Ultrason. Ferroelectr. Freq. Control.* **1997**, *44*, 1018–1026. [CrossRef]
9. Sherrit, S.; Mukherjee, B.K. Characterization of piezoelectric materials for transducers. *arXiv* **2007**, arXiv:0711.2657.
10. Galliere, J.M.; Latorre, L.; Papet, P. A 2-D KLM model for disk-shape piezoelectric transducers. In Proceedings of the 2009 Second International Conference on Advances in Circuits, Electronics and Micro-Electronics, Sliema, Malta, 11–16 October 2009; pp. 40–43.
11. Smyth, K.; Kim, S.G. Experiment and simulation validated analytical equivalent circuit model for piezoelectric micromachined ultrasonic transducers. *IEEE Trans. Ultrason. Ferroelectr. Freq. Control* **2015**, *62*, 744–765. [CrossRef] [PubMed]
12. Chen, D.; Zhao, J.; Fei, C.; Li, D.; Zhu, Y.; Li, Z.; Guo, R.; Lou, L.; Feng, W.; Yang, Y. Particle Swarm Optimization Algorithm-Based Design Method for Ultrasonic Transducers. *Micromachines* **2020**, *11*, 715. [CrossRef] [PubMed]
13. Narvaez, F.; Hosseini, S.; Farkhani, H.; Moradi, F. Analysis and Design of a PMUT-based transducer for Powering Brain Implants. *arXiv* **2021**, arXiv:2101.04443.
14. Mason, W.P. *Electromechnical Transducers and Wave Filters*; D. Van Nostrand Co.: New York, NY, USA, 1948.
15. Coates, D.; Maguire, P.T. Multiple-Mode Acoustic Transducer Calculations. *IEEE Trans. Ultrason. Ferroelectr. Freq. Control* **1989**, *36*, 471–473. [CrossRef] [PubMed]
16. Ramesh, R.; Ebenezer, D.D. Equivalent Circuit for Broadband Underwater Transducer. *IEEE Trans. Ultrason. Ferroelectr. Freq.Control* **2008**, *55*, 2079–2083. [CrossRef] [PubMed]
17. Kennedy, J.; Eberhart, R. Particle swarm optimization. In Proceedings of the International Conference on Neural Networks, Perth, Australia, 27 November–1 December 1995; pp. 1942–1948.
18. Bai, Q. Analysis of particle swarm optimization algorithm. *Comput. Inf. Sci.* **2010**, *3*, 180–184. [CrossRef]
19. Shi, Y.; Eberhart, R. A modified particle swarm optimizer. In Proceedings of the 1998 IEEE International Conference on Evolutionary Computation Proceedings, IEEEWorld Congress on Computational Intelligence (Cat. No.98TH8360), Anchorage, AK, USA, 4–9 May 1998; pp. 69–73.

Communication

Guided Acoustic Waves in Polymer Rods with Varying Immersion Depth in Liquid

Klaus Lutter *,†, Alexander Backer † and Klaus Stefan Drese

Institute for Sensor and Actuator Technology, Coburg University of Applied Sciences and Arts, Am Hofbräuhaus 1B, 96450 Coburg, Germany; alexander.backer@hs-coburg.de (A.B.); klaus.drese@hs-coburg.de (K.S.D.)
* Correspondence: klaus.lutter@hs-coburg.de; Tel.: +49-9561-317-8116
† These authors contributed equally to this work.

Abstract: Monitoring tanks and vessels play an important part in public infrastructure and several industrial processes. The goal of this work is to propose a new kind of guided acoustic wave sensor for measuring immersion depth. Common sensor types such as pressure sensors and airborne ultrasonic sensors are often limited to non-corrosive media, and can fail to distinguish between the media they reflect on or are submerged in. Motivated by this limitation, we developed a guided acoustic wave sensor made from polyethylene using piezoceramics. In contrast to existing sensors, low-frequency Hanning-windowed sine bursts were used to excite the L(0,1) mode within a solid polyethylene rod. The acoustic velocity within these rods changes with the immersion depth in the surrounding fluid. Thus, it is possible to detect changes in the surrounding media by measuring the time shifts of zero crossings through the rod after being reflected on the opposite end. The change in time of zero crossings is monotonically related to the immersion depth. This relative measurement method can be used in different kinds of liquids, including strong acids or bases.

Keywords: guided acoustic waves; piezo transducer; high density polyethylene; L(0,1) mode

Citation: Lutter, K.; Backer, A.; Drese, K.S. Guided Acoustic Waves in Polymer Rods with Varying Immersion Depth in Liquid. *Sensors* 2023, 23, 9892. https://doi.org/10.3390/s23249092

Academic Editors: Farook Sattar, Niladri Bihari Puhan and Reza Fazel-Rezai

Received: 11 October 2023
Revised: 14 December 2023
Accepted: 16 December 2023
Published: 18 December 2023

1. Introduction

Monitoring fluids in vessels is an indispensable necessity in various applications in the industrial, medical, and environmental fields. Traditional sensor concepts such as pressure sensors offer high precision and are widely used. While they are independent of the vessel they are submerged in, they are limited to water or other non-corrosive media. In addition, they are influenced by sedimentation at the bottom of a liquid tank. On the other hand, airborne ultrasonic echo sensors may be used for all types of liquids, or even solids; however, foam building up inside the vessel can distort the measurements. To find a solution for all of the aforementioned problems, sensors using guided acoustic waves may present an adequate approach. In previous papers, aluminum rods have been used for the L(0,2) mode as a waveguide [1]. Factors such as the immersion depth have been correlated with the signal energy attenuation because of leakages of waves into the surrounding fluid. Measuring attenuation to determine changes in fluid media is applicable to steel tubes using the aforementioned L(0,2) mode [2]. One sensor concept has used the L(0,1) mode within a nickel and iron wire by measuring the time of flight to determine the surface coverage of the waveguide [3]. All of the mentioned works used metals to propagate acoustic waves, which restricts their use in acidic or basic media. While there are metals resistant to certain acids and bases, these are expensive compared to polyethylene. Here, we present a related approach using polyethylene rods as waveguides. The possibility of exciting acoustic waves within polyethylene and other polymers has been broadly discussed [4–14]. In these papers, dispersion is a crucial feature that needs to be considered when using guided ultrasonic waves, as it distorts wave packets, in turn leading to nonuniform propagation. Therefore, the goal is to minimise both dispersion and attenuation.

---

2. Materials and Methods

2.1. Acoustic Wave Propagation in Polyethylene Rods

Acoustic waves in rods propagate in three different types of modes: longitudinal, flexural, and torsional [15–17]. The individual modes are influenced differently and in a frequency-dependent manner when liquid is present. A suitable mode and working frequency can be determined on the basis of dispersion diagrams [18,19]. In this work, Disperse simulation software [19] was used to calculate the dispersion diagrams. Because two different solid high-density polyethylene (HD-PE) rods (15 mm and 40 mm in diameter) are investigated in the measurement section of this paper, dispersion graphs of these two configurations are discussed in detail. These rod diameters were chosen because they are common sizes that can easily be ordered in large quantities. The thicker rod has increased stability compared to the smaller one which yields a conceptual advantage in industrial applications. The material parameters of the rods are $E = 2.2\,\text{GPa}$ as the Young modulus, $\rho = 960\,\text{kg}\,\text{m}^{-3}$ as the density, and $\nu = 0.38$ as the Poisson ratio, with the following parameters for the water: $\rho = 1000\,\text{kg}\,\text{m}^{-3}$ and $v = 1500\,\text{m}\,\text{s}^{-1}$ as the sound velocity.

The simulation was set up to have the rod fully immersed in either water or vacuum. It can be seen that there is a difference between the phase velocity depending on the surrounding medium (cp Figure 1). The phase velocities were decreased when submerged in water, leading to the the time shift of the zero crossing positions appearing to be caused by the immersion depth. Another feature to be extracted from these graphs is the desired frequency to be used for exciting the L(0,1) mode. The higher the frequency, the larger the difference in phase velocity due to the surrounding media. Therefore, it was important to find the most adequate frequency that shows a significant difference in phase velocity when submerged in water while analysing the signals and maintaining low dispersion. The dispersion characteristics can be extracted from the graphs in Figure 2.

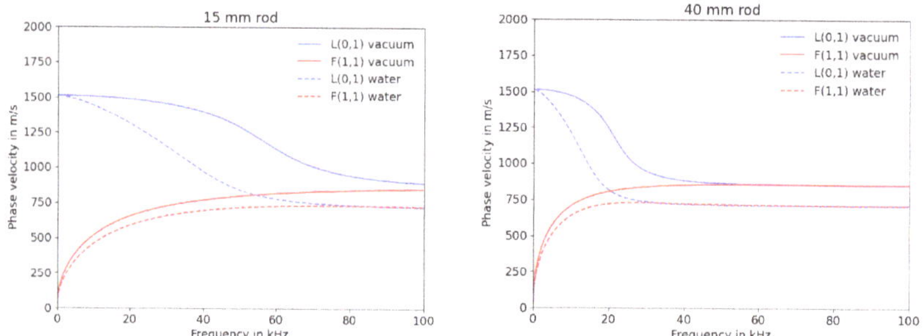

Figure 1. Simulated phase velocities of longitudinal (L) and flexural (F) modes for HD-PE rods with varying diameters.

Sections of the group velocity diagram with high gradients indicate increased dispersion in burst signals, as mentioned in [20]. Looking at the graphs in Figure 2, the group velocity for the 15 mm rod is quite constant for the vacuum scenario at low frequencies, but declines much faster when submerged into water. The 40 mm rod shows a higher gradient even at lower frequencies. This leads to the assumption that the range of frequencies that can be used for measuring the liquid level is significantly smaller for rods with larger diameter, which is comparable to the frequency and plate thickness product for Lamb waves [21].

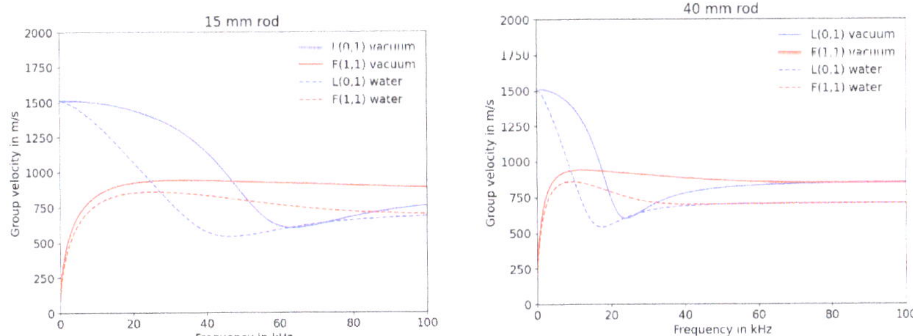

Figure 2. Simulated group velocities of longitudinal (L) and flexural (F) modes for HD-PE rods with varying diameters.

2.2. Signal Excitation

Piezoceramics are widely used for excitation of acoustic signals [22]. Depending on their geometry and polarisation, they can be used for longitudinal, torsional, or flexural waves. As the first concept for the approach of a guided acoustic wave sensor for measuring changes in surrounding media was to use polyethylene tubes instead of rods as a way to save both materials and costs, FEM simulations were performed in COMSOL Multiphysics (Version 6.0 [23]) to evaluate the reflected signal of the piezo transducers in multiple configurations. The simluation environment consisted of three different regimes, as can be seen in Figure 3. The model was set up as a two-dimensional rotationally symmetrical model that used the physics interfaces for solid mechanics and electrostatics and the multiphysics interface for piezoelectricity. The HD-PE tube had an outer diameter of 16 mm, a wall thickness of 1.8 mm, and a length of 500 mm. The material parameters for HP-PE matched the ones used in Disperse. To simulate the propagation of acoustic waves within the tube, a five-period Hanning-windowed sine burst with a frequency of 14 kHz was applied to the piezo. The acoustic wave coupled to the HD-PE tube was reflected on the opposite end and coupled back into the piezo transducer. Thus, the change of electric potential was recorded to evaluate the impulse echo propagation of this model. The piezo transducer used the material parameters of the PIC 255 Material by PI Ceramic.

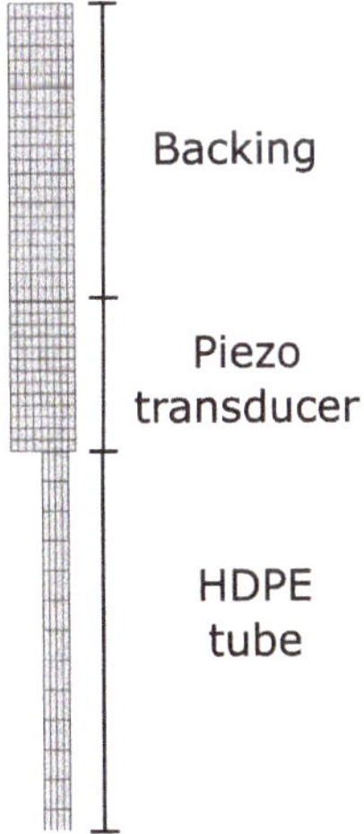

Figure 3. Simulation environment.

In order to find the ideal piezo geometry, a number of parameter sweep simulations were carried out. To match the shape of the tube, a ring shape was selected. Following parameters were varied:

- inner diameter
- thickness
- backing thickness

In following figures, the amplitude of the reflected acoustic signal was examined with varying piezo dimensions. The highest detected voltage (peak to peak, V_{pp}) of a wavelet is displayed on the y-axis as a target value. Higher amplitudes are preferred in this scenario. On one hand, this increases the signal to noise ratio and thus helps to stabilize any evaluation algorithms. On the other hand, higher amplitudes allow the use of longer rods due to high acoustic attenuation in HD-PE. In Figure 4, the influence of the inner diameter of the ring piezo is displayed. The lower the "wall thickness" of the ring, the higher the reflected amplitude. This can be explained by the decrease in mass that is to be moved by the reflected wave coupling from the HD-PE body into the piezoceramic.

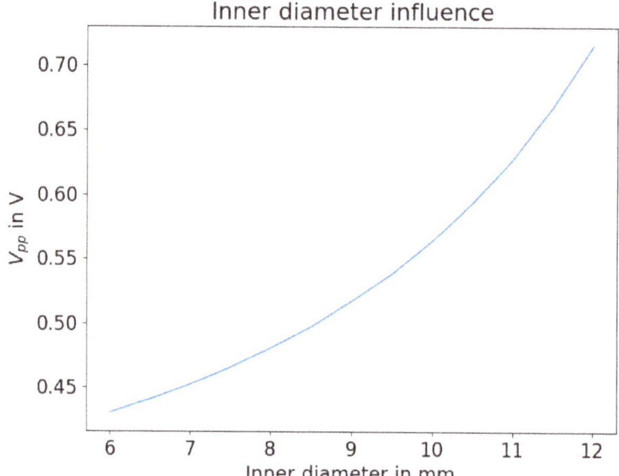

Figure 4. Influence of inner diameter of the piezo ring.

The influence of the piezo thickness is shown in Figure 5. There, it can be seen that the height positively correlates to the detected signal amplitude. Therefore, thicker piezo rings are to be preferred. The nonlinearity at 1 mm piezo height is caused by the simulation environment. In this parameter configuration, the simulation lattice is too coarse to satisfy the Courant–Friedrichs–Lewy condition [24].

Backings for piezoelectric transducers were used to increase the amount of energy coupled into the HD-PE body from the piezo transducer. The effect of different backing materials has been discussed in [25–27]. In this simulation, stainless steel with $E = 200\,\text{GPa}$, $\rho = 7850\,\text{kg}\,\text{m}^{-3}$ and $\nu = 0.3$ was used. The backing was varied in thickness and compared for two different piezo heights, with the results shown in Figure 6. An increase in backing height leads to increased signal amplitudes. The simulated piezo transducer with a height of 10 mm benefits more significantly from increased backing height.

2.3. Experimental Setup

To measure the change in wave propagation due to varying surrounding media, an experimental setup with a measurement vessel connected to a pump and a surge tank was assembled. This setup schematic is displayed in Figure 7. Six temperature sensors were used to monitor the temperature inside the vessel at four heights near the sensor and in the surge tank. A measurement software was built in Python (3.8, PyQt GUI) to control the valves connecting the two tanks, start and stop the pump, and control the immersion heater. The software was connected to an oscilloscope (LeCroy Waverunner 604 Zi) as well as a waveform generator (Agilent 33500B Series). To be able to measure reflected signals through long solid HD-PE rods, the signal coming from the generator and the reflection picked up by the piezo transducer needed to be amplified. An amplifier with a built-in multiplexer was used to boost the excitation signal by approximately 37 dB and the measured reflection by approximately 30 dB, leading to 190 V peak-to-peak applied to the piezo transducer. The vessel featured a pressure sensor (fluid.iO HD-100) to compare the acoustic measurements to those provided by the water pressure inside the vessel.

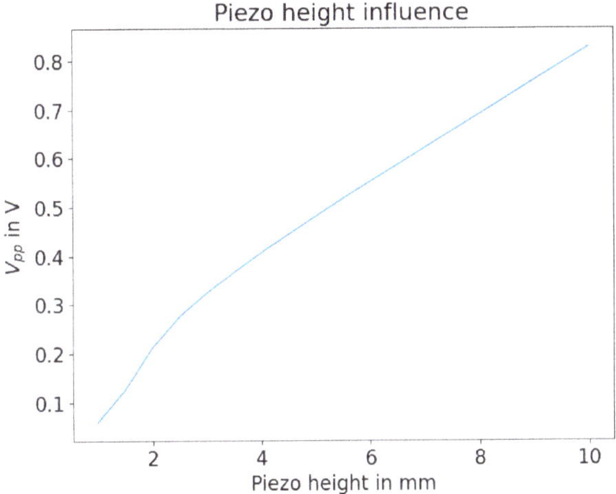

Figure 5. Influence of piezo height.

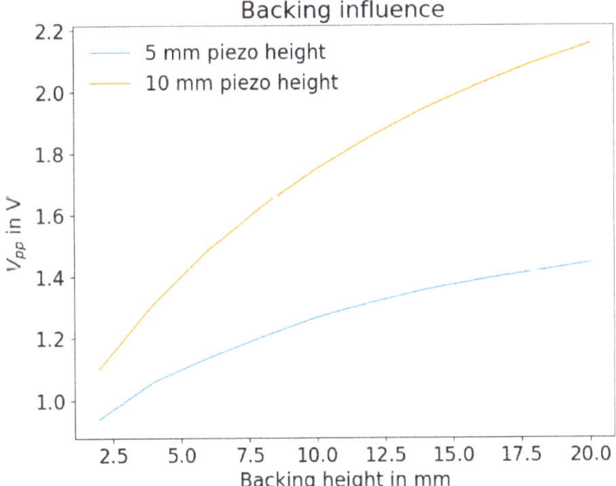

Figure 6. Influence of backing height with different piezo heights.

2.4. Zero Tracing

In order to relate the measured signals to the part of the rod immersed in the fluid, it is necessary to track the shift of zero crossing times due to the medium surrounding the rod. This can be achieved by tracing the zero crossings in the reflection of the L(0,1) mode. An algorithm to track these zero crossings was implemented and is explained below. At the beginning, the algorithm determines indices within the time and signal array centered around the maximum of the Hilbert transform of the signal in order to start tracing. To discard the excitation signal, a minimum timestamp needs to be passed. After setting the start indices, the algorithm finds points within the data array where a sign switch happens. The two data points around the sign switch are used to interpolate the time at which the zero crossing occurred. This algorithm is visualized in Figure 8. It can be seen that there is a difference between the rising and falling flanks. Because the excitation or signal frequency is known, phase jumps from one zero crossing to another within the evaluated wavelet can be detected and compensated. Taking the first traced timestamp into account, all following zero crosses can be determined by searching within a range of multiple of signal periods, determined by an input parameter ϵ. These traced signals can then be compared to a reference sensor. This type of tracing algorithm differs from pure time-of-flight measurements. The zero crossing timestamps for minimum and maximum immersion depth need to be known. Every measured timestamp between these values can then be interpolated to display the current immersion depth.

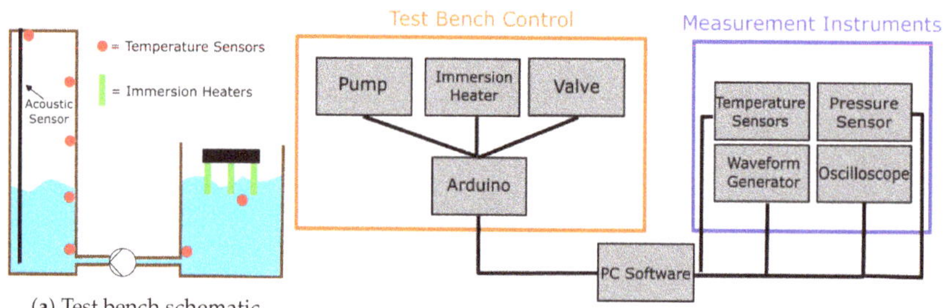

(**a**) Test bench schematic

(**b**) Electronics schematic

Figure 7. Schematics of the experimental setup.

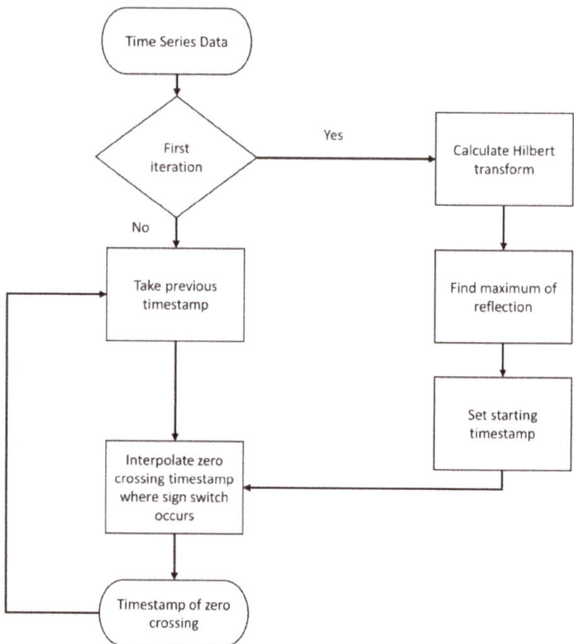

Figure 8. Flow chart of the zero tracing algorithm.

3. Results

3.1. Piezo Dimensions

The simulated piezo parameters were used to order ring shaped piezos (PIC 151, PI Ceramic) according to the simulated dimensions. These piezo transducers were mounted on the solid HD-PE rods and tubes using epoxy glue (UHU plus Schnellfest). A stainless steel backing cylinder was glued to the opposite site of the transducer to improve the energy coupled from the transducer to the HD-PE rod. In order to examine the influence of the backing (stainless steel, height 20 mm) on excited signals, a frequency sweep was measured to compare the maximum of the envelope of the reflected signal in a tube (16 mm diameter, 0.9 m length) without being submerged in water. The resulting signals are visualised in Figure 9.

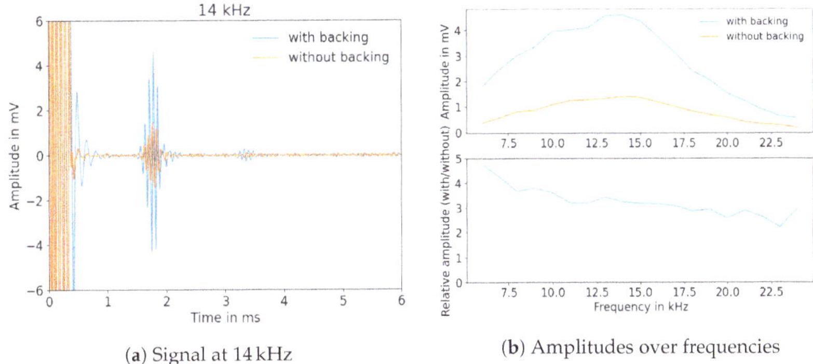

(**a**) Signal at 14 kHz   (**b**) Amplitudes over frequencies

Figure 9. Comparison of signal amplitudes with backing (20 dB amplification).

As can be seen, the amplitude increases significantly within a HD-PE tube when using a backing for the piezo transducer. In addition, it is possible to extract an ideal frequency to use for highest signal amplitudes. During our experiments, it turned out that a large part of the energy coupled out into the water, meaning that the signals could no longer be evaluated when the tube was immersed more than a few centimeters. Thus, these tubes could not be used for measuring variations in sensor surface coverage. Thus, the experiment was transferred to HD-PE rods, and the ideal excitation frequency was determined similarly. The measurements are shown in Figure 10. It can be seen that the overall signal amplitude is significantly higher for the 40 mm rod. Similar behaviour for amorphous media was investigated in [28].

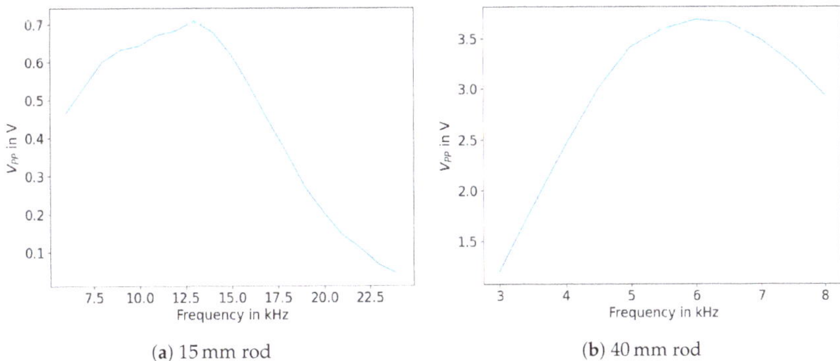

(**a**) 15 mm rod   (**b**) 40 mm rod

Figure 10. Amplitudes (peak to peak) over various signal frequencies.

The ideal excitation frequency for the solid 15 mm rod was around 13 kHz, whereas the 40 mm rod yielded the largest reflection amplitudes at 6 kHz. High amplitudes are to be preferred due to high attenuation in HD-PE. While these excitation frequencies should be chosen to optimize the signal to noise ratio for measuring liquid levels, the frequencies used in the following sections are predominantly lower, as mounting mechanisms for the rods lead to additional reflections in the signal at these frequencies. The mount applies mechanical stress to the top of the rod near the piezo transducer by clamping onto the rod; one example can be found in Figure 11. There, the signal frequency of 3 kHz was stable enough to use for immersion depth measurements.

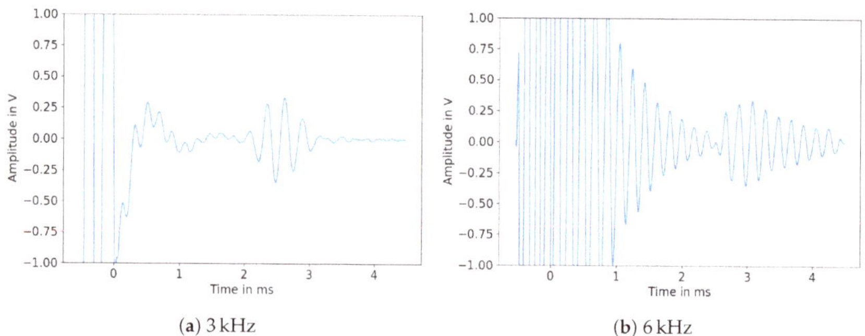

(a) 3 kHz   (b) 6 kHz

Figure 11. Measured signals of a 40 mm rod (1.7 m length).

3.2. Waveforms

To examine whether the L(0,1) mode could in fact be excited with the aforementioned sensor design, we measured the wave propagation within a 15 mm rod with a laser Doppler vibrometer. We then calculated the 2D-FFT transforms of these signals and compared them to the simulated dispersion graphs. The two graphs overlapped in terms of the excited modes. An example of this procedure can be seen in Figure 12.

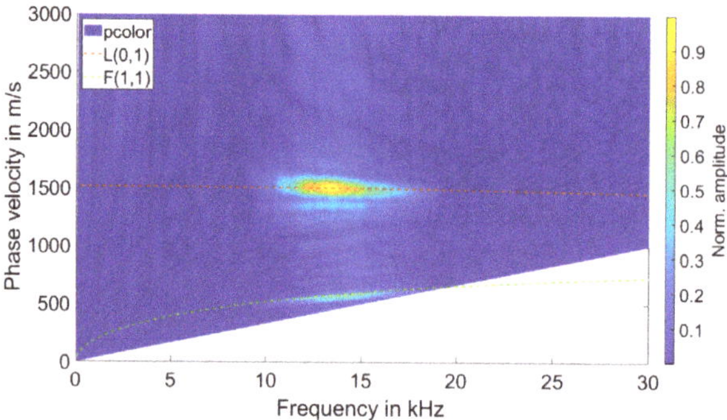

Figure 12. Overlap of dispersion graphs and 2D-FFT transform from laser Doppler vibrometer measurements.

Looking at the measured waveforms, it was possible to see both the main reflection propagating through the rod as well as the reflection at the fluid surface. One of these signals is shown in Figure 13.

It can additionally be seen that the reflected signal coming from the rod is dispersive and that there are multiple wave groups after the reflection of the L(0,1) mode. To further visualize the effect of changing immersion depths on the propagated acoustic signal within the rod, signals with two different immersion depths are shown in Figure 14.

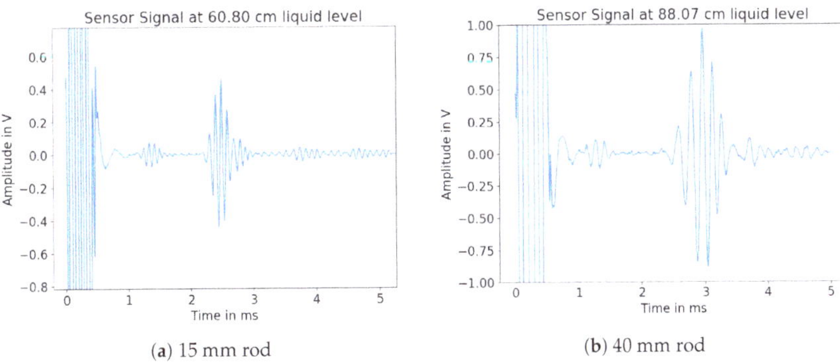

(**a**) 15 mm rod   (**b**) 40 mm rod

Figure 13. Measured signals of different HD-PE rods.

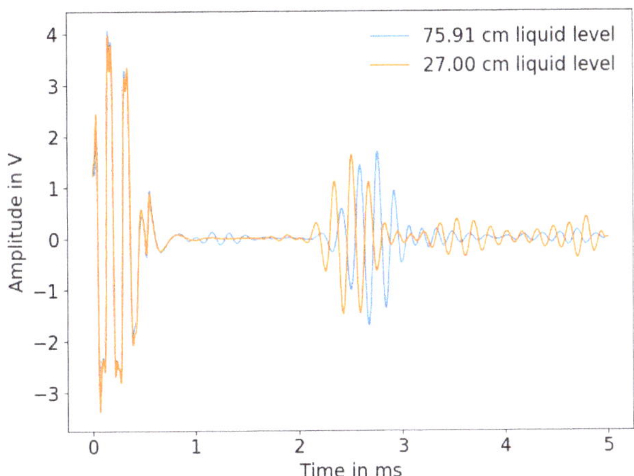

Figure 14. Acoustic signal at two different immersion depths.

3.3. Zero Tracing

The algorithm used to trace zero crossings was then applied on signals of two HD-PE rods. Figure 15 shows one measurement cycle of emptying and refilling the vessel. Both the guided acoustic wave sensor and the pressure sensor tracked the falling and rising liquid levels through their respective data points. The x-axis shows the number of measurements that were carried out during one test cycle in the vessel. In contrast to the simulations above, stable detection of reflections of the propagating waves was only possible with much lower excitation frequencies (13 kHz and 6 kHz respectively). This was due to high attenuation within long polyethylene rods (approximately 2 m). In early experiments, small HD-PE probes were examined. There, reflections of excited waves with 50 kHz could still be detected.

These traces already show one possible cause of error for using this principle as an immersion depth. At low levels, the reflection of the liquid surface overlaps with the reflection within the rod, leading to nonlinear tracing of the signal's zero crossings. Figure 16 displays this phenomenon in detail.

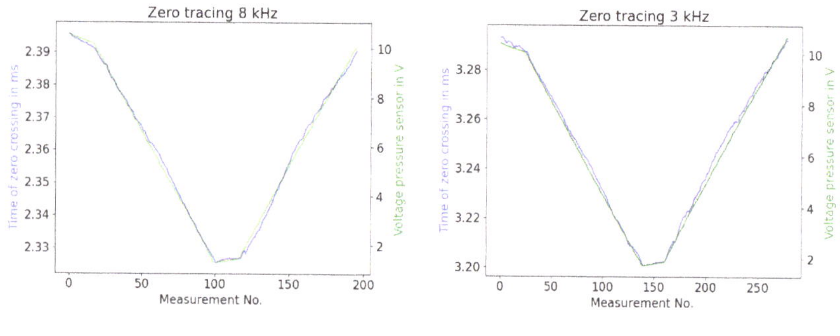

(a) 15 mm rod   (b) 40 mm rod

Figure 15. Zero tracing compared to pressure sensor voltages for rods of different diameter.

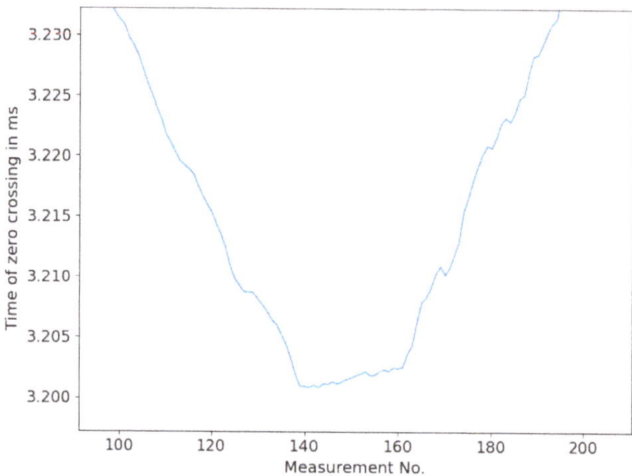

Figure 16. Nonlinear behaviour of traced zero crossing times at low immersion depths, showing a zoomed-in view of Figure 15b.

3.4. Temperature Dependency

Material parameters vary greatly due to changing temperatures. The elastic modulus of HD-PE decreases from 760 MPa to 407 MPa when the temperature increases from 23 °C to 40 °C [29]. Therefore, acoustic wave propagation changes accordingly [30,31]. One sample of a 40 mm rod was tested multiple times for water temperatures of 20 °C, 30 °C, 40 °C, and 50 °C. Two effects were predominantly apparent with temperature change, namely, signal attenuation and dilation of time of flight. With rising temperatures, signal amplitudes decrease and the dilation of the time of flight increases. These phenomena are displayed in Figure 17.

Thus, it is important to compare the overall dilation in the time of flight caused by the change in the surrounding medium to the one caused by temperature. The latter shows a larger amount of dilation. For a 40 mm rod with a three-period 3 kHz sine burst, the average range of zero crossing positions from fully submerged in water to not submerged at all spanned about 80 μs, while the range of the time shift of the wave group due to changes in temperature covered approximately 900 μs. The amplitude of the signal decreased by 42% at 50 °C compared to 20 °C. To compensate these variations in acoustic velocities, it is possible to either set up measurements to thoroughly examine the sensor behaviour within a specific temperature range and calculate compensation curves, or to try to excite additional acoustic modes in order to use multiple propagation times to create a compensation mechanism [32].

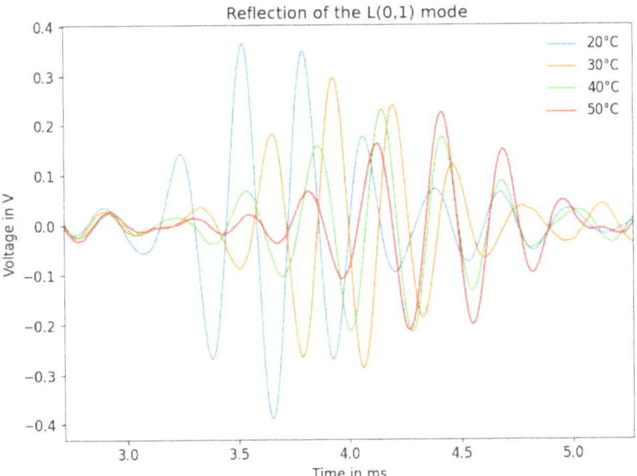

Figure 17. L(0,1) reflection at different temperatures.

4. Discussion

In this work, we present a novel approach to monitor changing fluid media in vessels using polyethylene rods. Compared to similar articles, we used lower frequencies to excite the acoustic waves due to attenuation and dispersion within amorphous media. Even though the concept worked in a laboratory environment, there are significant challenges that need to be overcome before this technique can be used in industrial applications. First, because the dilation caused by temperature exceeds the measured effect of the changing fluid media, the temperature drift needs to be thoroughly examined in order to calculate a compensation curve. The mounting mechanism poses a challenge as well, as reflections or multiples thereof can interfere with the desired pulse echo. In several of our experiments this led to the reflection at the surface not being traceable. Another aspect to be investigated is the possibility of amplifying both the excited signal and the received signal, as plastics such as polyethylene have much higher attenuation compared to most common metals used for guided acoustic waves. To conclude, this approach might bring advantages to certain industrial applications where using metal rods as waveguides or different sensors is generally not possible.

Author Contributions: Conceptualization, A.B. and K.L.; methodology, K.L.; software, K.L.; validation, K.L., A.B. and K.S.D.; formal analysis, K.S.D.; investigation, K.L.; resources, A.B.; data curation, K.L.; writing—original draft preparation, K.L.; writing—review and editing, A.B. and K.S.D.; visualization, K.L.; supervision, K.S.D.; project administration, K.S.D.; funding acquisition, K.S.D. All authors have read and agreed to the published version of the manuscript.

Funding: This project was supported by the Federal Ministry for Economic Affairs and Climate Action (BMWK) on the basis of a decision by the German Bundestag (KK 5048301DB0).

Institutional Review Board Statement: Not applicable.

Informed Consent Statement: Not applicable.

Data Availability Statement: The data presented in this study are available on request from the corresponding author. The data are not publicly available due to secrecy.

Acknowledgments: Special thanks belong to our project partner for providing necessary components to construct the measurement setup and the sensor rods as well as providing the amplification electronics.

Conflicts of Interest: The authors declare no conflict of interest.

Sensors **2023**, *23*, 9892

Abbreviations

The following abbreviations are used in this manuscript:

HD-PE   High Density Polyethylene

References

1. Subhash, N.N.; Krishnan, B. Fluid level sensing using ultrasonic waveguides. *Insight* **2014**, *56*, 607–612 [CrossRef]
2. Dhayalan, R.; Saravanan, S.; Manivannan, S.; Purna Chandra Rao, B. Developement of ultrasonic waveguide sensor for liquid level measurement in loop system. *Electron. Lett.* **2020**, *56*, 1120–1122. [CrossRef]
3. Royer, D.; Levin, L.; Legras, O. A Liquid Level Sensor Using the Absorption of Guided Acoustic Waves. *IEEE Trans. Ultrason. Ferroelectr. Freq. Control* **1993**, *40*, 418–421. [CrossRef] [PubMed]
4. Shah, J.; El-Hawwat, S.; Wang, H. Guided Wave Ultrasonic Testing for Crack Detection in Polyethylene Pipes: Laboratory Experiments and Numerical Modeling. *Sensors* **2023**, *23*, 5131. [CrossRef] [PubMed]
5. Sinha, M.; Buckley, D.J. Acoustic Properties of Polymers. In *Physical Properties of Polymers Handbook*; Springer: New York, NY, USA, 2007; pp. 1021–1031.
6. Ozaki, R.; Kadowaki, K. Analysis of Attenuation and Dispersion of Acoustic Waves in Low-Density Polyethylene. *IEEE Trans. Dielectr. Electr. Insul.* **2020**, *27*, 2007–2013. [CrossRef]
7. Qi, G.; Li, Y.; Ding, N. Measurement of Acoustic Basic Parameters of Polyethylene Pipe. In Proceedings of the IOP Conference Series: Materials Science and Engineering, Melbourne, Australia, 12–13 October 2019.
8. Pylaev, A.E.; Kostikova, E.A.; Yurkov, A.L. Velocity and Attenuation of Acoustic Waves in Polymers and Polymer Composites. *Polym. Sci. Ser. D* **2018**, *11*, 272–276. [CrossRef]
9. Mazeika, L.; Sliteris, R.; Vladisauskas, A. Measurement of velocity and attenuation for ultrasonic longitudinal waves in the polyethylene samples. *Ultragarsas J.* **2011**, *16*, 12–15.
10. Jordan, J.L.; Rowl, ; R.L.; Greenhall, J.; Moss, E.K.; Huber, R.C.; Willis, E.C.; Hrubiak, R.; Kenney-Benson, C.; Bartram, B.; Sturtevant, B.T. Elastic properties of polyethylene from high pressure sound speed measurements. *Polymer* **2021**, *212*, 123–164. [CrossRef]
11. Egerton, J.S.; Lowe, M.J.S.; Huthwaite, P.; Halai, H.V. Ultrasonic attenuation and phase velocity of high-density polyethylene pipe material. *J. Acoust. Soc. Am.* **2017**, *141*, 1535–1545. [CrossRef]
12. Rautenberg, J.; Bause, F.; Henning, B. Utilizing guided acoustic waves to measure dispersive material properties of polymers. In Proceedings of the SENSOR 2015, Nürnberg, Germany, 19–21 May 2015.
13. Kwun, H.; Bartels, K.A.; Dynes, C. Dispersion of longitudinal waves propagating in liquid-filled cylindrical shells. *J. Acoust. Soc. Am.* **1999**, *105*, 2601–2611. [CrossRef]
14. Wilcox, P.D. A rapid signal processing technique to remove the effect of dispersion from guided wave signals. *IEEE Trans. Ultrason. Ferroelectr. Freq. Control* **2003**, *50*, 419–427. [CrossRef] [PubMed]
15. Gazis, D.C. Three-Dimensional Investigation of the Propagation of Waves in Hollow Circular Cylinders. I. Analytical Foundation. *J. Acoust. Soc. Am.* **1959**, *31*, 568–573. [CrossRef]
16. Gazis, D.C. Three-Dimensional Investigation of the Propagation of Waves in Hollow Circular Cylinders. II. Numerical Results. *J. Acoust. Soc. Am.* **1959**, *31*, 573–578. [CrossRef]
17. Silk, M.G.; Bainton, K.F. The propagation in metal tubing of ultrasonic wave modes equivalent to Lamb waves. *Ultrasonics* **1979**, *17*, 11–19. [CrossRef]
18. The Dispersion Calculator: An Open Source Software for Calculating Dispersion Curves and Mode Shapes of Guided Waves. Available online: https://www.dlr.de/zlp/en/desktopdefault.aspx/tabid-14332/24874_read-61142/ (accessed on 24 August 2023).
19. Pavlakovic, B.; Lowe, M.; Alleyne, D.; Cawley, P. Disperse: A General Purpose Program for Creating Dispersion Curves. In *Review of Progress in Quantitative Nondestructive Evaluation: Volume 16*; Springer: Boston, MA, USA, 1997; pp. 185–192.
20. Wilcox, P. Long Range Lamb Wave Inspection: The Effect of Dispersion and Modal Selectivity. In *Review of Progress in Quantitative Nondestructive Evaluation: Volume 18A–18B*; Springer US: Boston, MA, USA, 1999; pp. 151–158.
21. Lowe, M.J.S. WAVE PROPAGATION | Guided Waves in Structures. In *Encyclopedia of Vibration*; Elsevier: Amsterdam, The Netherlands, 2001.
22. Duquennnoy, M.; Ouaftouh, M.; Qian, M.L.; Jenot, F.; Ourak, M. Ultrasonic characterization of residual stresses in steel rods using a laser line source and piezoelectric transducers. *NDT E Int.* **2001**, *34*, 355–362. [CrossRef]
23. Comsol Multiphysics: Simulate Real-World Designs, Devices, and Processes with Multiphysics Software from COMSOL. Available online: www.comsol.com (accessed on 24 August 2023).
24. Courant, R.; Friedrichs, K.; Lewy, H. On the Partial Difference Equations of Mathematical Physics. *IBM J. Res. Dev.* **1967**, *11*, 215–234. [CrossRef]
25. Moreno, E.; Pereira, W.; von Kruger, M.A.; Leija, L.; Ramos, A. FEM modeling and simulation of broadband ultrasonic transducers with randomized inhomogeneous backing material. *arXiv* **2022**, arXiv:2203.08785.
26. Kossoff, G. The Effects of Backing and Matching on the Performance of Piezoelectric Ceramic Transducers. *IEEE Trans. Sonics Ultrason.* **1966**, *13*, 20–30. [CrossRef]

27. Subki, M.S.H.M.; Ahmad, K.A.; Osman, M.K.; Boudville, R.; Yahya, S.Z.; Rahman, M.F.A.; Hussain, Z. Characterization of Backing Layer Piezoelectric Ultrasonic Transducers for Underwater Communication. In Proceedings of the IEEE International Conference on Control System, Computing and Engineering, Penang, Malaysia, 27–28 August 2021.
28. Benatar, A.; Rittel, D.; Yarin, A.L. A Theoretical and experimental analysis of longitudinal wave propagation in cylindrical viscoelastic rods. *J. Mech. Phys. Solids* **2003**, *51*, 1413–1431. [CrossRef]
29. Nomura, R.; Yoneyama, K.; Ogasawara, F.; Ueno, M.; Okuda, Y.; Yamanaka, A. Temperature Dependence of Sound Velocity in High-Strength Fiber-Reinforced Plastics. *Jpn. J. Appl. Phys.* **2003**, *42*, 5205. [CrossRef]
30. Wada, Y.; Yamamoto, K. Temperature Dependence of Velocity and Attenuation of Ultrasonic Waves in High Polymers. *J. Phys. Soc. Jpn.* **1956**, *11*, 887–892. [CrossRef]
31. Merah, N.; Saghir, F.; Khan, Z.; Bazoune, A. Effect of temperature on tensile properties of HDPE pipe material. *Plast. Rubber Compos.* **2006**, *35*, 226–230. [CrossRef]
32. Landskron, J.; Dötzer, F.; Benkert, A.; Mayle, M.; Drese, K.S. Acoustic Limescale Layer and Temperature Measurement in Ultrasonic Flow Meters. *Sensors* **2022**, *22*, 6648. [CrossRef]

MDPI
St. Alban-Anlage 66
4052 Basel
Switzerland
www.mdpi.com

Sensors Editorial Office
E-mail: sensors@mdpi.com
www.mdpi.com/journal/sensors